LANDOLT-BÖRNSTEIN

Numerical Data and Functional Relationships
in Science and Technology

New Series

Editors in Chief: K.-H. Hellwege · O. Madelung

Group III: Crystal and Solid State Physics

Volume 17
Semiconductors

Editors: O. Madelung · M. Schulz · H. Weiss †

Subvolume i

Special Systems and Topics
Comprehensive Index for III/17a ··· i

D. Bimberg · I. Eisele · W. Fuhs
H. Kahlert · N. Karl

Edited by O. Madelung · M. Schulz · H. Weiss †

Springer-Verlag Berlin · Heidelberg · New York · Tokyo

LANDOLT-BÖRNSTEIN

Zahlenwerte und Funktionen
aus Naturwissenschaften und Technik

Neue Serie

Gesamtherausgabe: K.-H. Hellwege · O. Madelung

Gruppe III: Kristall- und Festkörperphysik

Band 17
Halbleiter

Herausgeber: O. Madelung · M. Schulz · H. Weiss †

Teilband i

Spezielle Systeme und Themen
Gesamtregister für III/17a ··· i

D. Bimberg · I. Eisele · W. Fuhs
H. Kahlert · N. Karl

Herausgegeben von O. Madelung · M. Schulz · H. Weiss †

Springer-Verlag Berlin · Heidelberg · New York · Tokyo

ISBN 3-540-15072-2 Berlin, Heidelberg, New York, Tokyo
ISBN 0-387-15072-2 New York, Heidelberg, Berlin, Tokyo

CIP-Kurztitelaufnahme der Deutschen Bibliothek

Zahlenwerte und Funktionen aus Naturwissenschaften und Technik / Landolt-Börnstein. – Berlin; Heidelberg; New York; Tokyo:
Springer. Teilw. mit d. Erscheinungsorten Berlin, Heidelberg, New York. – Parallelt.: Numerical data and functional relationships
in science and technology

NE: Landolt, Hans [Begr.]; PT Landolt-Börnstein, … N.S./Gesamthrsg.: K.-H. Hellwege; O. Madelung. Gruppe 3, Kristall-
und Festkörperphysik. Bd. 17. Halbleiter/Hrsg.: O. Madelung … Teilbd. i. Spezielle Systeme und Themen. Generalregister/
D. Bimberg … Hrsg. von O. Madelung … – 1985

ISBN 3-540-15072-2 Berlin, Heidelberg, New York, Tokyo
ISBN 0-387-15072-2 New York, Heidelberg, Berlin, Tokyo

NE: Hellwege, Karl-Heinz [Hrsg.]; Madelung, Otfried [Hrsg.]; Bimberg, Dieter [Mitverf.]

Typesetting: Universitätsdruckerei H. Stürtz AG, Würzburg; printing: Druckhaus Langen-
scheidt KG, Berlin; bookbinding: Lüderitz & Bauer-GmbH, Berlin

2163/3020-543210

Contributors

D. Bimberg, Institut für Festkörperphysik (Fachbereich 4) der Technischen Universität Berlin, D 1000 Berlin

I. Eisele, Institut für Physik (Fakultät für Elektrotechnik) der Universität der Bundeswehr München, 8014 Neubiberg, FRG

W. Fuhs, Fachbereich Physik der Universität Marburg, 3550 Marburg, FRG

H. Kahlert, Institut für Festkörperphysik der Technischen Universität Graz, 8010 Graz, Austria

N. Karl, Physikalisches Institut der Universität Stuttgart, 7000 Stuttgart, FRG

O. Madelung, Fachbereich Physik der Universität Marburg, 3550 Marburg, FRG

M. Schulz, Institut für angewandte Physik der Universität Erlangen-Nürnberg, 8520 Erlangen, FRG

Vorwort

Der vorliegende Teilband 17i des Landolt-Börnstein schließt die Sammlung der wichtigsten physikalischen und technologischen Daten der halbleitenden Elemente und Verbindungen ab. Hierfür waren neun Bände mit insgesamt fast 5000 Seiten notwendig – ein Umfang, der bei der ersten Planung des Bandes 17 vor etwa sechs Jahren nicht abzusehen war.

Die physikalischen Eigenschaften der kristallinen Halbleiter wurden, geordnet nach Verbindungsklassen, in den Teilbänden 17a, b, e...h dargestellt. Die Technologie der in der Anwendung genutzten Halbleitermaterialien wurde in den Teilbänden 17c und d behandelt.

Der hier vorliegende Teilband 17i enthält nun Themen, die sich in die Ordnung der vorhergehenden Bände nicht einfügen ließen, deren Bedeutung aber ihre Aufnahme in das gesamte Werk erforderte. Der Abschnitt "Special systems" enthält Beiträge über amorphe und organische Halbleiter. Der Abschnitt "Special topics" enthält drei Beiträge über "Space charge layers at surfaces and interfaces", "Hot electrons", und "Electron-hole liquids", die als physikalische Spezialgebiete nicht in das Datenmaterial der vorhergehenden Bände eingeordnet werden konnten. In diesen Teilkapiteln ist die Information in geschlossener Form dargestellt worden und daher leicht auffindbar. Die zunächst vorgesehene Aufnahme von Kapiteln über Chalkogenid-Gläser, flüssige Halbleiter und über die elektronische Struktur von Halbleiteroberflächen war aus technischen Gründen leider nicht möglich. Es ist geplant, diese Themen in später erscheinenden Landolt-Börnstein-Bänden darzustellen.

Die Kapitel über "Special systems" und "Special topics" unterscheiden sich deutlich in der Art der Darstellung von den früheren Bänden. Jedes Kapitel ist in sich abgeschlossen und hat eine eigene Einführung mit eigener Symbolliste, die Figuren sind jeweils direkt nach den Tabellen der einzelnen Kapitel angeordnet.

Bei dem Umfang des Gesamtwerkes schien es angebracht, diesem letzten Band ein Gesamtregister beizufügen. Dieses wurde aus Gründen der Zweckmäßigkeit und leichteren Benutzung unterteilt in vier Register für die mehr physikalisch ausgerichteten Teilbände 17a, b, e...i und ein Register für die beiden Technologiebände 17c, d. Bei den physikalischen Bänden sind die Daten bereits straff geordnet. Die vorliegenden Register beschränken sich deshalb auf Materiallisten mit Seitenangaben. Für die Technologiebände war eine Darstellung der Daten ohne Überlappung verschiedener Teilinformationen nicht möglich. Es schien uns daher notwendig, ein Suchwortregister bereitzustellen, mit dem spezielle Begriffe, Daten und Informationen in verschiedenen Kapiteln lokalisiert werden können. In diesem Register ist für jedes Suchwort der Kontext, in dem es auftritt, zusammen mit der Seitenangabe angegeben.

Es ist den Herausgebern ein Bedürfnis, an dieser Stelle den Autoren und denen zu danken, die sich in der Landolt-Börnstein Redaktion um die Herausgabe dieses Bandes verdient gemacht haben: Herrn Dr. W. Polzin, Herrn Dr. H. Seemüller, Frau E. Hofmann, Frau I. Lenhart, Frau R. Lettmann. – Unser Dank gilt darüber hinaus zusammenfassend allen Autoren und Mitgliedern der Redaktion, die sich in den letzten sechs Jahren für das Zustandekommen der neun Teilbände eingesetzt haben. Ihrem Engagement ist es zu verdanken, wenn auch hier eines der wichtigsten Ziele des Landolt-Börnstein erreicht wurde: Eine zuverlässige und übersichtliche Darstellung von gesichtetem und ausgewertetem Datenmaterial über Halbleiter.

Erlangen und Marburg, August 1985   **Die Herausgeber**

Preface

The present subvolume 17i concludes the Landolt-Börnstein series on the most important physical and technological data of the elemental and compound semiconductor materials. Nine subvolumes totalling almost 5000 pages were necessary to present this survey of semiconductor data. This large extent was not expected when the editors started the concept six years ago.

The physical data of crystalline semiconductor materials are presented in the subvolumes 17a, b, e…h. The materials are sequentially ordered according to their composition. Information on the technology of semiconductor materials used in applications or devices is presented in the subvolumes 17c, d.

This present subvolume 17i now contains subjects which did not fit into the previous subvolumes. These subjects, however, contain important semiconductor data which should be represented in a concise form. The section on "Special systems" includes data on amorphous and organic semiconductors. The section on "Special topics" contains three contributions, dealing with physical data and information on "Space charge layers at surfaces and interfaces", "Hot electrons", and "Electron-hole liquids". In these chapters, semiconductor data inherent to physical areas are presented in a concise form so that they can be easily traced. The originally planned chapters on chalcogenide glasses, liquid semiconductors, and on the electronic structure of semiconductor surfaces could not be included because of technical reasons. It is planned to cover these systems in forthcoming volumes of this Landolt-Börnstein series.

The chapters on "Special systems" and "Special topics" differ from the previous subvolumes in the form of the data presentation. The clear and consistent ordering of the data used therein was not possible for these special subjects. Each chapter therefore has its own introduction and a list of symbols, and the figures are presented directly after the tables of the respective section.

Considering the total amount of data of this Landolt-Börnstein series on semiconductors, it seemed necessary to include a comprehensive index in this last subvolume. This index is subdivided into four listings for the more physically oriented subvolumes 17a, b, e…i and one for the subvolumes 17c, d presenting technology data. Since the physical data are already clearly ordered, the corresponding listings give only the materials and the pages on which the data are presented. As to the subvolumes presenting information on technology, a clearcut classification and ordering was not possible. It was therefore necessary to prepare a further listing (subject index) to retrieve special information or data spread in different subsections. For each subject, the context is listed together with the pages and the subvolume in which it appears.

At the end of this large effort to compile a vast amount of information on semiconductors, the editors wish to express their gratitude to all persons who contributed and worked on the nine subvolumes. The excellent collaboration with the contributing authors and all the members of the editorial office of the Springer-Verlag — for subvolume 17i, especially Dr. W. Polzin, Dr. H. Seemüller, Mrs. E. Hofmann, Mrs. I. Lenhart, and Mrs. R. Lettmann — is gratefully acknowledged. Thanks to the common effort of all persons involved, the main goal of the Landolt-Börnstein could be achieved: a reliable and clear presentation of an examined and selected semiconductor data compilation.

Erlangen and Marburg, August 1985   **The Editors**

Table of contents

Semiconductors

Subvolume i: Special systems and topics

A. Special systems

11 Amorphous semiconductors

11.0 Lists of frequently used symbols and abbreviations

a) List of symbols

Symbol	Unit	Property
$\boldsymbol{B}$	G	magnetic induction
c_{lm}	$\mathrm{dyn\,cm^{-2}}$	elastic moduli (stiffnesses)
$c_{H(O...)}$	at%	hydrogen (oxygen...) concentration
$c_{PH_3(B_2H_6)}$	–	concentration of dopant gas $PH_3(B_2H_6)$ (ratio of dopant gas $PH_3(B_2H_6)$ to SiH_4 or GeH_4 for a-Si or a-Ge, respectively)
C_j	–	coordination number
d	cm	separation, distance
d	$\mathrm{g\,cm^{-3}}$	density
D	$\mathrm{cm^2\,s^{-1}}$	diffusion coefficient
e	C	elementary charge ($|e| = 1.6021892\,(46)\cdot10^{-19}$ C)
$\boldsymbol{E}$	$\mathrm{V\,m^{-1}}$	electric field strength
E	eV	energy
E_b	eV	binding energy
$E_{c(v)}$	eV	band edge of conduction (valence) band
E_{ex}	eV	excitation energy (for luminescence)
E_F	eV	Fermi energy
E_g	eV	(optical) energy gap
E_p	eV	position of luminescence maximum
E_σ	eV	activation energy for conductivity
f	Hz	frequency
g	–	g-factor
$g(E)$	$\mathrm{cm^{-3}\,eV^{-1}}$	density of states
h	Js	Planck constant ($h = 6.626176\cdot10^{-34}$ Js)
$\boldsymbol{H}$	G	magnetic field
ΔH_{pp}	G	peak to peak linewidth
$\Delta H_{pp}(0)$	G	T-independent part of ΔH_{pp}
δH_{pp}	G	T-dependent part of ΔH_{pp}
$I_{(R)}$	$\mathrm{W\,cm^{-2}}$	intensity (Raman intensity)
I_{ph}	A	photocurrent
k	$\mathrm{J\,K^{-1}}$	Boltzmann constant (k $= 1.380662\,(44)\cdot10^{-23}\,\mathrm{J\,K^{-1}}$)
$\boldsymbol{k}$	$\mathrm{cm^{-1}}$	wavevector of electrons
K	$\mathrm{cm^{-1}}$	absorption coefficient
n	–	(real) refractive index
n_{eff}	$\dfrac{\text{electrons}}{\text{atom}}$	effective number of electrons participating in optical transitions
n_H	–	hydrogen content $\left(c_H = \dfrac{n_H}{n_{Si}}\right)$
n_I	$\mathrm{cm^{-3}}$	impurity concentration
n_s	$\mathrm{cm^{-3}}$	spin density
N	–	number (of atoms, of photons etc.)
p_{H_2}	bar	partial pressure of hydrogen
$P_{(rf)}$	W	rf power
r	Å	radius
r_j	Å	correlation distance ($r_{1,2}$: first, second neighbor distance of atoms)
R	–	reflectance

Symbol	Unit	Property
S	$V\,K^{-1}$	thermoelectric power, Seebeck coefficient
t	s	time
T_1	s	spin lattice relaxation time
T	–	transmission
T	K, °C	temperature
T_a	K, °C	annealing temperature
T_{cr}	K, °C	crystallization temperature
T_s	K, °C	substrate temperature
v	$cm\,s^{-1}$	sound velocity
v	cm^3	volume
X	bar	stress
α	$\text{Å}\,min^{-1}$	deposition rate, growth rate
$\varepsilon(0)$	–	low frequency dielectric constant
$\varepsilon_{1(2)}$	–	real (imaginary) part of dielectric constant
Θ_p	K	paramagnetic Curie temperature
κ	$W\,cm^{-1}\,K^{-1}$	thermal conductivity
λ	cm	wavelength
μ_{dr}	$cm^2\,V^{-1}\,s^{-1}$	drift mobility
$\mu_{n(p)}$	$cm^2\,V^{-1}\,s^{-1}$	electron (hole) mobility
μ_H	$cm^2\,V^{-1}\,s^{-1}$	Hall mobility
η	–	luminescence efficiency
$\bar{v}$	cm^{-1}	wavenumber
$\bar{v}_{TA(LA)}$	cm^{-1}	wavenumber of transverse (longitudinal) acoustic vibrational mode
$\Delta\varrho/\varrho$	–	magnetoresistance
σ	$\Omega^{-1}\,cm^{-1}$	electrical conductivity
σ_0	$\Omega^{-1}\,cm^{-1}$	preexponential factor of conductivity
σ_d	$\Omega^{-1}\,cm^{-1}$	dark conductivity
σ_{ph}	$\Omega^{-1}\,cm^{-1}$	photoconductivity
$\tau_{(n)}$	s	decay time, delay time, relaxation time, lifetime (of electrons)
χ	$cm^3\,g^{-1}$	magnetic mass susceptibility
ω	$rad\,s^{-1}$	circular frequency
$\hbar\omega$	eV	photon energy

b) List of abbreviations

a	amorphous
ac	alternating current
arb.	arbitrary
c	crystalline
CVD	chemical vapor deposition
dc	direct current
DLTS	deep level transient spectoscopy
e	electron
eff	effective
ESR	electron spin resonance
EXAFS	extended X-ray absorption fine structure
FWHM	full width half maximum
GD	glow discharge
IR	infrared
LA	longitudinal acoustic
LO	longitudinal optical
LESR	light induced ESR
max	maximum
ODMR	optical detected magnetic resonance
ph	photo-(subscript)
PC	photoconductivity

PL	photoluminescence
rel	relative
rf	radio frequency
RDF	radial distribution function
RT	room temperature
SIMS	secondary ionic mass spectroscopy
tr	transition, transit (subscript)
TA	transverse acoustic
TO	transverse optical
UHV	ultra high vacuum
UPS	UV photoelectron spectroscopy
UV	ultraviolet
W, B, S	wagging, bending, stretching mode (subscript)
XPS	X-ray photoelectron spectroscopy

11.1 Silicon (a-Si)

Amorphous silicon can be prepared by various methods: thermal evaporation, sputtering in inert gas, chemical vapor deposition, glow discharge decomposition of silane, ion bombardment. The properties of these films sensitively depend on the method of preparation and the details of the preparation process. The difficulty in comparing quantitatively data from different laboratories, consists in part in the incomplete control and specification of the deposition parameters. Many of the numerous qualitative and quantitative differences in the literature arise moreover from contamination during or after the deposition. For the data presented in this chapter the most important deposition parameters are given unless they are not available from the original papers or unimportant for the physical quantity under consideration.

Films prepared by evaporation, chemical vapor deposition and sputtering contain large defect densities which create a high density of localized gap states. Defect rich a-Si is also obtained by ion bombardment at high ion doses [80M2]. The glow discharge material and the hydrogenated sputtered a-Si have to date superior semiconductor quality [74S5, 75S, 76S1, 79L2, 81P1]. The remarkable properties of these films are associated with the fact that they contain a high amount of hydrogen (up to 50 at%) which saturates defects and thus reduces the density of gap states. Due to the high hydrogen content it is justified to refer to this material as a binary alloy, a-Si:H. Hydrogenation can also be achieved by reactive sputtering in an inert gas containing hydrogen [77M1, 79M1, 81P1, 81K2], and by diffusing hydrogen into films of a-Si prepared by evaporation [78K1], chemical vapor deposition [80S1, 81H2, 83S1, 83E] or ion bombardment of crystals [80P1]. Fluorinated amorphous silicon (a-Si:F:H) is obtained by using mixtures of SiF_4 and H_2 in the glow discharge technique [79M3, 80M9, 82M2, 83J1, 83M1] and by sputtering [82M2]. The deposition rate in the glow discharge technique using SiH_4 is relatively low (few Å/s). Higher deposition rates are obtained using silanes like Si_2H_6 or Si_3H_8 [82H1]. A detailed description of the glow discharge deposition method is given in [80K3, 79K2, 77B2, 83S2, 81V1, 83T].

Reviews: [79M5, 77B7, 80F3, 82H1, 84F, 84J].

11.1.1 Structural characterization

The radial distribution function (RDF) obtained by X-ray and electron diffraction is consistent with a 3-dimensional random network of tetrahedrally bonded atoms [69M, 70M]. The first and second neighbor distances (r_1, r_2) and coordination numbers (C_1, C_2) are almost identical in crystalline and amorphous silicon (Table 1). The width of the second neighbor peak indicates distortions in the tetrahedral bond angles of $\pm 9°$. The main features of the RDF do not depend on the method of preparation. This includes amorphisation by ion bombardment. However, due to incorporation of large amounts of hydrogen the structural parameters of glow discharge films are somewhat different [79G].

The densities of the amorphous films are 5···30% less than the densities of crystalline material [69M, 70M, 72B1, 77B1, 79F2, 79G] depending on the method and conditions of preparation. Structural inhomogeneities are observed by a number of techniques particularly in evaporated and sputtered films [69M, 74S1, 76C1, 77B3, 79F1, 80L3, 81L1, 83C, 83B4]. Small angle X-ray scattering results are used in particular to deduce the size and number of voids. Associated with the presence of these microvoids is the large defect density which is typical for evaporated and sputtered films (see section 11.1.2, Defect states).

In glow discharge deposited a-Si:H films under a variety of deposition conditions columnar growth morphology is found [79K1, 80K1, 84R2, 84H1]. Columns of 100···300 Å lateral dimensions are observed growing vertically to the substrate in particular at high rf power. Small deposition rates, high SiH_4 concentrations and large negative bias of the substrate with respect to the plasma potential tend to suppress the columnar structure [79K1]. In films of high semiconductor quality no microstructure is visible in the electron micrographs. However, proton magnetic resonance studies [80R1, 81R2, 82C1], small angle X-ray and neutron scattering [79L1] provide evidence of structural inhomogeneities in these films, too. These results present strong evidence that most of the hydrogen in these films is located at intergrain boundaries. These inhomogeneities, which very sensitively depend on the preparation conditions, are of major influence on the mechanical, electrical and optical properties of the films.

The crystallization of amorphous films is a gradual process with an exothermic heat of crystallization between 1.7 and 2.2 kcal/mol [79T1]. A wide range of crystallization temperatures is reported in the literature depending on sample preparation, impurities and in particular on the substrate (Table 2). In contact with metals the crystallization temperature is strongly reduced, it amounts roughly to 0.72 of the eutectic melting temperature [72H]. For further studies, see [74O1, 76G, 79J1, 79T1].

Under hydrostatic pressure of about 100 kbar a semiconductor-metal transition is observed which manifests itself in a decrease of the resistance by a factor of 10^6 [74S9, 80M3, 81M2].

All amorphous films are more or less contaminated. In evaporated films oxygen is the dominating impurity [74L1, 74M1]. In sputtered material oxygen and argon seem to be present in the %-range [78P, 79M1]. Films prepared in high vacuum or glow discharge are O-contaminated after exposure to air [79B2] and even contain high amounts of oxygen and nitrogen after deposition [81D2, 84K1, 84T1, 83T]. However, surface oxidation in high quality films proceeds at a rate which is smaller by a factor of 2 than in crystalline Si [80F3]. The films prepared by glow discharge decomposition and CVD contain large amounts of hydrogen. At high substrate temperatures hydrogen is incorporated mainly in SiH configuration. At lower substrate temperatures polyhydrides (SiH_2, SiH_3-configurations) seem to be preferred (see IR spectra in section 11.1.3, Vibrational properties). The same trend exists in reactively sputtered Si [77R1]. By heating hydrogen can be evolved. The H-evolution rate shows two peaks, one around 320···350°C and a second one between 400 and 600°C [77T1, 79B2, 83B1]. Both peaks follow first order kinetics. A free energy of activation is deduced to 46 kcal/mol and 61 kcal/mol, respectively [79M2]. In samples prepared at high substrate temperature the low temperature peak is strongly suppressed. Hydrogen analysis is performed by IR spectroscopy [77B4, 79L3, 80F1], gas evolution [77T1, 79B2, 83B1], nuclear resonance reaction [77B1, 77C1, 79M4, 80M1]. The latter method indicates that the hydrogen is inhomogeneously distributed in the film [83M4, 80M1].

Only few data exist on diffusion of impurities in a-Si (Table 3). By SIMS-analysis in a-Si:H the diffusion coefficient of deuterium is determined to $D = 1.17 \cdot 10^{-2} \exp(-1.53 \text{ eV}/kT)$ [cm^2/s] [78C1], see also [80S1, 82B, 83R1].

See Figs. 1···20 (p. 24ff.).

Table 1. a-Si, a-Si:H. Correlation distances r_j (in Å), coordination numbers C_j and densities d obtained from RDF-studies. Sample A: evaporated (no deposition parameters given) [70M, 69M], samples B, C, D: glow discharge films deposited at temperature T_s [79G]. The densities are smaller than those reported on thick films [77B1]. For additional data, see [80M3].

	A	B ($T_s = 520$ K)	C ($T_s = 300$ K)	D ($T_s = 300$ K)	c-Si
r_1	2.35	2.375 (50)	2.375 (50)	2.375 (50)	2.35
C_1	4.0 (1)	3.55 (10)	3.45 (10)	3.25 (10)	4.0
r_2	3.86	3.69 (5)	3.69 (5)	3.69 (5)	3.86
C_2	11.6 (5)	11.8 (1)	11.4 (1)	12.4 (1)	12
r_3		4.56	4.55	4.29	4.48
C_3					12
r_4		4.98	4.99	4.97	5.42
C_4					6
$\dfrac{d(\text{a-Si})}{d(\text{c-Si})}$		0.7866	0.7863	0.707	1.0
c_H [at%]		20	25	35···40	

Table 2. a-Si. Crystallization temperatures T_{cr}.

Sample preparation	T_{cr} °C	Method of observation	Remarks	Ref.
evaporated, NaCl	800···850	electron microscopy		72H
glow discharge, quartz	620	discontinuity in $\sigma(T)$	undoped	78R
CVD	680	X-ray	alloyed 18 at% C	79J1
CVD	950	X-ray	alloyed 18 at% C	80B1
evaporated, Si/Ag	540	electron microscopy		72H
evaporated, Al	335	electron microscopy		
evaporated. Au	186	electron microscopy		
evaporated, Cu	410	electron microscopy	formation of β-CuSi at 175°C	
evaporated, Cr	>900	electron microscopy	formation of Cr_5Si_3 at 500°C	
evaporated, Mo	>800	electron microscopy		
evaporated, Pd	640	electron microscopy	formation of Pd_2Si at 220°C	
evaporated, Pt	590	electron microscopy	formation of Pt_2Si at 280°C	

For additional data, see [83M2].

Table 3. a-Si:H. Diffusion coefficients in glow discharge deposited material.

Diffusing species	Temperature of measurement T [K]	D [cm^2/s]	Ref.
Mo	723	$<10^{-18}$	78C2
Pd	453	$3 \cdot 10^{-15}$	
O	723	$6 \cdot 10^{-18}$	
Fe	673	$2 \cdot 10^{-15}$	
	623	10^{-17}	
	573	10^{-18}	
Al	673	$>3 \cdot 10^{-15}$	
B	673	$<3 \cdot 10^{-17}$	
P	723	$<10^{-17}$	
Cr	623	$<10^{-18}$	
Nb	623	$<10^{-18}$	
Ta	623	$<10^{-18}$	
K	673	10^{-18}	81R4
Rb	673	$<3 \cdot 10^{-20}$	
Cs	673	$<1 \cdot 10^{-20}$	

Similar data on diffusion of B and Sb [83M4].

11.1.2 Defect states, characterization

Important techniques used for film characterization and study of defect states have been IR spectroscopy (see section 11.1.3, Vibrational properties), photoluminescence (PL), photo yield spectroscopy, electron spin resonance (ESR), and various kinds of space charge spectroscopy (see also section 11.1.8.3, Density of localized gap states).

According to the IR spectra in hydrogenated films, hydrogen is bonded in different local environments depending on the conditions of preparation. In a-Si:H deposited under ideal conditions (e.g. in case of glow discharge deposition at low pressure, low power, pure silane and high substrate temperature) bonding in SiH-configuration seems to predominate.

In photoluminescence (PL) of all kinds of hydrogenated films the dominant transition is between states near the band edges giving a broad peak between 1.2 and 1.4 eV (see section 11.1.6, Photoluminescence). The quantum efficiency is more or less determined by dangling bond (db) type defects. At low defect concentration (below $\approx 10^{16}$ cm^{-3}) the efficiency has been estimated to be between 1 and 100% at low temperature [77N3, 80S9, 82W1, 82C2, 81S4, 81S5, 83J2]. The efficiency correlates with the density of db-defects. Introduction of defects by electron- or ion bombardment [75E, 79S6, 80V1], non-ideal preparation conditions [78S4], dehydrogenation [80B6, 80V1, 80O2] or doping [76R, 77N3, 79A2, 80F5, 80S7] leads to quenching of the PL signal and to appearance of an additional band near 0.9 eV with a width between 0.3 and 0.4 eV. These changes are correlated with changes in the spin resonance. For review on photoluminescence properties, see [81S5].

The electron spin resonance (ESR) signal in undoped amorphous Si consists of a single line at $g = 2.0055$ which is attributed to a localized neutral Si-dangling bond defect. The energy level associated with the neutral state of this defect lies below the center of the band gap, whereas the negative state (doubly occupied) lies above midgap [81D1, 81D3]. The spin density n_s in undoped material for which the Fermi level is near midgap therefore is a measure of the density of dangling bonds. The spin concentration n_s depends on the deposition parameters. It is near $10^{19} \cdots 10^{20}$ cm^{-3} in evaporated a-Si [69B, 70B2, 77S4, 78T3] and varies from 10^{20} to 10^{15} cm^{-3} (level or detectability) depending on the deposition conditions [76C1, 76H1, 77S4, 78F3, 78S4, 79P2, 80C7, 80K7] in hydrogenated a-Si due to the saturation of dangling bonds. The linewidth ΔH_{pp} can be described by the superposition of a temperature independent contribution $\Delta H_{pp}(0)$ and a temperature dependent part $\delta H_{pp}(T)$. $\delta H_{pp}(T)$ is related to the hopping conductivity $\sigma_h(T)$ [76V1, 77S4]. There exist arguments for clustering of spins [79K3]. Even at low n_s antiferromagnetic interaction is reported in all kinds of a-Si [75F1, 76B2] (see section 11.1.7). In high quality glow-discharge films n_s is reported to be around 10^{15} cm^{-3} in the bulk and between $10^{12} \cdots 10^{13}$ cm^{-2} at the surface of the film [77K1]. Hydrogenation leads to a reduction of n_s also in CVD-deposited material [80S1] and evaporated Si [78K3, 78M4]. Dehydrogenation [77T1, 79B2, 80B6, 80V1, 80Z2] as well as particle bombardment [79S6, 80V1] raise the spin resonance signal appreciably. In a-Si:H-films when doped either n- or p-type the dangling bond signal disappears due to the decreasing number of singly occupied dangling bonds and new lines appear: $g = 2.013$ of width around 20 G in p-type and $g = 2.004$ of width around 6 G in n-type films [77S4, 80S7, 81D1, 81H1]. These signals are attributed to electrons and holes localized in the band tails. Lines at the same g-values are observed in light-induced electron spin resonance (LESR) which reveals states which are paramagnetic before recombination [77B1, 77K1, 77K3, 77P1, 78S4, 80S5, 82S1, 83B2]. Optically detected magnetic resonance (ODMR) experiments where resonance signals are detected by monitoring the intensity or polarization of luminescence have been studied in [78B, 78M2, 80M6, 81M3, 82S2, 83Y]. These effects are closely related to the dependence of luminescence and photoconductivity on magnetic field (see below). Such studies indicate that the Si-dangling bond defects are the most important recombination centers for non-radiative recombination. In undoped films the dangling bond spin density is a reliable measure of the film quality and is directly related with most of the physical properties of the films (see figures).

Alloying of a-Si:H has been studied with B [79T4], As [77K2, 80N3], C [78E2, 81S1, 81S6, 81T1, 83N2], Ge [80C2, 80H2, 80P5, 81A1, 81S1, 83M3, 83N1], O [78P, 79H2, 80K2, 80S6, 83J1], N [82N1, 84C3, 83K1], and Sn [83J2, 84M1, 84R1]. In practically all cases a more or less pronounced increase in the defect density is reported which manifests itself in a degradation of the electrical properties, a decrease of luminescence efficiency and modification of the ESR spectra. A reduction in defect density is reported in fluorinated samples a-Si:F [78M3, 80M4]. Data in [80K6] indicate a higher defect density in a-Si:F:H and a-Si:Cl:H films.

See Figs. 21 ⋯ 41 (p. 29 ff.).

11.1.3 Vibrational properties

Infrared (IR) and Raman spectroscopy have provided valuable information about all of the vibrational modes, because of the disorder induced breakdown of momentum conservation rules. IR and Raman spectra are qualitatively very similar to each other and have been demonstrated to reflect the main features of a one phonon density of vibrational states. Differences arise because of variations in the matrix elements and of statistical factors [74L2, 75A1, 72C1, 77B6]. Upon hydrogenation the IR absorption decreases as a whole except for a hydrogen induced structure near $215\,cm^{-1}$. At higher energy hydrogen and fluorine induced local modes are observed, which can be attributed to stretching and bending modes of SiH and SiF bonds in different local structures (Table 4) [79L3, 80K3]. The IR absorption of B and P doped films is studied in [81S2]. Local modes of B-H and B-Si are observed. Remarkably, the intrinsic IR absorption of the Si network is enhanced by doping. The details of the spectra depend on the conditions of preparation and on the annealing state of the films. IR spectroscopy is therefore one of the most important techniques for sample characterization.

Only very few investigations exist on elastic properties. The sound velocities reported differ widely and depend appreciably on the microstructure of the films. In Brillouin scattering [78G1, 78S1] photoelastic coupling seems to be much enhanced by the presence of hydrogen (Table 5). The attenuation of acoustic surface waves exhibits a strong peak near 270 K in sputtered films [81B1]. Heat capacity has been studied at low temperature in [83L, 82L4].

See Figs. 42⋯51 (p. 35 ff.).

Table 4. a-Si, a-Si:H. Wavenumbers (in cm^{-1}) of vibrational modes.

Vibrational mode	Numerical value	Experimental method	Remarks	Ref.
$\bar{\nu}_{TA}$	150	Raman	sputtered film	79B1
	130	IR	sputtered film	80S8
$\bar{\nu}_{LA}$	310	Raman	sputtered film	79B1
	300	IR	sputtered film	80S8
	310	IR	sputtered film, a-Si:H	
$\bar{\nu}_{LO}$	380	Raman	sputtered film	79B1
	392	IR	sputtered film	80S8
$\bar{\nu}_{TO}$	480	Raman	sputtered film	79B1
	465	IR	sputtered film	80S8
	460	IR	sputtered film, a-Si:H	80S8
$\bar{\nu}$ (local modes)	215	IR	sputtered film, a-Si:H	80S8
	211	Raman	sputtered film, a-Si:H	79B1
			For similar data on IR investigations, see [74B2], and on Raman studies, see [71S2, 75A1, 72S1, 71S1, 77B5]	
$\bar{\nu}$ (local modes)	900⋯1100	IR, a-Si sputtered	O-induced	71T1
	850	IR	N-induced	71T1
			H-induced modes are assigned to local structural groups	79L3
	2000	IR, a-Si:H glow discharge	SiH stretch	
	2090		SiH stretch (clustered [83W1])	
	630		SiH bend (rock or wag)	
	2090		SiH_2 stretch	
	880		SiH_2 bend (scissors)	
	630		SiH_2 rock	
	2090⋯2100		$(SiH_2)_n$ stretch	
	890		$(SiH_2)_n$ bend (scissors)	
	845		$(SiH_2)_n$ wag	
	630		$(SiH_2)_n$ rock	

(continued)

Table 4 (continued)

Vibrational mode	Numerical value	Experimental method	Remarks	Ref.
$\bar{\nu}$ (local modes)	2140		SiH_3 stretch	79L3
	907		SiH_3 stretch (degenerate)	
	862		SiH_3 bend (symmetric)	
	630		SiH_3 rock	
	2030	Raman, a-Si:H sputtered	H-bond stretch	79B1
	2100			
	660		H-bond bend (rock or wag)	
	1010	IR, a-Si:F:H sputtered	SiF_4 stretch	80F2
	930		SiF_2 SiF_3 } stretch	
	828		SiF stretch	
	510		Si—TO mode, induced by F	
	380		SiF_4 bend	
	300		SiF SiF_2 } wag	

For additional data, see [80K5].

The frequencies depend weakly on the preparation parameters as do the relative contribution of the various modes to the spectra [76C1, 77T1, 77B5, 77B4, 78K2, 78F1, 79K3, 79T1, 80F2]. The assignment of the stretching modes ($2150\cdots2000$ cm^{-1}) has been disputed in [83W1, 78F1, 80P6, 80S2].

Table 5. a-Si:H. Elastic properties.

	Numerical value	Experimental method	Remarks	Ref.
sound velocity (in 10^5 cm s^{-1}):				
v (longitudinal, bulk)	7.4	Brillouin scattering	$T_s = 300$ K, $c_H \approx 35$ at%	78G1
v (surface)	3.59		(T_s; substrate temperature, c_H: hydrogen concentration)	
v (longitudinal, bulk)	8.4		$T_s = 525$ K, $c_H \approx 15$ at%	
v (surface)	4.28		different results [77T2, 72T]	
elastic moduli (in 10^{11} dyn cm^{-2}):				
c_{11}	9	Brillouin scattering	$T_s = 300$ K, $c_H \approx 35$ at%	78G1
c_{11}	15		$T_s = 525$ K, $c_H \approx 15$ at%	
c_{44}	4.7	Brillouin scattering	$T_s = 525$ K, $c_H = 15$ at%	78S1
c_{44}	2.4		$T_s = 300$ K, $c_H = 35$ at%	

11.1.4 Density of states

The density of states in the valence- and conduction band are investigated by UV and X-ray excited photoemission spectroscopy (UPS, XPS) and soft X-ray absorption. It is observed that due to the loss of k-conservation the spectra are relatively featureless. The total width of the valence bands remains unchanged as well as the valley between the region of predominantly s-derived states and the p-like peak at the top of the valence band. The most intensive part of the UPS spectra is related to the p-like states in the upper valence bands. The valence band edge appears steeper than in crystalline spectra. Reviews: [74S2, 74S3, 84L1]. In hydrogenated material the UPS-spectra reveal features which have been identified as states arising from hydrogen 1s/silicon 3p-3s bonding orbitals which sensitively depend on the bonding configuration of hydrogen [77R1, 79R4, 79R5]. In the XPS spectra no comparable structure is found due to the low cross-section of the H1s states for X-rays. The density of states of the conduction band has been studied by yield spectroscopy [72G, 83R3]. The spectra of hydrogenated samples show a recession of the top of the valence band by as much as 0.7 eV which indicates redistribution of states [79R5]. The incorporation of boron and phosphorus in a-Si:H is studied by photoemission spectroscopy leading to an incorporation efficiency of 80% for P and 70% for B [79R5]. It is observed that upon doping the shift in the activation energy of the conductivity E_σ follows the shift of the core levels, which points to a rather low density of surface states [79R4, 79R5, 79W]. Measurements on oxygen exposed samples gave an upper limit to the surface density of states of $4 \cdot 10^{13}$ cm^{-2} [81M1]. The photoemission spectra of a-Si:F films has been studied in [81G1, 81L2] and the influence of oxygen and surface states has been adressed in [81M1, 82K1, 81L3].

See Figs. 52···56 (p. 38 ff.).

11.1.5 Absorption edge and optical spectra

The optical spectra of a-Si differ from the crystalline spectra mainly in the three aspects: (1) The fine structure disappears in a-Si; (2) the ε_2-spectrum is shifted to lower energy; (3) the height of the ε_2-peak has decreased. The height as well as the position of the ε_2-peak depend on the method of preparation and the preparation parameters. These trends are explained by the loss of the k-selection rule and by changes in the matrix elements due to localization of electron states and a weakening of the bonds in a-Si [73B, 74B1, 79E].

The dielectric function has also been investigated by electron energy loss spectroscopy. The results deviate in part from those obtained from optical spectroscopy. The bulk plasma energy varies between 16.1 and 16.3 eV depending on the method of preparation [66Z, 81R1].

The absorption edge consists roughly of three parts: (1) At high absorption $K > 10^4$ cm^{-1} the absorption coefficient K can be described by $K \cdot \hbar\omega = b(\hbar\omega - E_g)^2$ defining an optical gap E_g the value of which varies from 1.1 to 1.8 eV depending on sample preparation and annealing history. (2) In the range $10^2 < K < 10^4$ cm^{-1} the absorption coefficient depends exponentially on photon energy as $K = K_0 \exp(\hbar\omega/E_1)$ with the parameter E_1 typically between 0.05 and 0.08 eV. (3) A weak absorption tail at low absorption constants $K < 10^2$ cm^{-1}, the shape and magnitude of which depends on sample history is assigned to defect absorption. In this range much experimental uncertainty exists. For this low energy range different techniques have been used: photoconductivity measurements assuming a constant $_{\mu\tau}$-product [72F1, 73L1, 80C2, 81S3, 81W3], photoacoustic spectroscopy [81Y1, 81Y2], photothermal deflection spectroscopy [81J, 82J]. Hydrogenated sputtered material has quite similar properties as glow discharge deposited a-Si:H. Hydrogen widens the gap [79F3, 79J1, 79P1, 79T1, 80B2, 80M3]. The absorption edge shifts to lower energy with rising temperature with a coefficient between $4···4.4 \cdot 10^{-4}$ eV/K [79F3, 79J3]. With pressure the gap decreases with a coefficient $-0.7···-2.0 \cdot 10^{-6}$ eV/bar depending on the pressure range [77W1]. However, also positive values have been reported in the low pressure range [72C1]. Impurities, in particular O, As, B, Li may have a profound influence on the absorption [76K1, 79F3, 80K2]. Electroabsorption reveals a simple structure near the absorption edge [80A1, 81N]. Photoinduced absorption is reported in [79A5, 80O3, 80V2, 81V2, 82K3, 82O, 82T1, 83V1] and has been used to study excess carrier relaxation and recombination.

See Figs. 57···82 (p. 44 ff.) and Figs. 95 (p. 51), 232 (p. 90).

11.1.6 Photoluminescence

Photoluminescence of high quantum efficiency is found in hydrogenated a-Si in particular in glow discharge deposited a-Si:H (for review, see [79F4, 81S5]). The form of the spectra and the time decay of the luminescence depend on the details of the preparation process [74E2, 78S4].

In glow discharge deposited a-Si:H prepared under optimum conditions (low pressure, low power, high substrate temperature, pure silane) the spectrum (at RT) consists of a single peak between 1.2 and 1.4 eV with a full width of half maximum of $0.25 \cdots 0.3$ eV [77E3, 78S4, 79A2] (cf. section 11.1.2). At low temperature the quantum yield is considered to be near unity in high quality films [81S5, 81S4, 83J2]. Above 50 K the photoluminescence drops with an activation energy of $0.04 \cdots 0.05$ eV below 120 K and $0.15 \cdots 0.23$ eV above 200 K [74E2, 77N3, 79A2, 79T3]. The photoluminescence intensity is proportional to the intensity of the excitation light at low excitation levels [76E1, 77E3, 79T3]. For high intensities the behavior is sublinear and a square root behavior is reported in [79R2]. Electric fields near 10^5 V/cm effectively quench the photoluminescence [76E1, 77N2, 79R2]. In the luminescence decay besides a fast component of $10 \cdots 20$ ns a long nonexponential dependence is found [78T2, 79A2, 79R2, 79S5, 79T3]. The decay time as well as the temperature dependence vary with the photon energy of the exciting light [79R2, 79S4]. The main features can be explained by radiative tunneling. It is believed that the excited electron-hole pairs do not diffuse to great separations but retain a high degree of correlation or that the carriers are trapped at random in band tail states. Corresponding to a spectrum of electron-hole separations a broad distribution of photoluminescence decay times exists $(10^{-8} < \tau < 10^{-2}$ s$)$. With increasing decay time the luminescence spectrum shifts by up to 0.2 eV to lower energy. A substantial shift to energies below the band gap has already occurred on a time scale below 10 ns [82W1, 82C3, 81H3]. Time resolved measurements with 250 ps-resolution are reported in [83W2, 83W3].

A pronounced similarity exists between photoluminescence and electroluminescence [76P1, 82L1, 82N2, 83R4]. Thickness dependence of the lifetimes points to a critical influence of surface quality [80R2]. Fatigue effects have been reported [80M5]. The photoluminescence is found to depend on magnetic field in resonant (ODMR-measurements [78M2, 81M3, 82S2]) and nonresonant [78B, 78S3] experiments. Only a weak influence of pressure is reported, the coefficient of the luminescence peak being $-2.0\,(5) \cdot 10^{-6}$ eV/bar at RT [81W1].

Hydrogenated sputtered material has quite similar luminescence properties as glow discharged deposited a-Si:H [77B7, 79A3, 79A4, 80C5, 80C6, 80N2, 80P3]. Photoluminescence in evaporated films is reported in [77E2]. The photoluminescence intensity is much reduced and the spectrum is shifted to lower energy by defect creation either by deposition at nonideal conditions or by particle bombardment. In less perfect films an additional defect induced band is found near 0.9 eV with a width between 0.3 and 0.4 eV. This defect luminescence correlates with the spin density of the singly occupied Si-dangling bonds (see section 11.1.2).

See Figs. 83 ··· 106 (p. 48 ff.).

11.1.7 Magnetic properties

The lattice diamagnetic susceptibility of a-Si is enhanced by a factor of between 2.5 and 4.6 as compared to the value of the crystalline material. It depends on the method of preparation: sputtered $\chi = -2.74 \cdot 10^{-7}$ cm^3/g, glow discharge $\chi = -4.82 \cdot 10^{-7}$ cm^3/g, reactively sputtered $\chi = -5.2 \cdot 10^{-7}$ cm^3/g [75E] (χ in CGS-emu). The influence of annealing and hydrogen content on the diamagnetic enhancement of a-Si:H has been studied in [78C3]. The origin of this effect is still unclear.

The paramagnetic susceptibility is attributed to dangling bond type defects and thus depends on the preparation and annealing history of the films like the ESR spectrum (see section 11.1.2). The spin susceptibility follows a Curie-Weiss relation with the Curie-Weiss temperature of 1.1 K [76P3]. Antiferromagnetic interaction is observed in all kinds of a-Si [75F1, 76B2, 76C1, 76D].

Spin-lattice relaxation of paramagnetic states [83B3] indicates the existence of low energy excitations of the disordered network (localized two-level systems). Measurements of the ^{1}H spin-lattice relaxation time point to the existence of trapped H_2 molecules in a-Si:H films [82C1, 82C4, 84T2].

See Figs. 33 ··· 40 (p. 33 ff.).

11.1.8 Transport properties

The amorphous films contain structural inhomogeneities (see sections 11.1.1, 11.1.2). It is not surprising, therefore, that the electronic properties depend strongly on the method of preparation and thus differ widely also between different authors. The results may be obscured by non-linear contacts, space charges due to adsorbates [78T1, 80T2, 82T2, 83A, 80F3, 84A] at the surface of the films or at the film-substrate interface [78S2]. Furthermore thickness dependent conductivity has been reported [80A2, 80A3, 82H2, 83M5] and metastable conductance states can be produced by illumination with light [77S2, 80S3], by rapid cooling and high electric fields [79A1]. For reviews on transport properties, see [74M1, 74S5, 77S1, 80F3, 84B2, 84F].

11.1.8.1 Electrical conductivity

In evaporated and sputtered material contamination of the films during deposition or by exposure to air [74L1, 74M1, 74B4, 75B1, 75B2, 76B1] leads to large differences in the reported results. One observes two ranges in the temperature dependence of the conductivity: At high temperatures $\sigma = \sigma_{01} \exp(-E_\sigma/kT)$, generally ascribed to conduction in extended states above the mobility edges; at lower temperatures due to the high density of defect states hopping conduction at the Fermi level, $\sigma = \sigma_{02} \exp(-T/T_0)^{1/4}$. The relative contribution depends on details of the preparation and annealing history. Ion bombardment has been shown to increase or decrease the hopping contribution depending on whether the number of defects is increased (He) or decreased by bond saturation [75B3, 78A1]. Hopping parameters have been extracted from experimental $\sigma(T)$ curves in [72B4, 73P3, 75K]. Tunneling through an oxide barrier into localized gap states leads to T_0-values which are consistent with planar conductivity measurements [70O, 72S4]. In the hopping range the conductivity depends on frequency and magnetic field.

Hydrogenation reduces the density of localized gap states by saturation of dangling bonds. Films prepared by reactive sputtering and by glow discharge decomposition behave quite similarly. They are, if undoped, highly resistive and show activated conduction. Only those films can effectively be doped from the gas phase by addition of controlled amounts of gases like PH_3 and B_2H_6 or by ion implantation. In case of the glow discharge technique widely different values are given for the incorporation ratio (=ratio of the concentrations in the film and in the gas phase): 0.5 [76S1]···5.2 [81S4] for phosphorus, 0.5 [76S1]···4.8 [80Z3] for boron. Only part of the incorporated species is electronically active. For arsenic doping only about 20% of the incorporated As-atoms are reported to be fourfold coordinated with Si-atoms [77K2]. The doping effect in a-Si:H is closely related to the low density of gap states. Quite a number of techniques indicate that doping leads to an increase of the density of defects: transport studies [77B8], DLTS [82L2, 83B5], photothermal deflection spectroscopy [82J], photoacoustic spectroscopy [82T3], photoconductivity [82W2] and ESR [81S4, 81D1]. In compensated films the increase of the defect density by doping seems to be much less pronounced [82S3, 81S4]. See also sections 11.1.2, 11.1.6.

Different interpretations have been given to explain details of the temperature dependences of conductivity and thermoelectric power particularly in doped samples [80S4, 77J, 77F, 79B3, 79B4, 79D2, 81B2, 83O1, 84B2]. The detailed investigations of n- and p-type films show that the mobility gap of a-Si:H is near 1.8 eV and the preexponential factor of the conductivity σ_0 amounts to about $2000\,\Omega^{-1}\,cm^{-1}$ independent of doping level and preparation conditions. σ_0 is supposed to be practically the same for electron- and hole-conduction. The apparent prefactor σ_{01} obtained from an extrapolation of the $\sigma(T)$-data in a $\log\sigma$ vs. $1/T$-plot to $1/T=0$ may vary appreciably due to the statistical shift of the Fermi level with temperature [83O1]. It is most remarkable, that for all n- and p-type films σ_{01} obeys over about 5 orders of magnetitude the Meyer-Neldel rule: $\sigma_{01} = C \exp(AE_\sigma)$ [79C]. The difference in the activation energies of conductivity and thermoelectric power, which is found quite generally, has been interpreted by long range spacial fluctuations of the mobility edges due to various kinds of inhomogeneities [83O1, 84B2].

See Figs. 107···134 (p. 55ff.) and Fig. 11 (p. 26).

11.1.8.2 Drift mobility

The drift mobility in glow discharge deposited a-Si:H films has been studied by time-of-flight techniques for electrons [72L3, 77A2, 77M3, 80T3, 81T3, 83S4, 84S] and holes [78A2, 77M3, 81S8, 81T3]. For results on sputtered films, see [81T4, 82K5]. Dispersive current transients are found in some cases [77A2, 78A2, 78F2] for electrons and generally for holes. Transient and steady state photoconductivity have been used to determine drift mobilities for glow discharge deposited films [78F2] and hydrogenated sputtered films [79M1]. The drift mobilities depend exponentially on $1/T$ with an activation energy of $0.12\cdots0.18$ eV and $0.35\cdots0.45$ eV for electrons and holes, respectively. Room-temperature values for glow discharge material are reported in the range $0.3\cdot10^{-2}\cdots0.8$ cm^2/V s for electrons and 10^{-4} cm^2/V s for holes. The behavior is attributed to transport in extended states and multiple trapping. At low temperature a change of the transport mechanism to hopping among tail states is proposed to occur in some cases [72L3]. In highly phosphorus doped films transport in the phosphorus donor band is observed [77A2]. The free carrier mobilities are estimated from such experiments to $1\cdots10$ cm^2/Vs for electrons and 0.67 cm^2/Vs for holes. For review, see [83S5].

Different results are obtained for compensated a-Si:H-films [84M3, 84K2]. It is found that with increasing degree of compensation the drift mobilities decrease and are field dependent.

See Figs. 144···152 (p. 65ff.).

11.1.8.3 Density of localized gap states

The main features of the $g(E)$-distribution are localized band tails which decay roughly exponentially from either band edge, $g(E) = N_0 \exp(-E/kT_c)$, and defect states near mid-gap, which predominantly originate from Si-dangling bonds. From time-of-flight transport (see section 11.1.8.2) kT_c is deduced to 0.027 eV for the conduction band tail and to 0.043 eV for the valence band tail [80T3, 81T3].

The density of localized states $g(E)$ is mainly studied by field effect [72S2, 74M2, 75N, 76M1, 79J4, 80G1, 80G2, 81P2, 82W3], tunneling [79B6], capacitance as a function of temperature and/or frequency [80B5, 80T5, 80V3, 83G1], deep-level-transient-spectroscopy [80C3, 80C4, 81T2, 82C5, 82L2, 83B5, 83O2, 84L2, 84T3], capacitance-voltage measurements [79H1], and space charge limited currents [84O]. Indirect information is obtained from photoconductivity [73L1, 77A3, 77R2, 78F2], photoluminescence [79F4, 81S5] or electron spin resonance [77K1, 81D1]. More recently the distribution of gap states is studied by photoelectric yield spectroscopy [83G2]. Field effect studies give in particular in hydrogenated a-Si the gross features of $g(E)$. However, disagreement exists as to the reliability of some of the experiments and to the significance of specific structure in $g(E)$. It is furthermore not clear to how far an extent this method probes surface states or the bulk states of a-Si. For a-Si:H $g(E)$ decreases considerably by hydrogenation and in glow discharge deposited films with increasing substrate temperature. $g(E)$ is found to depend on doping. Therefore one and the same $g(E)$-curve cannot be used for interpreting observations in undoped and doped films. In glow discharge material deposited under optimum conditions in the gap center $g(E)$ values between 10^{16} cm^{-3} eV^{-1} and 10^{17} cm^{-3} eV^{-1} have been obtained (see also [79S4, 80B5]). Is is reported that in fluorinated films (a-Si:F:H) the density of states is lower than in a-Si:H-films [80M4]. In evaporated a-Si the density of gap states attains much higher values $10^{19}\cdots10^{20}$ cm^{-3} eV^{-1} [75N, 74M2]. The $g(E)$ derived from DLTS studies differs appreciably from the field effect data. It is characterized by a deep minimum $0.3\cdots0.6$ eV below the conduction band. The energy scale is controversal [84T3, 84L2], different assumptions are made for the prefactor of the attempt-to-escape frequency i.e. the capture cross-section.

See Figs. 125···128 (p. 60).

11.1.8.4 Hall mobility and magnetoresistance

Hall effect and magnetoresistance are far less informative than they are in crystalline materials. The Hall effect in glow discharge deposited films is reported to exhibit n-p-anomaly: n-type samples have a positive Hall potential as expected for holes, the opposite applies to p-type samples [77B8, 77L1, 80D3]. Explanations have been given in terms of random phase theory [77E1] but so far many questions are still open. Magnetoresistance has been studied predominantly on evaporated films. A rather complex behavior is found with positive and negative contributions, the relative magnitude of which depends on temperature. Particularly the positive contribution depends on sample preparation [74M1]. This behavior is connected with hopping transport. It has been explained by the modification of the spin-flip relaxation time in the external magnetic field [78M1]. Studies on sputtered a-Si:H [82K4], on glow discharge deposited a-Si:H [80M7, 81W4].

See Figs. 131···133 (p. 61 f.).

11.1.8.5 Photoconductivity

The photoconductivity strongly depends on the preparation and annealing history of the films. As in case of the photoluminescence the photoconductivity is the higher, the lower the defect density is [80V1, 83D]. Hence only hydrogenated films have high photoconductivity. For reviews, see [76S2, 79L2, 80F3, 84F, 84C2].

The details of the recombination processes are not yet clear and different models have been proposed to explain special dependences on temperature and intensity [74S4, 76S2, 77A3, 77R2, 78F2, 80F5, 81W2]. Band bending effects have been shown to be important [80F3, 83J3, 83J4]. The influence of magnetic fields has been used to study the excess carrier recombination [77S3, 78M3, 82S4, 83D]. Si-dangling bonds have been shown to be the dominant recombination centers. There is a pronounced influence of doping on the photoconductivity [77A3, 77R2, 80F5, 81V3]. Transient photoconductivity is studied in [78F2, 81S5, 81H4]. From time-of-flight studies electron- and hole lifetimes and mobilities have been determined [83S4, 83S5, 83S6, 83S7] (see section 11.1.8.2). Si-dangling bonds are found to act as deep traps both for electrons and holes. It is reported that in undoped glow discharge deposited films the product of $\mu\tau$ and the spin density n_s is constant and amounts to $2.5 \cdot 10^8 \, \mathrm{cm^{-1} \, V^{-1}}$ for electrons and $4 \cdot 10^7 \, \mathrm{cm^{-1} \, V^{-1}}$ for holes [83S7]. Weak doping drastically reduces the $\mu\tau$-product of the minority carriers. The diffusion length in solar cell structures is studied in [77W2, 80M8].

The spectral dependence of the photoconductivity has been investigated in [73L1, 80C2, 80R3, 81W3]. In the subbandgap region this has been used to determine the absorption coefficient K (see section 11.1.5). With subbandgap light in undoped films quenching of the photoconductivity can be found [82P, 82V, 83F). Thermally stimulated currents point to structure in the density of localized gap states [80F4, 84J] as does the influence of impurities on $\sigma_{ph}(T)$ [80G3].

See Figs. 135···152 (p. 63 ff.) and Fig. 223 (p. 87).

11.2 Germanium (a-Ge)

Amorphous Ge can be prepared by evaporation in high vacuum, by ion bombardment, by sputtering in argon, by electrolytic deposition [51S, 79P2] or decomposition of germane in a glow discharge [69C]. The properties of the films depend very sensitively on details of the preparation process. They are very often dominated more or less by defects introduced during preparation and by contamination. Hydrogenation leads to partial saturation of defects although less perfect than in case of a-Si [76C1, 76L1, 79J5, 77M5, 82H3, 83S9]. Therefore glow discharge deposited films and hydrogenated sputtered films have similar properties as the corresponding a-Si:H films.

Reviews: [79M5, 79B7, 72A].

11.2.1 Structural characterization

The radial distribution function (RDF) does not depend in a systematic way on the method of preparation [72B5, 72S3, 73G1, 76K2, 69G, 73T]. It is in accordance with a random network of tetrahedrally bonded Ge atoms. The mean square relative displacement σ^2 of nearest neighbors is different in crystalline and amorphous Ge [79R3]. For structural data of sputtered films, see Table 1.

The density varies between 3.8 and 5.3 g/cm^3 dependent on the preparation parameters [69L, 72B1, 72R2, 70D2, 79V, 73P1]. Only very few investigations exist where deposition parameters have been varied in a systematic manner (see Table 2).

The films contain a large number of microvoids. For sputtered a-Ge and a-Ge:H a void density is reported in the order of 10^{11} voids/cm^2 with a size of 50···300 Å [80K8]. The total volume of voids estimated from small angle scattering varies widely with the method and details of preparation [69M, 72S3, 72C2, 74S1]. Different void shapes have been observed: crack-like voids [71D] or rod-like voids, which are oriented perpendicular to the substrate [73H1]. The voids are often found to be aligned parallel to the direction of the vapor stream [72B6, 73R]. The number of voids and their size depend on substrate temperature and the kind of substrate [74H1, 71D]. According to these experiences void-free i.e. nearly ideal a-Ge is obtained by evaporation under the following conditions: low deposition rate, high substrate temperature, low base pressure. It is not astonishing that under these circumstances the film properties strongly depend on the annealing history.

The crystallization temperatures T_{cr} quoted in the literature differ widely (Table 3). The values do not only depend on the preparation techniques but also on the method used for the measurement: differential thermal analysis [72R1, 74L3, 81F], electron micrographs [72H], discontinuities in the conductivity etc. [73G2, 76C2, 77C3, 76B3, 74J, 74H2] and laser beam annealing [80F7]. Values for T_{cr} are found between 200 °C and 400 °C. In oxide-free films surface crystallization starts at 380 °C, oxygen contamination leads to an increase of T_{cr} [72B6]. It is also observed that in a-Ge:H T_{cr} is raised by about 60 °C. In contact with metals T_{cr} is considerably reduced and amounts to roughly 0.65 of the eutectic melting temperature of the system [72H]. The crystallization process, nucleation and crystal growth, is reported to include an activation energy of 3···3.5 eV [76C2]. The heat of crystallization is determined from DTA-analysis to 2.6(4) kcal/mol [74L3, 81F].

Chemical analysis is published only in few cases. It is however clear, that most of the films contain larger amounts of oxygen (up to percent) (Table 4) [73P1, 74M1, 80N4]. This contamination is of particular importance for the transport properties [74M1]. Hydrogen analysis is performed by gas evolution, nuclear resonance reaction or IR spectroscopy [80F1].

A pressure induced semiconductor-metal transition is reported to occur near 60 kbar in a-Ge [74S9, 81M2].

See Figs. 153···160 (p. 68 f.) and Figs. 8 (p. 25), 167 (p. 71).

Table 1. a-Ge, c-Ge. Structural data for films sputtered at 425 and 625 K from RDF studies. r_i: neighbor distances (in Å), C_i: coordination number, σ_i: width of the RDF peaks (in Å), d_0: atomic density (in atoms/Å^3) [73T].

	a-Ge		c-Ge
	($T_s = 425$ K)	($T_s = 625$ K)	
r_1	2.47 (1)	2.47 (1)	2.45 (1)
C_1	3.79 (10)	3.91 (10)	3.95 (10)
σ_1	0.087 (8)	0.089 (8)	0.080 (8)
r_2	4.00 (4)	4.00 (4)	4.00 (1)
C_2	12.3 (3)	12.6 (3)	11.8 (3)
σ_2	0.29 (1)	0.27 (1)	0.12 (2)
r_3	4.71	(4.71)	4.69
C_3	(5.5)	(6.1)	11.5
d_0	$4.06 (4) \cdot 10^{-2}$	$4.28 (4) \cdot 10^{-2}$	$4.42 \cdot 10^{-2}$

Table 2. a-Ge. Variation of the density and room-temperature conductivity σ_{RT} for films sputtered at various substrate temperatures T_s. The density was determined by weighing unsupported films in air and toluene [73P1].

T_s [K]	d [g/cm^3]	σ_{RT} [Ω^{-1} cm^{-1}]
300	4.85 (5)	$7 \cdot 10^{-3}$
425	4.90 (5)	$6 \cdot 10^{-3}$
525	5.06 (3)	$(8 \cdots 16) \cdot 10^{-4}$
575	5.14 (3)	$2 \cdot 10^{-4}$
600	5.17 (3)	$5 \cdot 10^{-5}$
625	5.18 (3)	$5 \cdot 10^{-5}$

Table 3. a-Ge. Crystallization temperature T_{cr}.

Sample preparation	T_{cr} [°C]	Method of observation	Remarks	Ref.
evaporation onto CaF$_2$ (111) $150 < T_s < 200$ °C				
$\quad p = 1.33 \cdot 10^{-7}$ Pa, $\alpha = 5$ Å/min	150	X-ray diffraction		65S
$\quad 5.33 \cdot 10^{-5}$ Pa, $1.8 \cdot 10^4$ Å/min	250			
evaporation onto CaF$_2$				
$\quad p < 1.33 \cdot 10^{-6}$ Pa, $\alpha = 10$ Å/min	275	electron diffraction		66K1
evaporation onto Ge (111)				
$\quad p < 1.33 \cdot 10^{-6}$ Pa, $\alpha = 10$ Å/min	255	electron diffraction		66K1
sputtering onto				
$\quad$ CaF$_2$ (111)	330	electron diffraction		66K1
$\quad$ Ge (111)	330			
evaporation onto Ge -150°C $< T_s < 20$°C	$300 \cdots 350$	photoemission annealing	no influence of deposition rate	71R
evaporation near 20°C	240	differential thermal analysis	small piece of Ge	72R1
evaporation onto vitreous carbon $T_s = 120$°C	230	photoemission annealing		73L2
$\quad p < 2.67 \cdot 10^{-8}$ Pa, $\alpha = 1.2 \cdot 10^3$ Å/min				
evaporation onto SiO$_x$ $p = 1.33 \cdot 10^{-7}$ Pa, $T_s = 25$°C	380	electron microscopy	surface and volume nucleation, oxide-free	72B6
evaporation onto NaCl $p = 1.33 \cdot 10^{-4}$ Pa, $T_s = 25$°C	322	electron microscopy	Ag-covered	72H
	205	electron microscopy	Al-covered	
	125	electron microscopy	Au-covered	
	< 20	electron microscopy	Cu-covered	
	$300 \cdots 400$	electron microscopy		

Table 4. a-Ge. Impurity content of two sputtered a-Ge films in ppm prepared at substrate temperatures of 425 and 600 K from polycrystalline and hot-pressed targets, respectively [73P1]. Method: Ion source mass spectroscopy and electron microprobe analysis.

Impurity	$T_s = 425$ K	$T_s = 600$ K
H	40	130
Be	40	200
C	12	–
O	<300	3000
F	500	–
Na	50	400
Mg	14	140
Al	1100	2600
Si	50	80
Cl	–	100
P	–	–
K	6	18
Ar	$3.4\,(3) \cdot 10^4$	$1.8\,(2) \cdot 10^4$
Ti	–	7
Cr	4	90
Mn	–	20
Fe	7	80
Cu	40	200

11.2.2 Defect states, characterization

a-Ge films generally have a much lower density than c-Ge and contain a large number of microvoids (see section 11.2.1). Electron spin resonance measurements have been interpreted consistently in a model which identifies the majority of unpaired spins with dangling bond type defects on internal void surfaces [72B2, 73P1]. The ESR spectrum of evaporated and sputtered films consists of a single line, the g-value of which depends somewhat on preparation. In undoped films $g = 2.0190 \cdots 2.023$ has been reported. The spin density n_s depends on details of the preparation and amounts to $10^{17} \cdots 10^{20}$ cm^{-3} [69B, 73A, 76C1, 74A2]. n_s may decrease by order of magnitude upon annealing [73A]. Hydrogenation leads to saturation of defect states and thus to a reduction of n_s [74L4, 76C1, 77P1, 76P3, 83S9, 83S10, 83S11]. From the temperature and doping dependence of the ESR spectra it is concluded that the neutral singly occupied state of the dangling bond defects lies slightly above midgap [83S9]. The states of the doubly occupied negative dangling bonds are supposed to be higher by a correlation energy of 0.1 eV [83S10, 83S11]. In doped a-Ge:H-films lines are observed at $g = 2.0535$ (halfwidth 112 G) for boron doped samples and $g = 2.0120$ (halfwidth 33 G) for phosphorus doped films. These signals are attributed to valence band tail holes and conduction band tail electrons, respectively [83S10, 83S11].

The linewidth of the ESR signal can be described by a superposition of a temperature independent contribution $\Delta H_{pp}(0)$ and a temperature dependent part $\delta H_{pp}(T)$. The variation of $\delta H_{pp}(T)$ is correlated with the temperature dependence of the conductivity in the hopping regime [77P1, 76P3, 76C1]. Photoinduced changes in the ESR spectrum are reported in [77P1].

In undoped films the dangling bond spin density is a reliable measure for the film quality and is directly correlated with quantities like density, porosity, stress and others, which characterize the film structure and morphology and furthermore with most physical properties of the films. For review, see [77S4].

The bonding configurations of the hydrogen in a-Ge:H-films reveal in the IR absorption spectra and also in photoemission. Evolution of hydrogen starts above an annealing temperature of about 150°C. The changes in the IR spectra upon annealing are studied in [79B8, 80F1].

See Figs. 161 ⋯ 170 (p. 70 ff.).

11.2.3 Vibrational properties

Infrared (IR) and Raman spectroscopy give direct information about all of the vibrational modes because of the disorder induced break down of the *k*-selection rule (Table 5). Since the tetrahedral short range orders is preserved, the density of vibrational states is mainly a broadened version of the crystalline counterpart [72L4, 75A1]. The lowest energy peak in the phonon density of states of all kinds of film is broader and is located at a lower energy (80 cm^{-1}) than the TA phonon peak in the crystal (88 cm^{-1}). The infrared absorption spectra differ significantly between the different kinds of amorphous Ge. For hydrogenated material, a-Ge:H (Table 5), in IR and Raman spectroscopy local modes are observed which are attributed to Ge-H bonds in different local environments [76C1, 79B1, 79B8, 80S8, 80F1].

Thermal properties are studied in: [74K1, 82L4, 83L] heat capacity; [74N1] and [77L2] thermal conductivity.

For ultrasonic absorption on a-Ge:H with different hydrogen concentrations, see [81B3].

See Figs. 169, 171···182 (p. 72ff.).

Table 5. a-Ge, a-Ge:H. Wavenumbers (in cm^{-1}) of vibrational modes.

Vibrational mode	Numerical value	Experimental method	Remarks	Ref.
$\bar{\nu}_{TA}$	80	Raman		79B1
	75	IR		73S1
$\bar{\nu}_{LA}$	177	Raman		79B1
	165	IR		73S1
	170	IR		80S8
$\bar{\nu}_{LO}$	230	Raman		79B1
	240	IR		80S8
$\bar{\nu}_{TO}$	278	Raman		79B1
	280	IR		80S8
			H-induced modes in a-Ge:H assigned to	
$\bar{\nu}$ (local modes)	1890	Raman	Ge—H stretch	79B1
	1975	Raman	Ge—H stretch	
	1895	IR	Ge—H stretch	79B8
	1970	IR	Ge—H stretch	
	565	Raman	Ge—H wag	79B1
		IR		79B8
	755	IR	Ge—H bend	79B8
	820	IR	Ge—H bend	
	125	Raman	H-induced local mode	79B1
	116	IR		80S8
	720	IR	O-induced	73C

11.2.4 Density of states

Information about the electronic density of states is obtained from UPS [71R] and XPS [72L2, 74E3] studies. Like in case of amorphous silicon the sharp features of the crystalline spectra associated with van Hoove singularities are lost for all kinds of a-Ge. The uppermost p-like valence band density of states peak is narrower in a-Ge, the edge appears steeper and the maximum is slightly shifted towards the gap. Also in the conduction band the structure is lost. A pronounced influence of preparation conditions is reported on the valence band spectra [74O2]. In hydrogenated a-Ge:H different bonding configurations of hydrogen are identified in the photoemission spectra [80G4].

Reviews: [74S2, 74S3].

See Figs. 183···187 (p. 75ff.).

11.2.5 Absorption edge and optical spectra

The energetic position and shape of the absorption edge depend on the method of preparation and on the preparation parameters. By raising the substrate temperature or by annealing, the edge becomes steeper and shifts to higher energy [67C, 70T2, 71T2, 72C1, 74T]. With ion bombardment a shift to lower energy is attained [74O3]. Two ranges may be distinguished: Below about $3 \cdot 10^3 \, \text{cm}^{-1}$ the absorption co-efficient K often varies as $\exp(\hbar\omega/E_1)$ with $E_1 \approx 0.1 \, \text{eV}$ [72C1, 70C1, 73C]; at higher values of K, $\hbar\omega \cdot K = B(\hbar\omega - E_g)^2$ which defines an optical gap E_g. E_g typically varies between 0.6 and 1 eV. The value of 1 eV is ascribed to near ideal void free amorphous films. Such films are obtained by deposition in UHV at low rates and at high substrate temperatures. Films prepared in high vacuum at a low rate, high density films, exhibit sharp edges like in the crystal [70D2, 70D3, 69D]. However, there may be large uncertainties in the measurements of such low K-values [74T]. From photoconduction spectra in UHV deposited a-Ge a gap of 0.6 eV and an upper limit to the density of localized gap states of $10^{17} \, \text{cm}^{-3}$ are derived. After exposition to air the edge shifts by 0.4 eV to higher energy [73K1]. Like in case of a-Si hydrogenation leads to a pronounced increase of the optical gap [76C1, 84P2]. Photoinduced absorption of a-Ge:H is reported in [84P1, 82O].

In the reflectance spectrum the structure of the crystalline spectra is lost. The ε_2-spectrum of evaporated films has a single maximum. The peak height and energetic position depend on preparation and film history. The peak position varies between 2.6 and 3 eV, the peak height from 18 to 23 [74T]. In void free a-Ge, $\varepsilon_2(\omega)$ falls off more rapidly at high energies than in the crystal. Therefore the effective number of electrons contributing to the absorption is lower in these amorphous films. The influence of preparation parameters is studied in [71J, 72B7, 70D3, 73C, and 72B8]. The optical data are summarized in Table 6. For a review, see [74T].

See Figs. 188···199 (p. 77f.) and Fig. 232 (p. 90).

Table 6. a-Ge. Optical data of a-Ge films sputtered at $T_s = 300 \, \text{K}$ and $625 \, \text{K}$. E_1: steepness of edge in exponential range, $n(0.1)$: refraction index at long wavelengths, E_g: optical gap, E_2: maximum in $\varepsilon_2(\omega)$, E_g^P: Penn gap, Nn_{eff}: total electron density and n_{eff}: effective number of electrons per atom which are partizipating in optical transitions in the energy range up to 4.5 eV; also compiled are pressure and temperature coefficients of n and E_g [73C].

	$T_s = 300 \, \text{K}$	$T_s = 625 \, \text{K}$
E_1 [eV]	0.13	0.10
$n \, (0.1 \, \text{eV})$	4.33 (5)	4.13 (5)
E_g [eV]	0.70 (2)	0.90 (2)
E_2 [eV]	2.74 (5)	2.90 (5)
E_g^P [eV]	3.13	3.34
$Nn_{\text{eff}} \, (4.5)$ [electrons/cm^3]	$7.40 \cdot 10^{22}$	$7.84 \cdot 10^{22}$
$n_{\text{eff}} \, (4.5)$ [electrons/atom]	1.83 (10)	1.82 (10)
$1/n(\partial n/\partial p)_T$ [bar^{-1}]	$-0.8 \, (2) \cdot 10^{-6}$	
$(\partial E_g/\partial p)_T$ [eV/bar]	$3.5 \, (5) \cdot 10^{-6}$	
$(\partial E_g/\partial T)_p$ [eV/K]	$-4.5 \, (2) \cdot 10^{-4}$	
$(\partial E_g/\partial T)_v$ [eV/K]	$-4.0 \, (3) \cdot 10^{-4}$	

11.2.6 Photoluminescence

Photoluminescence has only been observed in hydrogenated glow discharge deposited a-Ge:H [80H2]. The spectrum consists of a single peak at 0.6 eV with a width of 0.25 eV.

11.2.7 Magnetic properties

The diamagnetic susceptibility of a-Ge is larger by about a factor of 2.5 than in crystalline Ge. The value depends on the preparation method: evaporated $\chi = -3.02 \cdot 10^{-7}$ cm^3/g, sputtered: $\chi = -2.29 \cdot 10^{-7}$ cm^3/g, glow discharge: $\chi = -2.22 \cdot 10^{-7}$ cm^3/g, sputtered in Ar/H$_2$: $\chi = -2.10 \cdot 10^{-7}$ cm^3/g [75F1] (χ in CGS-emu). This quantity does not depend on annealing up to the crystallization temperature [75F1, 76D, 74P1, 73H2].

The paramagnetic susceptibility depends considerably on annealing temperature, hydrogenation and other parameters. The paramagnetic susceptibility is reported to follow a Curie-Weiss law with a temperature $\Theta_p = 0.26$ K [76P3]. Antiferromagnetic spin interaction has been observed below 10 K [75F1, 76D]. In samples with the same spin concentration this interaction is found to be quite differently pronounced [75F1].

See Figs. 161 ⋯ 165 (p. 69 ff.).

11.2.8 Transport properties

The electrical properties drastically depend on the method of preparation as well as on the annealing history. Representative results are given to emphasize the main features.

11.2.8.1 Electrical conductivity

A general feature is that in undoped evaporated and sputtered films at high temperatures the conductivity is activated with an energy E_σ between 0.35 and 0.55 eV and a preexponential factor of $\sigma_0 = 10^3 \cdots 10^4\,\Omega^{-1}$ cm^{-1} (see e.g. [74M1, 75P, 76C3, 75B2, 74S8, 74B3]). In UHV deposited films E_σ is found to be 0.25⋯0.32 eV and to increase upon exposure to air [73K1, 74D, 74K2, 74E4]. Thermoelectric power S has been found either positive or negative depending on details of the preparation. Films deposited in high vacuum (UHV) generally are p-type. Oxygen contamination is reported to lead to n-type conduction. Different attempts have been made to decompose $\sigma(T)$ and $S(T)$ curves into contributions from electrons, holes and hopping conduction at the Fermi level. Generally a difference in the activation energies of S and σ is observed near 0.2 eV [76L2, 82H3]. Undoped glow discharge deposited films are n-type and exhibit activated behavior down to lower temperatures [79J5, 76J]. Such films can effectively be doped by adding small amounts of PH$_3$ and B$_2$H$_6$ to the discharge gas [79J5, 82H3]. The influence of ion bombardment is studied in [73O and 74B3].

At low temperatures in particular in unhydrogenated films straight lines are often obtained in a plot of $\ln\sigma$ vs. $T^{-1/4}$ indicating variable range hopping near the Fermi level (Fig. 204). From the slope of such curves the density of states near the Fermi level $g(E_F)$ is obtained as $8 \cdot 10^{17}$ to $5 \cdot 10^{18}$ cm^{-3} eV^{-1} assuming a decay length of the wave functions $\alpha^{-1} = 10$ Å. Unreasonably large values, however, are deduced from the preexponential factor (e.g. [78G2, 73P3]). The dependence of the hopping conductivity on film thickness enabled the independent calculation of the decay length α^{-1} and $g(E_F)$ by applying 2-dimensional hopping theory. This leads to $\alpha^{-1} = 10$ Å and $g(E_F) = 10^{18}$ cm^{-3} eV^{-1} [74K2].

Frequency- and field dependence of the conductivity is observed in the hopping region, the results differing widely from one author to the other [74M3, 70C2, 74A3, 71T3, 76T, 80L1, 80L2]. Often $\sigma \propto f^n$ is obtained with an exponent n near 0.8. The density of states near the Fermi level deduced from these data is much higher than that one obtained from dc-measurements. As a function of hydrostatic pressure the conductivity is found to decrease [74K3].

See Figs. 200 ⋯ 220 (p. 80 ff.).

11.2.8.2 Density of localized gap states and drift mobility

From field effect, $g(E_F)$ of evaporated films is estimated to be in the range 10^{19} to 10^{21} cm^{-3} eV^{-1} [75M]. In films deposited by glow discharge technique, $g(E_F)$ is reduced down to 10^{18} cm^{-3} eV^{-1} [76J, 79J5]. A reduction of the density of gap states is also achieved by sputtering of Ge-films in a H-atmosphere [76C1, 76L2]. Only these hydrogenated films can effectively be doped substitutionally [79J5, 76P3, 82H3]. For doping studies on evaporated or sputtered films, see [76A2, 74A4, 79S8, 82H3].

Steady state and transient photoconductivity is investigated in hydrogenated sputtered films as a function of temperature. The deduced electron drift mobility is activated with 0.1 eV above 200 K and only weakly temperature dependent below. This behavior is interpreted by small polaron hopping [77M5].

See Fig. 217 (p. 85).

11.2.8.3 Hall mobility and magnetoresistance

The Hall effect exhibits sign anomaly, i.e. the sign is negative when the thermoelectric power is positive and vice versa [74M1, 74S8, 73L3]. Like in case of a-Si, magnetoresistance can be of either sign depending on the temperature and magnetic field. It is independent of the relative orientation of current and field [74M1, 74C3]. The same type of behavior is observed both for the dc- and ac-conductivity [74M3].

See Figs. 218···220 (p. 86).

11.2.8.4 Photoconductivity

There exist only few investigations of photoconductivity due to the small value of the photoresponse [76V2, 74D, 75F2]. Bolometric effects may be dominant in some of the experimental results. In evaporated a-Ge the $\mu\tau$-product is near 10^{-10} cm^2/V [76V2]. Hydrogenation leads to an increase of the $\mu\tau$-product, in undoped glow discharge films values up to 10^{-7} cm^2/V have been obtained. This effect is thus much less pronounced than in case of a-Si:H. Studies on sputtered a-Ge:H [77M5, 76M3, 83R2] and on glow discharge deposited films [79J5]. Doping with phosphorus is reported to considerably enhance the photo-response [79J5]. Different recombination models are proposed [75F2, 77M5, 79J5].

The spectral dependence of the photoconductivity is used to measure the absorption coefficient [73K1]. In evaporated material the photoconduction spectrum is reported to have an edge near 0.6 eV and to shift to higher energies by annealing and hydrogenation [77M3, 74H3, 76H2].

See Figs. 221···225 (p. 87f.).

11.3 III–V compounds

The data on III–V compounds are far less complete than in case of a-Ge and a-Si. Various methods are used for film preparation: thermal or e-gun evaporation, flash evaporation, sputtering in inert gas or hydrogen atmosphere, in case of GaAs glow discharge decomposition of gas mixtures of AsH$_3$ and Ga(CH$_3$)$_3$. The details of the preparation method strongly determine the film properties. Only very few systematic investigations exist, however, about the influence of preparation parameters and annealing which in case of a-Ge and a-Si have given valuable information about the type of disorder present in the material. That is why a comparison of results from different laboratories is rather difficult. In the following, therefore, only a brief survey will be given to show the major trends.

11.3.1 Structural characterization

Most of the information on structure is obtained from X-ray and electron diffraction [73T, 73S2, 73S3], see Table 1. The resultant radial distribution functions (RDF) are very similar to those of a-Ge and a-Si. The major difference is a slight shift of the 3rd RDF-peak to higher values of r [75C]. The structure is considered to be best described by a random network model like in case of a-Si and a-Ge. In addition, wrong bonds are expected to exist [74C2, 72C1] but their existence has not yet been proved conclusively. EXAFS investigations indicate different environments of Ga and As atoms in sputtered a-GaAs which was interpreted as evidence of wrong bonds [78D]. However, in flash evaporated material normal GaAs bonds are found [80T4, 81G2, 82T4]. Also photoemission studies [74S3, 74S7, 74P2], Raman spectroscopy [72W, 74L5] and infrared absorption spectra [73P2] did not give convincing evidence of wrong bonds. Structural and optical studies lead to the conclusion that wrong bonds might exist in flash-evaporated a-GaSb [81G2] and a-InP [83G3].

Hydrogenation by sputtering from GaAs targets in H$_2$/Ar atmospheres [77P2, 81A2, 80H3] and by glow discharge decomposition [82S5] from gas mixtures of AsH$_3$ and Ga(CH$_3$)$_3$ leads to a reduction of the defect density. However, bond saturation is much less effective than in case of a-Si:H and a-Ge:H [80H3, 80P4]. Hydrogen evolution in a-GaAs:H is taking place between 150 and 200 °C [82W4].

The density of amorphous films has been determined by weighting [73S2, 72S5] and from X-ray diffraction [73S2]. Like in case of a-Ge and a-Si the density is by up to 10% lower than in the crystalline state. Crystallization of a-GaAs is reported to set in near 280 °C [83K1, 82W4].

Table 1. III–V compounds. Structural data [73S2], first and second neighbor distances r_1 and r_2 (in Å), width of RDF peaks σ_1 and σ_2 (in Å), coordination numbers C_1 and C_2, tetrahedral bond angle θ and spread $\Delta\theta$, density deficit $\Delta d/d$ determined from X-ray data. The numbers in brackets refer to the crystalline values. The films were sputtered at a substrate temperature of $T_s = 253$ K.

	GaAs	GaP	GaSb	InSb
r_1	2.48 (3)	2.44 (10)	2.67 (3)	2.86 (3)
	[2.45]	[2.36]	[2.65]	[2.81]
σ_1	0.085 (10)	0.18 (1)	0.14 (1)	0.14 (2)
C_1	3.93 (10)	3.47 (20)	3.5 (2)	3.82 (30)
r_2	4.10 (5)	3.90 (5)	4.30 (5)	4.5 (5)
σ_2	0.30 (5)	0.35 (5)	0.40 (5)	0.45 (5)
C_2	12.8 (20)	12 (2)	13.2 (20)	12.8 (20)
θ	109 (2)°	107 (2)°	109 (2)°	106 (2)°
$\Delta\theta$	10°	12°	14°	15°
$\Delta d/d$	5%	5%	2%	2%

11.3.2 Vibrational properties

The IR and Raman spectra show great similarity with the spectra of a-Si and a-Ge and may be related similarly to a broadened crystalline one phonon density of states. In contrast to a-Ge a much weaker coupling to low frequency modes is observed in the IR spectra [71S1, 73P2, 73S1, 82W4]. Raman effect is studied in several III–V compounds [72S1, 72W, 74L5, 82W4] and in hydrogenated GaAs [80P4] where H-associated modes have been identified. From IR studies it is concluded that in sputtered a-GaAs:H hydrogen is incorporated mainly in bridging configuration between two Ga atoms [82W4]. The question of existence of wrong bonds has been adressed by Raman spectroscopy [72W, 74L5]. No As—As bonds could be detected in GaAs, but approximately 2···4% of Sb—Sb bonds were supposed to exist in a-GaSb. Other III–V compounds gave similar results.

See Figs. 226···229 (p. 88 f.).

Table 2. III–V compounds. Local modes (wavenumbers in cm^{-1}) from IR spectroscopy.

Local mode	Numerical value	Assignment	Ref.
a-GaAs:H (sputtered at 293 K):			
$\bar{v}$	2130	As—H$_2$ stretch	82W4;
	2040	As—H stretch	see also 80P4
	1870	Ga—H$_2$ stretch	
	1760	Ga—H stretch	
	1460	Ga—H—Ga stretch	
	990	As—H$_2$ bend	
	700	Ga—H$_2$ bend	
	620	As—H wag	
	565	Ga—H wag	
	530	Ga—H—Ga wag	
a-GaP:H (sputtered at 293 K):			
$\bar{v}$	2320	P—H$_2$ stretch	82W4
	2220	P—H stretch	
	1930	Ga—H$_2$ stretch	
	1800	Ga—H stretch	
	1480	Ga—H—Ga stretch	
	1010	P—H$_2$ bend	
	700	Ga—H$_2$ bend	
	630	P—H wag	
	570	Ga—H wag	
	540	Ga—H—Ga wag	(continued)

Table 2 (continued)

Local mode	Numerical value	Assignment	Ref.
a-GaSb:H (sputtered at 293 K):			
$\bar{\nu}$	1940	Sb—H_2 stretch	82W4
	1870	Sb—H stretch	
	1790	Ga—H_2 stretch	
	1690	Ga—H stretch	
	1400	Ga—H—Ga stretch	
	960	Sb—H_2 bend	
	700	Ga—H_2 bend	
	600	Sb—H wag	
	555	Ga—H wag	
	510	Ga—H—Ga wag	

In a-GaAs:H with c_H between 6.8 and 12.3 at% the most intensive modes are the Ga—H—Ga stretching and wagging modes [82W4].

11.3.3 Density of states

Photoelectron spectroscopy (XPS) has been investigated on all III–V compounds [74S3, 74S7, 74L5, 83K2] which had been prepared by sputtering. In contrast to the situation in a-Ge there is no shift in the top portions of the density of states of the valence band and the three-peak structure is retained. The density of valence band states can be generated merely by broadening that one of the corresponding crystals. The core level spectra observed in the amorphous films are surprisingly sharp and show no evidence of coulombic fluctuations which one should expect from a random network with wrong bonds [74S3]. Optical absorption experiments involving the highest lying d-levels show that the two-peak structure in the crystalline density of states in the conduction band is lost in the amorphous films [72G].

See Figs. 230, 231 (p. 89).

11.3.4 Optical properties

Like in case of a-Ge and a-Si the details of the optical spectra like absorption edges, height and position of the ε_2-peak depend on the preparation parameters and on the annealing state (Table 3). There are no systematic investigations to date where the films are sufficiently characterized.

The ε_2-spectra of all the amorphous III–V compounds consist of a single broad maximum which is considerably shifted to lower energy with respect to the crystalline spectra [72S5, 80G5, 81G3]. This shift is most pronounced for GaP. In the spectrum of the real part of the dielectric constant $\varepsilon_1(\omega)$, structure is completely lost. In all cases the low frequency value $\varepsilon_1(0)$ and thus the refractive index n is higher in the amorphous film than in the crystal [72S5, 80G5]. By annealing, the position of the ε_2-maximum is shifted to higher energy and $\varepsilon_1(0)$ decreases.

The absorption edges differ widely in form and position and strongly depend on the method of preparation [71E, 72C1, 72C3, 74N2, 79M6, 78K4, 80G5]. It is remarkable that in unannealed films relatively sharp edges are found too. A general trend is that the absorption edge shifts to higher energy by annealing [80G5]. The annealing process seems to be complete at 250 °C for a-GaAs and at 300 °C for a-GaP [80G5]. Hydrogenation leads to a pronounced shift of the edge to higher energy in GaAs [77P2] and GaP [79M6].

The influence of non-stoichiometry on the optical properties is discussed in [81G2, 81G3].

See Figs. 232···236 (p. 90 f.).

Table 3. a-GaP, a-GaAs. Survey of optical data.

	GaP	GaAs	Remarks	Ref.
optical gap E_g [eV]	0.42[1]	0.61[1]	sputtered films (various T_s, fully annealed)	72C1, 72C3
n (0.1 eV)	3.8 (2)	3.6 (2)		
$\dfrac{1}{n}\cdot\dfrac{dn}{dp}\left[\dfrac{10^{-3}}{\text{kbar}}\right]$	-0.25	-0.7		
$\dfrac{dE_g}{dp}\left[\dfrac{\text{eV}}{\text{kbar}}\right]$	$2\cdot10^{-4}$	$7\cdot10^{-4}$		
$\left(\dfrac{\partial E}{\partial T}\right)_p\left[\dfrac{\text{eV}}{\text{K}}\right]$	$-3.1\,(2)\cdot10^{-4}$	$-3.8\,(2)\cdot10^{-4}$		
E_g [eV]	1.55	1.15	flash evaporated ($T_s = 100$ K, annealed	81G2, 80G5
n (0.5 eV)	3.3	3.6	to steady state at	
ε_2-peak position [eV]	4.1	3.65	525 K)	
E_g [eV]		0.65[2] before anneal 0.95[2] after anneal at 535 K	flash evaporated ($T_s = 300$ K)	74N2
		1.45[1]	sputtered hydrogenated ($T_s = 277$ K)	77P2
	1.17[2]		sputtered ($T_s = 300$ K) non annealed	79M6
	1.8[2]		hydrogenated	
	1.8[2]		plasma-deposited ($T_s = 625$ K)	78K4

[1]) E_g defined by $\hbar\omega$ at $K = 10^4\,\text{cm}^{-1}$.
[2]) E_g defined from plot of $\hbar\omega\sqrt{\varepsilon_2}$ vs. $(E_g - \hbar\omega)$.

11.3.5 Transport properties

In principle the transport behavior is similar to a-Ge and a-Si. There is activated conduction at high temperature with an activation energy which often is about half the optical gap. At lower temperature variable range hopping near the Fermi level predominates where $\sigma = \sigma_0\exp(-T/T_0)^{1/4}$. The details strongly depend on the preparation parameters and on annealing. In contrast to the behavior of a-Si and a-Ge hopping at the Fermi level is much less pronounced and can be eliminated by annealing at relatively low temperature. It is suggested that the partial ionic character of the bonding in III–V compounds is the reason for a low density of dangling bond type defect states. Photoconductivity is relatively poor in all cases where it has been measured. The influence of hydrogenation has been studied in case of GaAs [77P2, 80H3] and GaP [79M6].

Further references: InSb [76B4, 73H3, 71E, 77H]. GaAs [77H, 76B4, 72B9, 80H3, 77P2, 74C1, 74B7]. GaSb [79N, 77D]. GaP [74Y, 79S7, 79M6, 79M7, 78K4].

See Figs. 233, 237 (p. 90f.).

Figures for 11

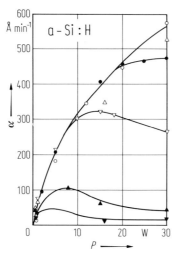

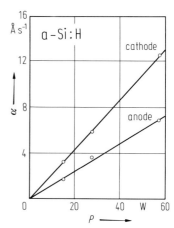

Fig. 1. a-Si:H. Deposition rate α of a-Si:H in a glow discharge at a substrate temperatur $T_s = 503$ K as a function of rf-power P. Parameter is the relative content of SiH_4 in SiH_4/Ar mixture: $\triangle$ 100%, $\circ$ 10%, $\bullet$ 5%, $\triangledown$ 3%, $\blacktriangle$ 1%, $\blacktriangledown$ 0.1% [79K2]. Somewhat different results: see [79P1].

Fig. 2. a-Si:H. Deposition rate α in a glow discharge on cathode and anode as a function of rf-power P for an electrode separation of $d = 2.5$ cm and a temperature of $T_s = 543$ K. The cathode is defined by a negative potential with respect to the plasma [79T1].

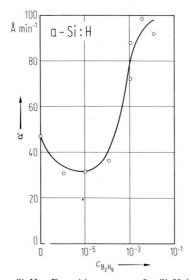

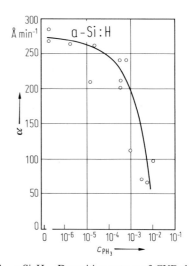

Fig. 3. a-Si:H. Deposition rate α of a-Si:H in a glow discharge at a substrate temperature $T_s = 503$ K vs. doping level defined by the concentration of B_2H_6 in SiH_4 with rf-power 1 W and pure silane. The pronounced influence of even small amounts of B_2H_6 underlines the complicated plasma chemistry of the process [80K1]. Additional data [81S4].

Fig. 4. a-Si:H. Deposition rate α of CVD-deposition of a-Si ($T_s = 900$ K) as a function of the gaseous concentration of the doping gas (phosphine to silane ratio c_{PH_3}) [79T2].

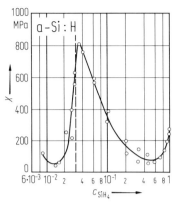

Fig. 5. a-Si:H. Compressive stress X in a-Si:H prepared by glow discharge decomposition at $T_s = 523$ K from SiH$_4$ diluted in argon as a function of the SiH$_4$ concentration c_{SiH_4}. Below $c_{SiH_4} \approx 10^{-2}$ columnar film growth is observed [84H1].

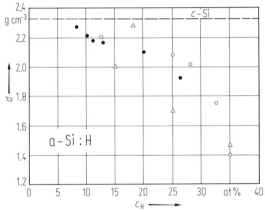

Fig. 7. a-Si:H. Density d of undoped glow discharge deposited films vs. H-concentration c_H. Data from [79F2, 76K1, 77B1]. Widely different deposition conditions.

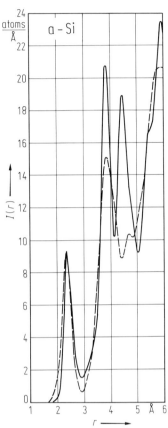

Fig. 6. a-Si. Radial distribution function $I(r)$ of an evaporated film of 100 Å thickness, as deposited (---) and after crystallization (———) [70M]. No preparation parameters given.

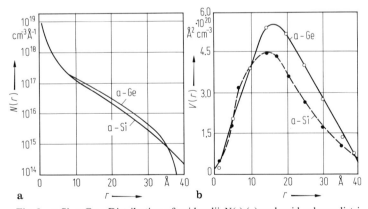

Fig. 8. a-Si, a-Ge. Distribution of void radii $N(r)$ (a) and void volume distribution $V(r) = \frac{4}{3} r^3 N(r)$ (b) for evaporated Si and Ge (no deposition parameters given) as deduced from small angle X-ray scattering. The density is by 10% lower than in the crystal [74S1]. Qualitatively similar data were reported for a-Si:H films prepared by glow discharge decomposition [79D1].

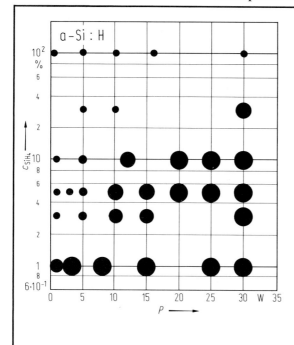

◀

Fig. 9. a-Si:H. Influence of preparation conditions on columnar structure in a-Si:H films deposited by glow discharge at $T_s = 503$ K. Increasing size of the dots indicates increasing visibility of the columnar structure in electron microscopy. Shown is the concentration ratio of SiH_4 in Ar, c_{SiH_4} vs. rf-power P [79K1].

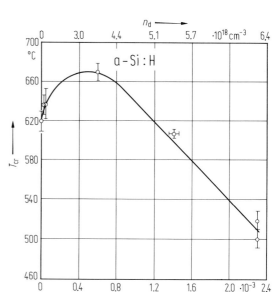

Fig. 10. a-Si:H. Crystallization temperature of glow discharge deposited a-Si:H ($T_s = 520$ K) as a function of the doping level defined by the concentration of the dopant gas PH_3 in silane. The density of donors is given in the top scale [78R].

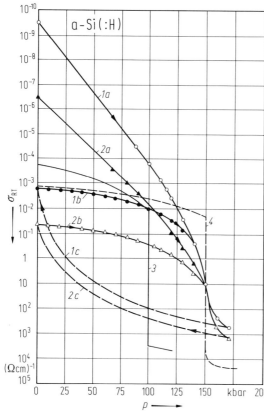

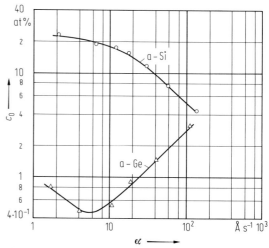

Fig. 11. a-Si. Pressure dependence of the room-temperature conductivity for different Si films sputtered at 570 K. Curve 1: a-$Si_{80}H_{20}$, 2: a-$Si_{90}H_{10}$, 3: evaporated a-Si, 4: crystalline Si [80M3]. After release hysteresis behavior occurs. See also [74S9, 77W1] and for elastoresistance studies, see [72F2, 70D1].

Fig. 12. a-Si, a-Ge. Oxygen content c_O in evaporated Ge and Si films as a function of deposition rate α determined by α-back scattering. Si: base pressure $7 \cdot 10^{-6}$ Torr, $T_s = 500$ K; Ge: $7 \cdot 10^{-6}$ Torr, 300 K, only relative values [74M1]. 1 Torr $\cong$ 133.3 Pa.

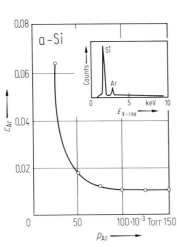

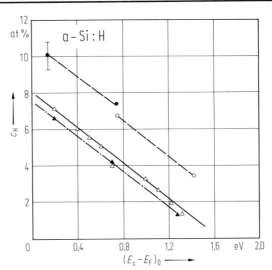

Fig. 13. a-Si. Ar-content c_{Ar} in rf-sputtered a-Si as a function of Ar pressure p_{Ar}. Argon content is estimated from the ratio of Ar to Si+Ar peak heights in e-beam excited X-ray fluorescence spectrum (see insert). Preparation parameters: substrate temperature 683 K, rf-power per target area 3 W/cm², cathode-anode distance 3.2 cm [79M1]. 1 Torr $\cong$ 133.3 Pa.

Fig. 15. a-Si:H. Hydrogen concentration c_H of glow discharge deposited a-Si:H films ($T_s = 520$ K) as a function of doping. $(E_c - E_F)_0$ denotes the activation energy of the conductivity. The data were obtained by profiling different types of junctions using nuclear resonance technique: ● in⁺i, ○ ip⁺i, △ graded pn, ▲ p⁺in⁺ [80M1].

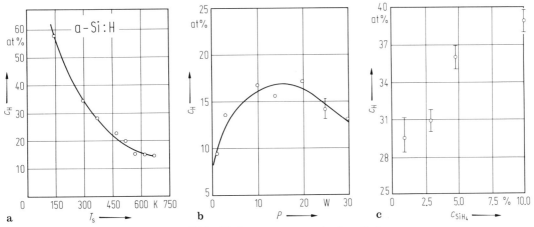

Fig. 14. a-Si:H. Hydrogen content c_H in glow discharge deposited films determined by IR spectroscopy as a function of (a) substrate temperature T_s, (b) rf-power P at $T_s = 503$ K and 5% SiH₄ in Ar, and (c) silane concentration c_{SiH_4} in Ar at $T_s = 298$ K and $P = 25$ W [79K2].

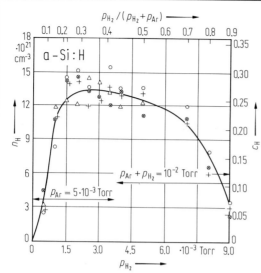

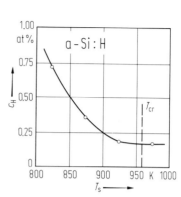

Fig. 16. a-Si:H. Hydrogen content n_H and c_H ($=n_H/n_{Si}$) in sputtered samples of a-Si:H (no T_s given) as a function of H$_2$ partial pressure p_{H_2}. The concentrations were determined in four different ways: IR spectroscopy stretching mode ($\otimes$) and wagging mode ($+$), gas evolution ($\circ$) and nuclear reaction ($\triangle$) [80F1]. 1 Torr $\cong$ 133.3 Pa. ($n_{Si} = 5 \cdot 10^{22}$ cm^{-3}).

Fig. 17. a-Si:H. Hydrogen content c_H in CVD deposited a-Si:H as a function of the substrate temperature T_s determined by SIMS. Near T_{cr} the films crystallize [80B1].

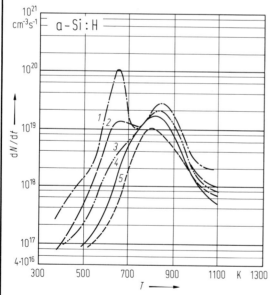

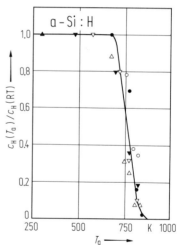

Fig. 18. a-Si:H. Hydrogen evolution rate dN/dt of a-Si:H films deposited by glow discharge at various temperatures T_s: curve 1: 300 K, 2: 373 K, 3: 473 K, 4: 573 K, 5: 673 K. Heating rate 20 K/min. Similar results are obtained using SiD$_4$ and Si$_2$H$_6$ [83B1]. See also [79B2].

Fig. 19. a-Si:H. Hydrogen content (relative to room-temperature value) in glow discharge deposited a-Si:H ($T_s = 520$ K) as a function of the annealing temperature T_a (annealing time: 15 min). The samples contained different amounts of hydrogen at 300 K [79J2].

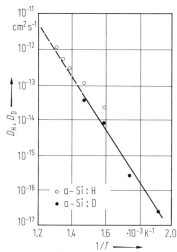

Fig. 20. a-Si:H. Diffusion coefficient D, of H and D in glow discharge deposited films ($T_s = 520$ K) vs. reciprocal temperature. H [80Z2], D [78C1].

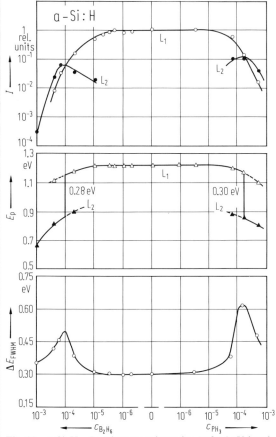

Fig. 23. a-Si:H. Luminescence intensity I of 1.3 eV band (L_1) and 0.9 eV band (L_2), position of the luminescence peak E_p and width ΔE_{FWHM} (full width half maximum) as a function of the doping level in glow discharge deposited films ($T_s = 520$ K). Excitation energy $E_{ex} = 1.92$ eV. On the abscissa the concentrations c_{PH_3} and $c_{B_2H_6}$ of the dopant gases PH_3 and B_2H_6 in SiH_4 are given [80F5]. Additional data also on compensated films, see [81S4] and Fig. 98.

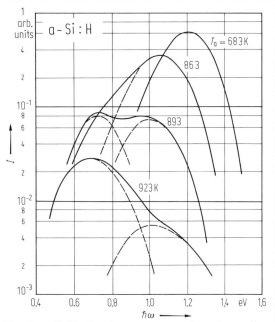

Fig. 21. a-Si:H. Luminescence spectra (intensity vs. photon energy) at 77 K of glow discharge deposited a-Si:H ($T_s = 670$ K) after annealing for 20 min at the indicated temperatures T_a [80V1]. (The appearance of a second band is clearly demonstrated.) Excitation by the 1 W line of a krypton laser, $E_{ex} = 1.92$ eV.

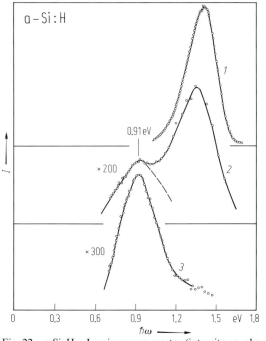

Fig. 22. a-Si:H. Luminescence spectra (intensity vs. photon energy) taken at 8 K of glow discharge deposited a-Si:H films ($T_s = 500$ K). Excitation energy $E_{ex} = 2.335$ eV. Curve 1: undoped, 2: P doped with c_{PH_3} in $SiH_4 = 3 \cdot 10^{-3}$ and 3: B doped with $c_{B_2H_6}$ in $SiH_4 = 10^{-3}$ [80S7]. Additional data [79A2, 80F5, 81S4, 84B1].

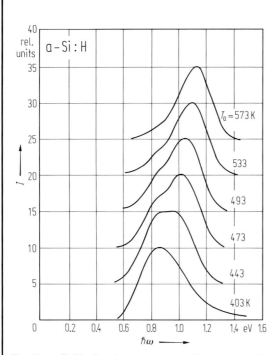

Fig. 24. a-Si:H. Luminescence spectra (intensity vs. photon energy) taken at 77 K of glow discharge deposited films ($T_s = 250\,°C$) after bombardment with Ar^+ ions. The films were annealed at the indicated temperatures T_a [75E]. Excitation energy $E_{ex} = 1.92$ eV. Similar data on electron bombarded films [79S6].

Fig. 25. a-Si:H. Electron spin density n_s and luminescence intensity I of the 1.3 eV band in a-Si:H deposited by glow discharge ($T = 500$ K) from different SiH_4/inert gas mixtures (in all cases 5%) [81K1]. The arrows indicate the trend of changes after exposure to air.

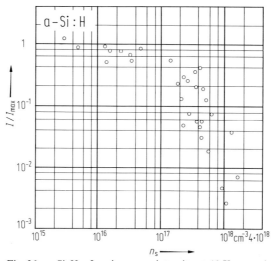

Fig. 26. a-Si:H. Luminescence intensity at 10 K normalized to I_{max} of the 1.3 eV band vs. spin density n_s at 300 K for glow discharge films deposited under a variety of conditions [78S4]. Excitation energy $E_{ex} = 2.41$ eV.

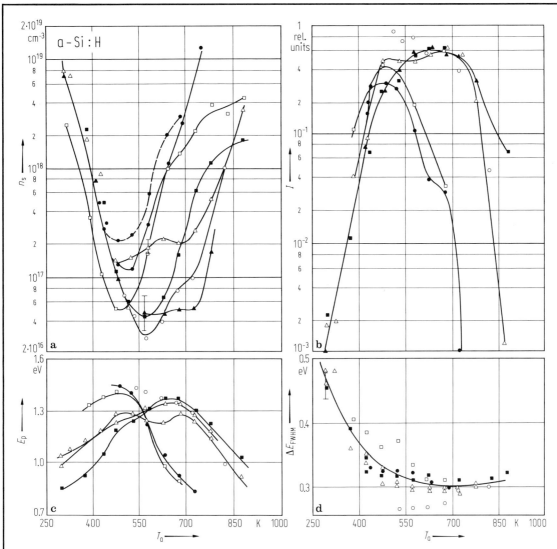

Fig. 27. a-Si:H. Spin density n_s (a), luminescence intensity I at E_p (b), position of the luminescence maximum E_p (c) and halfwidth ΔE_{FWHM} of several glow discharge a-Si:H-films as a function of annealing temperature T_a [80B6]. 10 minutes anneals at the given temperatures. Widely different deposition conditions: ▲, △ SiH_4, $P = 1\,W$, $T_s = 300\,K$; o SiH_4, $1\,W$, $500\,K$; ● $c_{SiH_4} = 5 \cdot 10^{-2}$ in Ar, $20\,W$, $425\,K$; □ $c_{SiH_4} = 5 \cdot 10^{-2}$, $25\,W$, $300\,K$; ■ $c_{SiH_4} = 5 \cdot 10^{-2}$, $25\,W$, $500\,K$. Excitation energy $E_{ex} = 2.335\,eV$. See [80O2] for sputtered material.

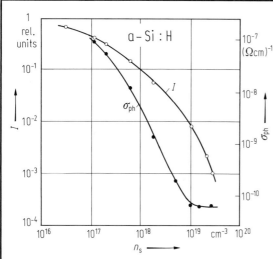

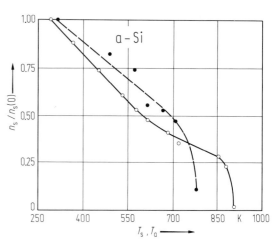

Fig. 28. a-Si:H. Dependence of luminescence intensity I (peak at 1.3 eV) and photoconductivity σ_{ph} at 200 K of a thick (100 μm) glow discharge deposited sample ($T_s = 520$ K) on spin density n_s. An initial spin density of $3 \cdot 10^{19}$ cm^{-3} was produced by electron bombardment. Annealing was performed in steps to 400 °C [80V1]. σ_{ph} was measured with IR glass filter between sample and tungsten-iodine light; excitation energy for luminescence $E_{ex} = 1.92$ eV.

Fig. 29. a-Si. Normalized variation of spin density n_s in evaporated a-Si (UHV conditions) as a function of deposition temperature T_s (●) and annealing temperature T_a (annealing time: 2 h) (○) [78T3]. RT evaporation and annealing at T_a leads to the same spin density as deposition at the same temperature. Different results [70B2].

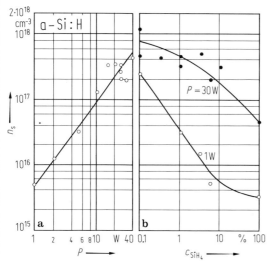

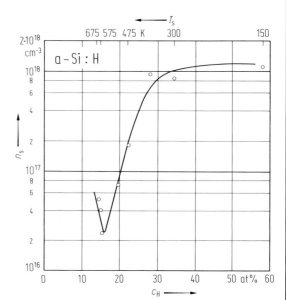

Fig. 30. a-Si:H. Dependence of room-temperature ESR spin density n_s of glow discharge deposited films ($T_s = 500$ K) on (a) rf-power at $c_{SiH_4} = 5\%$ in Ar and (b) on concentration of silane in Ar, c_{SiH_4}, at 1 W and 30 W [78S4].

Fig. 31. a-Si:H. Electron spin density n_s as a function of hydrogen content c_H for films deposited by glow discharge at different substrate temperatures at the following conditions: rf-power 25 W, 5% SiH$_4$ in Ar [79K3].

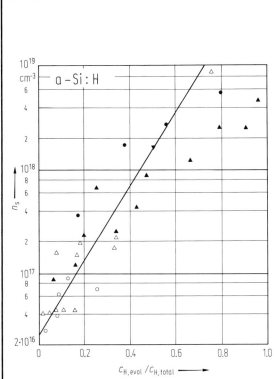

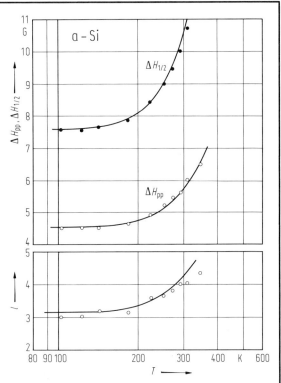

Fig. 32. a-Si:H. Spin density n_s of glow discharge deposited a-Si:H as a function of the ratio of evolved and the total hydrogen concentration. Preparation conditions: $\triangle$ pure SiH_4, $T_s = 300$ K; $\circ$ pure SiH_4, $T_s = 500$ K; $\blacktriangle$ $c_{SiH_4} = 5\%$ in Ar, $T_s = 300$ K; $\bullet$ $c_{SiH_4} = 5\%$ in Ar, $T_s = 420$ K [79B2]. See also [78F3].

Fig. 34. a-Si. ESR peak-to-peak linewidth ΔH_{pp}, half-width $\Delta H_{1/2}$ and lineshape factor l of evaporated a-Si ($T_s = 360$ K) as a function of temperature. Points experimental [76V1]. Solid curves are calculated using a Gaussian distribution of relaxation times [77M4].

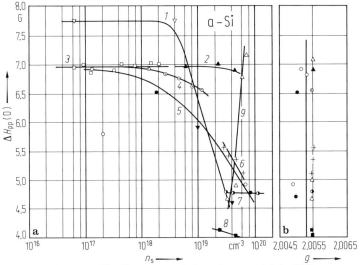

Fig. 33. a-Si. Temperature independent part of the ESR linewidth ΔH_{pp} (0) vs. spin concentration n_s (a) and g-values (b) for a number of different specimens [77S4]. _1_ and _5_ sputtered films, _2_ glow discharge deposited at low T_s, _3_ and _4_ electron bombarded glow discharge films, _6_ ion bombarded glow discharge films, _7_ films evaporated, _8_ sample amorphized by ion bombardment [76M2], _9_ films evaporated in UHV at different T_s [77T3].

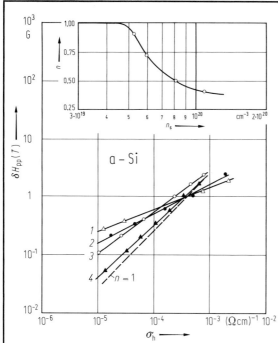

Fig. 35. a-Si. Temperature dependent part of the ESR linewidth $\delta H_{pp}(T) = \Delta H_{pp}(T) - \Delta H_{pp}(0)$ of evaporated a-Si ($T_s = 360$ K) as a function of the hopping conductivity σ_h. Curves 1, 2, 3 and 4 were obtained after annealing at 330, 450, 510 and 530 K, respectively. The insert shows the exponent n of the relation $\delta H_{pp}(T) = C\,\sigma_h(T)^n$ as a function of the spin concentration n_s [77S4, 76V1]. Similar results on sputtered films [77P1].

Fig. 38. a-Si:H. Spin lattice relaxation time of dangling bonds in undoped glow discharge deposited a-Si:H ($T_s = 520$ K) as a function of temperature after 30 min anneals: ○ as deposited near 520 K, △ 773 K, ▲ 820 K, ▽ 845 K, ● 880 K. The hydrogen evolution peaks at 620 K, all temperatures are below crystallization temperature [83B3].

For Fig. 36, see next page.

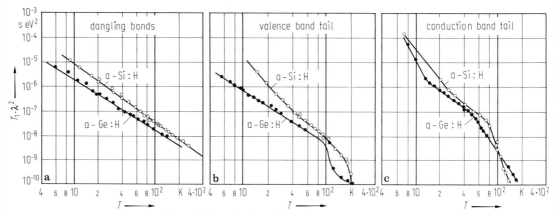

Fig. 37. a-Si:H, a-Ge:H. Spin lattice relaxation time T_1, of paramagnetic centers in glow discharge deposited a-Si:H and a-Ge:H as a function of temperature for dangling bonds (a), band tail holes (b) and band tail electrons [83B3]. No preparation parameters specified. In the figure T_1 is scaled by the spin orbit coupling factor ($\lambda_{Ge}/\lambda_{Si} = 7.3$). See also [83S11] for data on a-Ge:H.

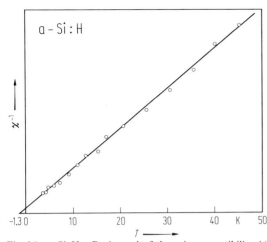

Fig. 36. a-Si:H. Reciprocal of the spin susceptibility $1/\chi$ vs. temperature for glow discharge deposited a-Si:H. Θ_p defined by the Curie-Weiss-Law $\chi = C/(T + \Theta_p)$ is derived to 1.3 (4) K independent of n_s [76B2]. Similar data in [75F1] and on sputtered material [76C1, 76D].

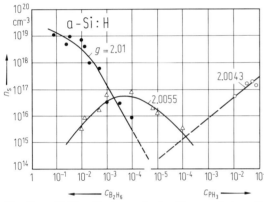

Fig. 40. a-Si:H. Electron spin densities n_s of glow discharge deposited films ($T_s = 520$ K) as a function of the doping level for three different lines of given g-value. On the abscissa the concentration c_{PH_3} and $c_{B_2H_6}$ of the dopant gas in silane are given [81D1]. Similar data on CVD-deposited a-Si:H: see [81H1, 78T4, 80F6].

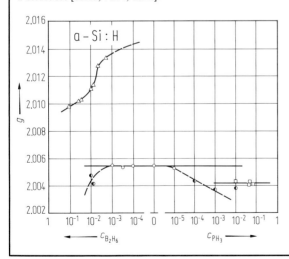

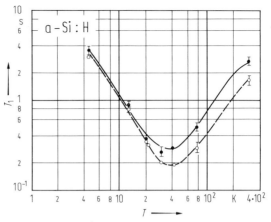

Fig. 39. a-Si:H. ^{1}H-spin lattice relaxation time, T_1, as a function of temperature in powdered (●) and thin (○) samples of a-Si:H prepared by glow discharge at low defect deposition conditions [84H2]. Similar data [84B3, 81R3].

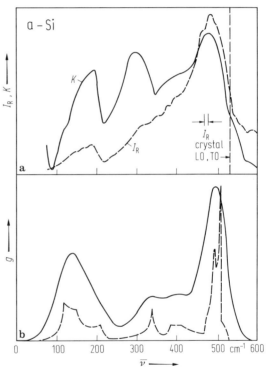

Fig. 42. a-Si. (a) IR spectrum (absorption coefficient vs. wavenumber, solid line) and reduced Raman spectrum (reduced Raman intensity vs. wavenumber, dashed line) of sputtered a-Si at 300 K [75A1]. (b) Density of states of crystalline Si (dashed line). The solid line is a broadened density of states [71S2].

◄

Fig. 41. a-Si:H. g-values of ESR signal of glow discharge deposited samples ($T_s = 520$ K) as a function of the doping level. c_{PH_3} and $c_{B_2H_6}$ denote the concentration of PH$_3$ and B$_2$H$_6$ in SiH$_4$. In moderately doped films two lines are present (○) [81D1]. For additional data, see [81H1].

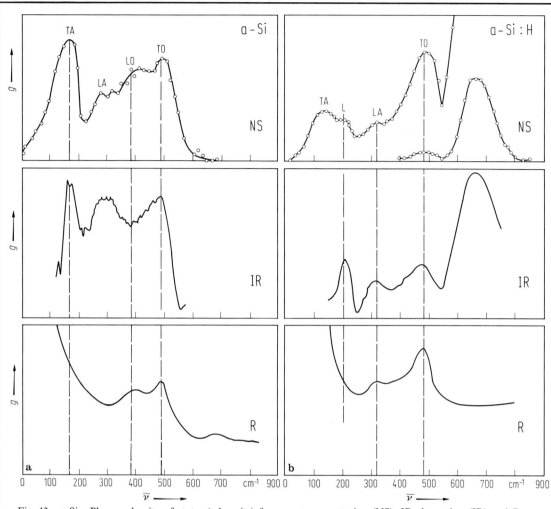

Fig. 43. a-Si. Phonon density of states (rel. units) from neutron scattering (NS), IR-absorption (IR) and Raman scattering (R) of two sputtered amorphous Si-samples; $T_s = 300$ K: (a) pure Ar, (b) $c_{H_2} = 0.66$ in Ar; H content in film 12 at% [83S3].

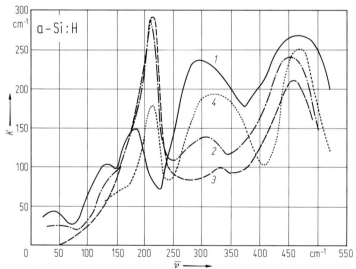

Fig. 44. a-Si:H. Vibrational absorption spectra (absorption coefficient vs. wavenumber) of a-Si:H deposited by rf-sputtering [80S8]. H concentrations: Curve 1: $c_H = 0$, 2: $c_H = 19$ at%, 3: $c_H = 24$ at%, 4: $c_H = 15$ at%.

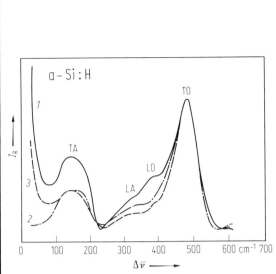

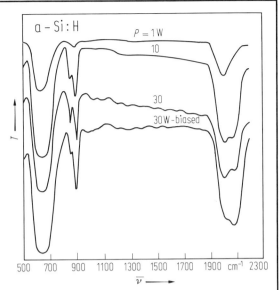

Fig. 45. a-Si:H. Stokes Raman spectra (Raman intensity vs. Raman shift, ∥ configuration) of sputtered a-Si, curve *1*, and a-Si:H film sputtered in Ar—H_2 mixture containing 1%, curve *2*, and 10% H_2, curve *3*; laser energy $\hbar\omega_L =$ 2.41 eV. Assignment of the structures is given in the figure [79B1].

Fig. 46. a-Si:H. Infrared transmission vs. wavenumber of glow discharge deposited a-Si:H films. Preparation conditions: $T_s = 500$ K, $p = 0.15$ Torr, $c_{SiH_4} = 0.05$ in Ar, flow rate 130 sccm, rf-power indicated. The lowest curve was taken on a sample deposited on a negatively biased substrate holder [79K3]. Note that the 1 W-curve indicates predominant bonding of hydrogen in SiH bonds. (sccm: standard cubic centimeter per minute, 1 Torr ≙ 133.3 Pa.)

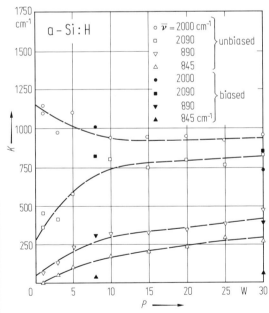

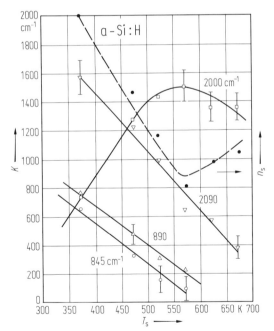

Fig. 47. a-Si:H. Absorption coefficient K of various Si—H modes vs. rf-power P of glow discharge deposited a-Si:H films [79K3]. Preparation conditions, see Fig. 46.

Fig. 48. a-Si:H. Absorption coefficient K of various Si—H modes vs. substrate temperature T_s. The films were deposited at the same conditions as the samples in Fig. 46 and with rf-power $P = 25$ W [79K3]. Included is the variation of the electron spin density (●, log scale). Parameter gives the wavenumber $\bar{\nu}$ of the Si—H modes.

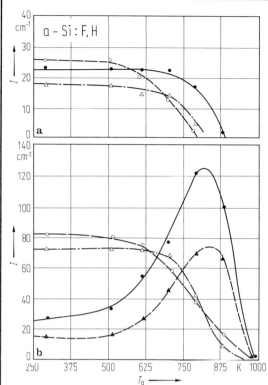

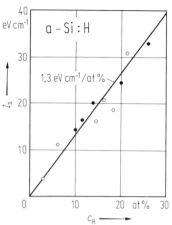

Fig. 50. a-Si:H. Area under the stretching mode absorption $I_s = \int K(\omega)d(\hbar\omega)$ for various glow discharge deposited films as a function of their total hydrogen content c_H, determined by H effusion [79T1] (open circles) and nuclear resonance reaction [80C1] (full circles).

Fig. 49. a-Si:F, H. Intensity $I = \int \dfrac{K(\omega)}{\omega} d\omega$ of infrared modes as a function of annealing temperature (annealing time: 20 min), (a) Si—H modes: 2090 cm^{-1} (—·—, $\triangle$), 1985 cm^{-1} (—, $\circ$), 630 cm^{-1} (—, $\bullet$) (b) Si—F modes: 1010 cm^{-1} (—, $\bullet$), 830 cm^{-1} (—·—, $\triangle$), 380 cm^{-1} (——, $\blacktriangle$) 510 cm^{-1} (——, $\circ$). The films were deposited by reactive sputtering in a gas mixture containing argon, hydrogen and SiF$_4$ onto substrates held at about 65 °C [80F2].

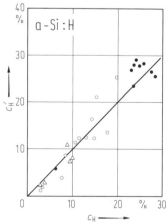

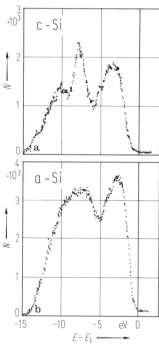

Fig. 51. a-Si:H. Hydrogen content $c'_H = I_w \cdot A_w / 5 \cdot 10^{22}$ cm^{-3} of reactively sputtered films determined from the integrated strength $I_w = \int \dfrac{K(\omega)}{\omega} d\omega$ of the IR absorption band near 630 cm^{-1} as a function of the hydrogen content $c_H (= n_H / 5 \cdot 10^{22}$ cm$^{-3})$ as obtained from nuclear reaction techniques. The proportionality factor is $A_w = 1.6 \cdot 10^{19}$ cm^{-2}. $\circ$ a-Si:H [80S2], $\triangle$ a-Si:F:H [80F2], $\bullet$ data from [80F1].

Fig. 52. a-Si. Corrected XPS valence band spectra (number of counts vs. binding energy) of crystalline (a) and evaporated amorphous Si ($T_s = 300$ K) (b) using monochromatized AlK$_\alpha$ X-rays. The binding energy is referred to the Fermi level E_F of a thin Au layer [72L2]. Additional data [68D, 79P2, 79R5].

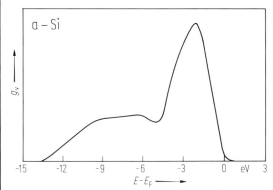

Fig. 53. a-Si. Valence band density of states of evaporated amorphous silicon ($T_s = 300$ K) calculated from the spectrum in Fig. 52 after corrections for variations in photoelectric cross-sections [84L1].

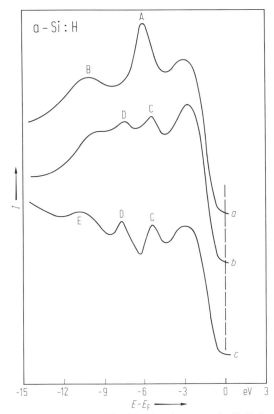

Fig. 54. a-Si:H. He II valence band spectra of a-Si:H (intensity vs. binding energy). Glow discharge sample deposited at 525 K: curve a: as deposited, b: after anneal at 625 K. c: sample prepared by reactive sputtering at $T_s = 625$ K [79R5]. See also [77R1, 81M1, 82K1, 84L1]. A, B, C, D and E denote H-induced structures.

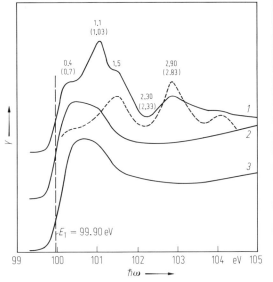

Fig. 55. a-Si/a-Si:H. The L_{III}(Si2p$_{3/2}$→conduction band) yield spectra (yield vs. photon energy) of c-Si (curve 1) and amorphous Si sputtered in pure Ar (2) and in an atmosphere containing hydrogen $c_{H_2} = 0.5$ in Ar (3) at $T_s = 300$ K [83R3]. In the spectrum of c-Si characteristic energies are given together with the corresponding numbers in paranthesis of [72B3]. The broken line is the density of conduction states calculated for c-Si [66K2]. E_T: threshold energy for L_{III} absorption. (Values correspond to $\hbar\omega - E_T$.)

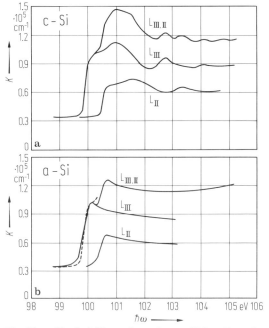

Fig. 56. a-Si. Soft X-ray absorption coefficient K vs. photon energy for crystalline (a) and evaporated amorphous Si (b). The spectra are resolved in L_{III} and L_{II} components separated by 0.6 eV. The dashed line in (b) is the crystalline spectrum near the threshold [72B3]. See also [79S1].

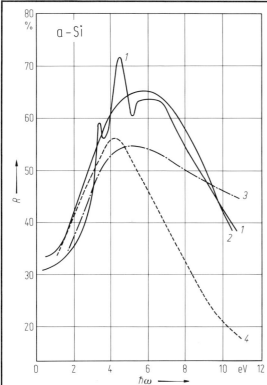

Fig. 57. a-Si. Reflectance of differently prepared films vs. photon energy [79E]. Curve 1: c-Si, 2: glow discharge a-Si deposited at 673 K, 3: slowly evaporated film [72P1], 4: fast-evaporated a-Si. For data on sputtered hydrogenated material, see [80P2].

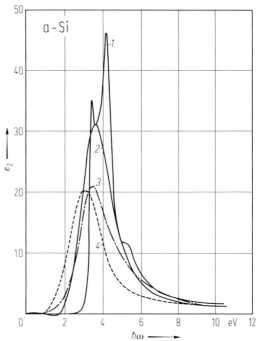

Fig. 58. a-Si. Imaginary part of the dielectric constant vs. photon energy for the samples in Fig. 57 [79E]. Similar data including influence of defects in evaporated a-Si [73J]. For Fig. 60, see next page.

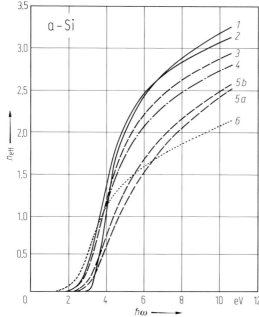

Fig. 61. a-Si. Number of electrons n_{eff} per atom contributing to the optical absorption below $\hbar\omega$ vs. $\hbar\omega$ [79E]. Curve 1: c-Si; curves 2···6: glow discharge deposited a-Si:H at different substrate temperatures; 2: 673 K; 3: 533 K; 4: 413 K; 5a: 300 K; 5b: 300 K annealed at 623 K; 6: evaporated a-Si [70B1]. See also [80W].

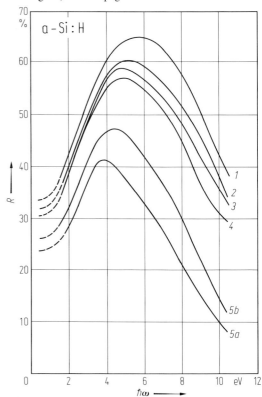

Fig. 59. a-Si:H. Reflectance of a-Si:H films deposited in a rf-glow discharge at various substrate temperatures vs. photon energy [79E]. Curve 1: 673 K; 2: 533 K; 3: 413 K; 4: 338 K; 5a: 300 K; 5b: 300 K after annealing at 623 K.

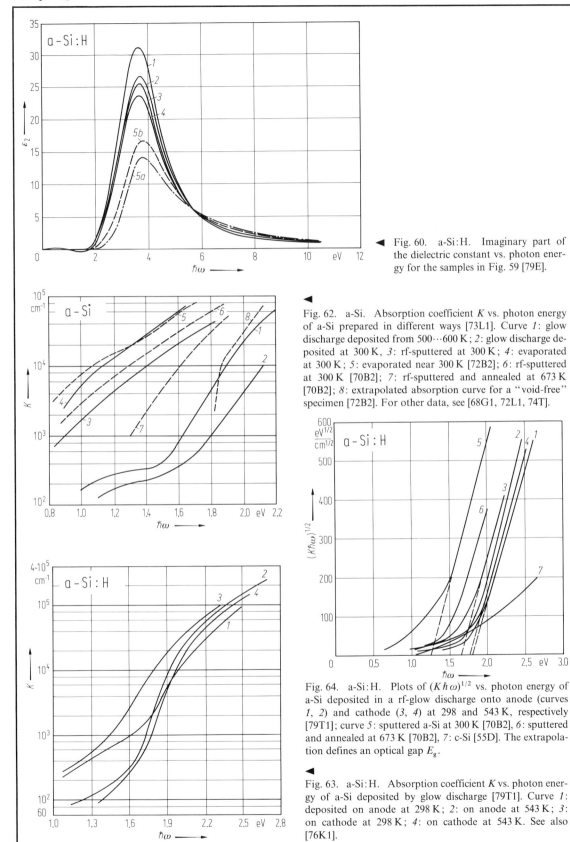

Fig. 60. a-Si:H. Imaginary part of the dielectric constant vs. photon energy for the samples in Fig. 59 [79E].

Fig. 62. a-Si. Absorption coefficient K vs. photon energy of a-Si prepared in different ways [73L1]. Curve 1: glow discharge deposited from 500···600 K; 2: glow discharge deposited at 300 K, 3: rf-sputtered at 300 K; 4: evaporated at 300 K; 5: evaporated near 300 K [72B2]; 6: rf-sputtered at 300 K [70B2]; 7: rf-sputtered and annealed at 673 K [70B2]; 8: extrapolated absorption curve for a "void-free" specimen [72B2]. For other data, see [68G1, 72L1, 74T].

Fig. 64. a-Si:H. Plots of $(K\hbar\omega)^{1/2}$ vs. photon energy of a-Si deposited in a rf-glow discharge onto anode (curves 1, 2) and cathode (3, 4) at 298 and 543 K, respectively [79T1]; curve 5: sputtered a-Si at 300 K [70B2], 6: sputtered and annealed at 673 K [70B2], 7: c-Si [55D]. The extrapolation defines an optical gap E_g.

Fig. 63. a-Si:H. Absorption coefficient K vs. photon energy of a-Si deposited by glow discharge [79T1]. Curve 1: deposited on anode at 298 K; 2: on anode at 543 K; 3: on cathode at 298 K; 4: on cathode at 543 K. See also [76K1].

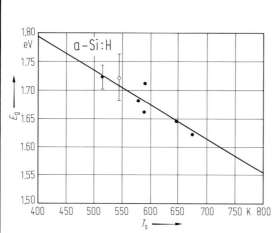

Fig. 65. a-Si:H. Optical gap E_g of glow discharge deposited films vs. substrate temperature T_s. The films were grown on the anode. ● [80C1], ○ [79T1], solid line [77Z].

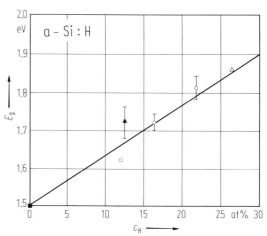

Fig. 66. a-Si:H. Optical gap E_g of glow discharge deposited films vs. hydrogen content c_H. ○ [80C1], △ [79T1], ▲ [79J1]. See also [80C8].

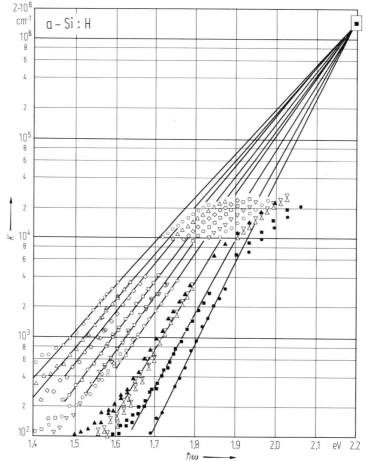

Fig. 67. a-Si:H. Absorption coefficient K vs. photon energy of glow discharge deposited a-Si:H (13 at% hydrogen) at different temperatures (● 12.7 K, ■ 151 K, ▲ 293 K). The other data are obtained at 293 K after 30 min annealing at different temperatures: ⵣ 293 K, ○ 773 K, ▽ 800 K, □ 823 K, ◇ 848 K, △ 873 K, ○ 900 K [81C]. Similar data: [80C8, 80C2, 77C2, 80M10] sputtered a-Si:H.

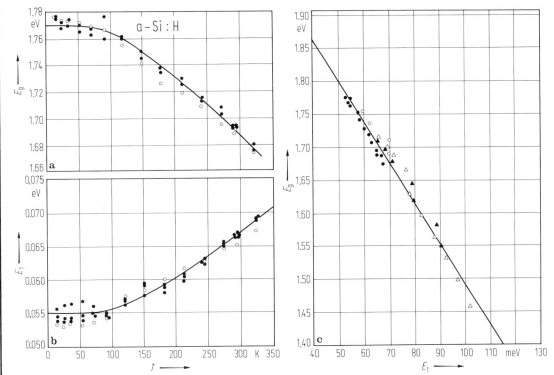

Fig. 68. a-Si:H. Optical gap E_g (a) and width of the Ur-bach tail E_1 (b) of glow discharge deposited a-Si:H (13 at% hydrogen) as a function of temperature. The solid points are from isoabsorption data, the open circles are from a fit to the $K(\omega)$-results in Fig. 67. (c) Optical gap E_g for three samples of a-Si:$H_{0.13}$ as a function of E_1. ● T-dependence, ▲ evolution; ○ T-dependence; △ evolution [81C].

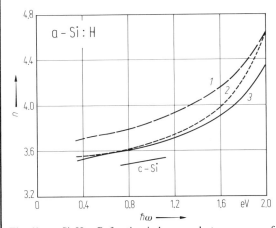

Fig. 69. a-Si:H. Refractive index vs. photon energy of glow discharge deposited a-Si:H. Curve *1*: undoped, *2*: p-doped ($c_{B_2H_6}$ in $SiH_4 = 10^{-2}$), *3*: n-doped (c_{PH_3} in $SiH_4 = 10^{-2}$); substrate temperature 525 K, low pressure growth conditions [80B3]. Crystalline Si for comparison. See also [80C1, 82K2].

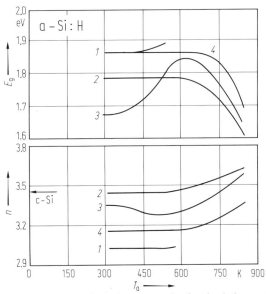

Fig. 70 a-Si:H. Optical gap E_g and refractive index n of glow discharge films measured at $\lambda = 2$ μm as a function of annealing temperature T_a (annealing time: 30 min) [79T1]. Same films as in Figs. 63 and 64.

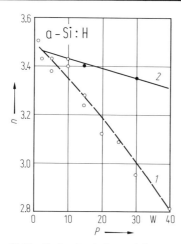

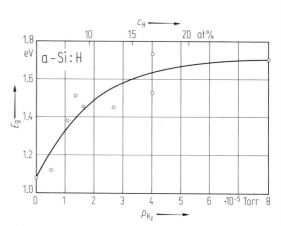

Fig. 71. a-Si:H. Refractive index at 1.5 μm vs. rf-power P of glow discharge deposited a-Si:H. Preparation conditions: $T_s = 500$ K, total pressure $p = 0.15$ Torr, c_{SiH_4} in Ar = 0.05, flow rate 130 sccm. Curve 1: unbiased substrate holder, 2: substrate holder negatively biased [79K3]. (sccm: standard cubic centimeter per minute; 1 Torr ≅ 133.3 Pa.)

Fig. 72. a-Si:H. Optical gap E_g vs. partial pressure of hydrogen p_{H_2} for sputtered a-Si:H films. Preparation conditions: $p_{Ar} = 4 \cdot 10^{-4}$ Torr, $T_s = 570$ K. In the upper scale the hydrogen content is given [80M3]. Similar data in [80Z1, 80D1, 80D2, 79F3].

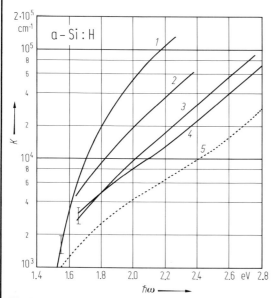

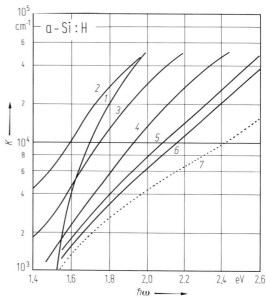

Fig. 73. a-Si:H. Absorption coefficient K as a function of photon energy $\hbar\omega$ of samples prepared by CVD at different substrate temperatures [79J1]. Curve 1: $820 < T_s < 920$ K, 2: 960 K, 3: 1010 K, 4: 1050 K, 5: c-Si [55D].

Fig. 74. a-Si:H. Absorption coefficient K as a function of photon energy of a sample prepared by CVD at 920 K after various annealing periods [79J1]. Curve 1: as deposited, 2: after 105 min at 920 K, 3: after 130 min at 920 K, 4: after 265 min at 920 K, 5: after 60 min at 1120 K, 6: after 60 min at 1300 K, 7: c-Si [55D].

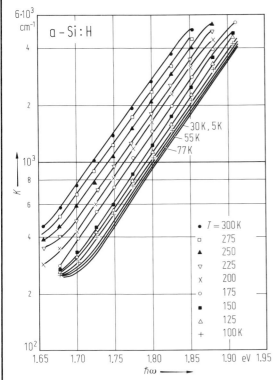

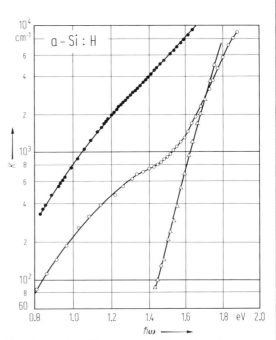

Fig. 75. a-Si:H. Absorption coefficient K vs. photon energy for sputtered a-Si:H ($p_H = 4 \cdot 10^{-4}$ Torr, $T_s = 520$ K) at various temperatures [79F3]. Results on CVD-film in [79J3]. 1 Torr $\cong 133.3$ Pa.

Fig. 76. a-Si:H. Absorption coefficient K vs. photon energy for differently doped sputtered samples of a-Si:H [79F3]. Preparation conditions: substrate temperature $T_s = 525$ K, $p_H = 4 \cdot 10^{-4}$ Torr, $\triangle$ undoped, $\circ$ $p_{PH_3} = 10^{-5}$ Torr, $\bullet$ $p_{B_2H_6} = 9 \cdot 10^{-6}$ Torr. Similar results on glow discharge films [76K1, 81S4]. 1 Torr $\cong 133.3$ Pa.

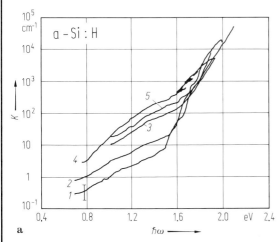

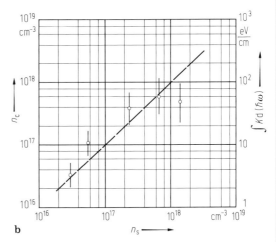

Fig. 77. a-Si:H. (a) Absorption coefficient vs. photon energy for films deposited by glow discharge using various rf-powers at a substrate temperature $T_s = 500$ K (photothermal deflection spectroscopy). Curve *1*: 2 W, *2*: 5 W, *3*: 15 W. *4*: 30 W, *5*: 40 W. (b) Integrated absorption $\int K(\omega) \mathrm{d}(\hbar\omega)$ of the subbandgap absorption and calculated defect density n_c as a function of the dangling bond spin density n_s, dashed line as guide for the eye [82J].

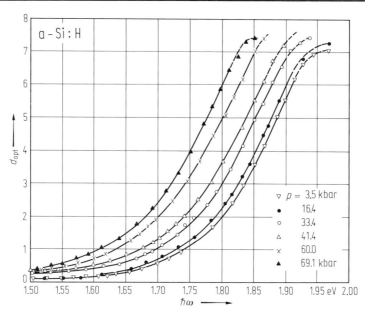

Fig. 78. a-Si:H. Optical density d_{opt} vs. photon energy at various pressures for glow discharge deposited a-Si:H ($T_s = 520$ K). Above 50 kbar irreversible red shifts indicate defect creation [77W1]. Different results [72C1, 80M3].

For Fig. 79, see next page.

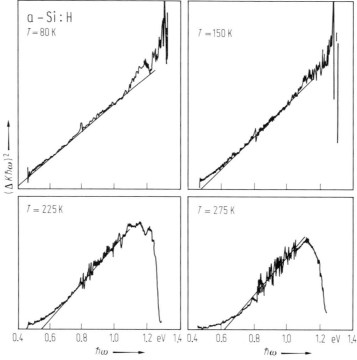

Fig. 80. a-Si:H. Photoinduced absorption spectra, $(\Delta K \hbar \omega)^2$ vs. photon energy, of sputtered a-Si:H at various temperatures. No preparation parameters given [80O3].

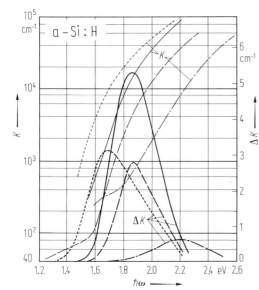

Fig. 79. a-Si:H. Electroabsorption ΔK (bold lines) and absorption coefficient K (thin lines) vs. photon energy for differently prepared films [80A1]. CVD-deposited near 870 K (-----); glow discharge deposited at 670 K (—); 410 K (– –) and 300 K (–·–). See also [84M2].

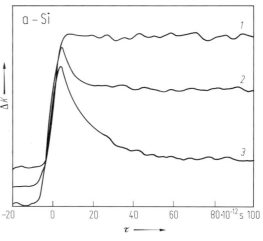

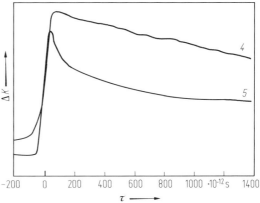

Fig. 81. a-Si. Decay of photoinduced absorption at 300 K in a-Si prepared by evaporation (curve 1), sputtering (curve 2), sputtering in gas containing SiF$_4$ (curve 3), glow discharge from SiH$_4$ (curve 4), and sputtering in a hydrogen-argon atmosphere on heated substrates (curve 5) [83V1, 81V2].

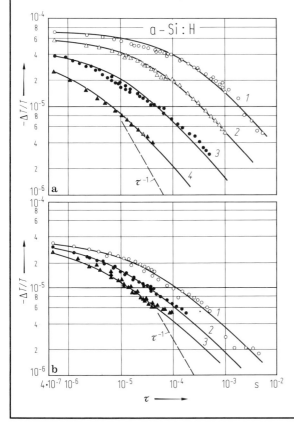

◄

Fig. 82. a-Si:H. Decay of photoinduced absorption (change of transmission $\Delta T/T$ vs. delay time τ) of undoped (a) and doped (b) a-Si:H at different temperatures. (a): reactively sputtered film ($p_{H_2} = 8 \cdot 10^{-4}$ Torr), curve 1: $T = 80$ K; 2: 105 K; 3: 225 K; 4: 306 K. (b): glow discharge film, $c_{PH_3} = 10^{-3}$, curve 1: $T = 80$ K; 2: 200 K; 3: 305 K. No preparation parameters given [80V2]. Similar data [82K3]. 1 Torr ≙ 133.3 Pa.

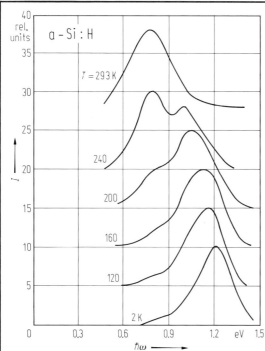

Fig. 83. a-Si:H. Luminescence spectra (intensity I vs. photon energy) of a-Si:H films deposited at 520 K in a rf-glow discharge. Measurements were taken at the indicated temperatures with a 0.5 W krypton laser ($\lambda = 647.1$ nm) [77E3]. Similar spectra [79T3, 79A2].

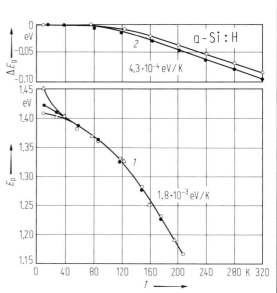

Fig. 84. a-Si:H. Peak energy of the photoluminescence E_p and variation of the optical energy gap ΔE_g as a function of temperature of glow discharge deposited films ($T_s = 500$ K). Curve 1: different samples and excitation energies; ● pure silane, $E_{ex} = 1.91$ eV; ○ pure silane, $E_{ex} = 2.335$ eV; △ 5% SiH_4 in Ar, $E_{ex} = 1.91$ eV. Curve 2: optical gap measured at $K = 10^5$ cm^{-1} (○) and $K = 10^3$ cm^{-1} (●) [79T3].

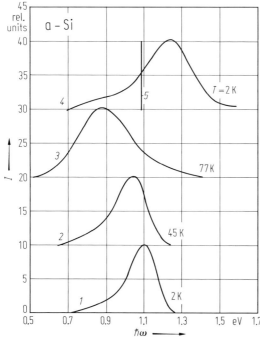

Fig. 85. a-Si. Luminescence spectrum (intensity I vs. photon energy) of a-Si evaporated in ultra high vacuum ($T_s = 670$ K) measured at the indicated temperatures (curves $1 \cdots 3$). Included are the spectra of glow discharge deposited a-Si ($T_s = 520$ K), (curve 4) and crystalline Si (curve 5) measured at 2 K [77E2]. $E_{ex} = 1.91$ eV.

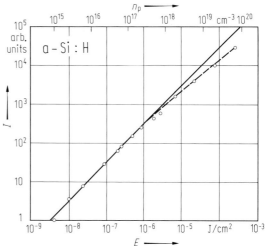

Fig. 86. a-Si:H. Luminescence intensity I at 1.35 eV of glow discharge deposited a-Si:H ($T_s = 520$ K, $c_H = 17$ at%) at 2 K as a function of the pulse energy per unit area E (excitation energy 2.41 eV, pulse width 10 ns, pulse frequency ≈ 100 Hz). In the upper scale the photoexcited pair density per pulse, n_p, is given [80S9].

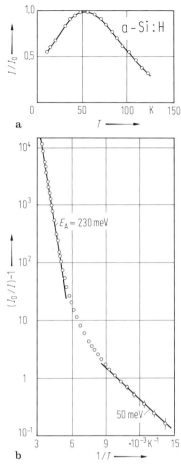

a

b

Fig. 87. a-Si:H. (a) Luminescence intensity I/I_0 at 1.35 eV normalized to the maximum value at 50 K vs. temperature. (b) Plot of (I_0/I)-1 vs. $1/T$. The film was glow discharge deposited from pure silane (power 1 W at 500 K). Excitation energy was 2.335 eV. The slope of straight line portions defines activation energies of 230 meV and 50 meV at high and low temperature, respectively [79T3]. For similar data, see [79A2, 77E3, 77N3], on sputtered a-Si:H [80C5].

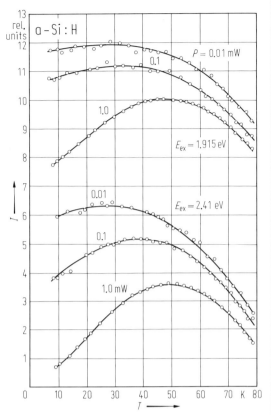

Fig. 88. a-Si:H. Relative luminescence intensity at 1.35 eV of glow discharge deposited a-Si:H ($T_s = 500$ K) as a function of temperature for two excitation energies. Parameter is the excitation power in mW [81S7].

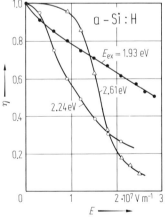

Fig. 89. a-Si:H. Luminescence efficiency η of glow discharge deposited a-Si:H ($T_s = 550$ K) as a function of the electric field E at various excitation energies measured at 77 K [77N2]. Similar data in [76E1], sputtered material [80P3].

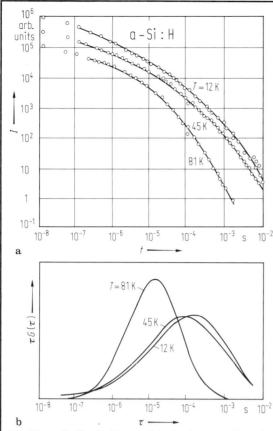

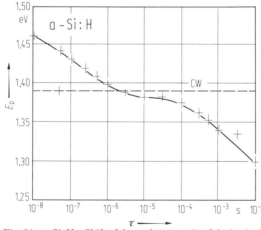

Fig. 90. a-Si:H. (a) Decay of the luminescence intensity I (arb. units) at 1.3 eV of glow discharge deposited a-Si:H (pure silane, $T_s = 500$ K) after a sharp excitation pulse (≈ 20 ns, $E_{ex} = 2.41$ eV) at the indicated temperatures. (b) Distribution of the decay times $G(\tau)$ normalized to the same value of $\int G(\tau) d\tau$. Plotted is the function $\tau G(\tau)$ because of the logarithmic time scale ($G(\tau) d\tau = \tau G(\tau) d\ln \tau$) [79T3]. Additional results [79S5, 79R2, 79K4, 80N5].

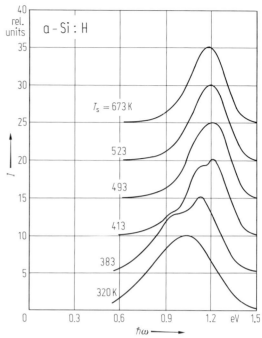

Fig. 92. a-Si:H. Photoluminescence spectra (intensity vs. photon energy) of a-Si:H films deposited in a rf-glow discharge at the given temperatures T_s. Measurements were taken at 77 K and $E_{ex} = 1.91$ eV [74E1].

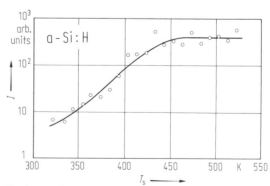

Fig. 91. a-Si:H. Shift of the peak energy E_p of the intrinsic luminescence band of glow discharge deposited a-Si:H (pure silane, $T_s = 500$ K) as a function of the delay time τ at 12 K. For cw-excitation the peak energy amounted to 1.39 eV. Excitation energy $E_{ex} = 2.41$ eV [79T3].

Fig. 93. a-Si:H. Integrated luminescence intensity I at 77 K of glow discharge deposited a-Si:H as a function of substrate temperature T_s with $E_{ex} = 1.91$ eV [74E1].

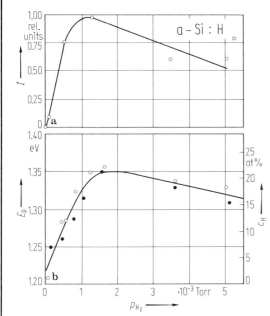

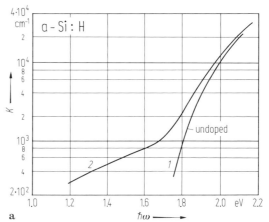

Fig. 94. a-Si:H. (a) Integrated photoluminescence intensity ($E_{ex} = 2.41$ eV, $T = 77$ K) of sputtered a-Si:H ($T_s = 470$ K), (b) peak position in the PL-spectrum E_p (●) and hydrogen concentration c_H (○) as a function of the partial pressure of hydrogen in sputtering gas [80P3]. 1 Torr $\cong 133.32$ Pa.

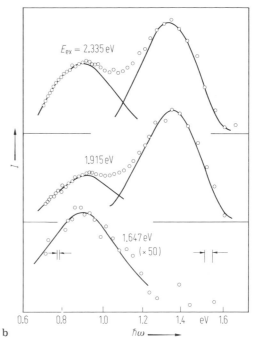

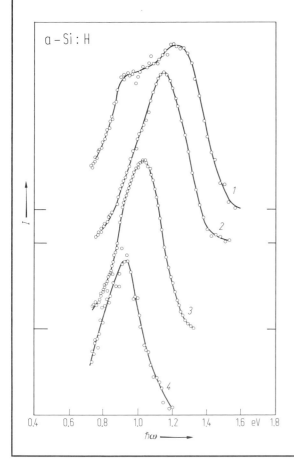

Fig. 95. a-Si:H. (a) Absorption coefficient K at 300 K as a function of photon energy of glow discharge deposited a-Si:H ($T_s = 500$ K). Curve 1: undoped and curve 2: doped with 10^4 ppm PH_3 [76K4]. (b) Luminescence spectra (intensity vs. photon energy) of a film doped with 300 vppm PH_3 taken at 10 K at different excitation energies [80S7]. Absorption in the extrinsic absorption tail quenches the intrinsic luminescence band but not the defect induced band.

For Fig. 96, see next page.

◀

Fig. 97. a-Si:H. Luminescence spectra (intensity vs. photon energy) at 8 K of glow discharge deposited compensated a-Si:H ($T_s = 500$ K, $P = 2$ W, SiH_4 with no carrier gas). The spectra are normalized to the same peak heights: curve 1: $c_{PH_3} = 10^{-3}$, 2: $c_{PH_3} = 10^{-3}$, $c_{B_2H_6} = 10^{-4}$, 3: $c_{PH_3} = 10^{-3}$, $c_{B_2H_6} = 10^{-3}$, 4: $c_{PH_3} = 10^{-3}$, $c_{B_2H_6} = 4 \cdot 10^{-3}$ (concentration of doping gases in SiH_4). $E_{ex} = 2.41$ eV [81S4]. Additional data [79A2, 84B1].

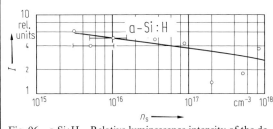

Fig. 96. a-Si:H. Relative luminescence intensity of the defect peak at 300 K of various glow discharge deposited samples vs. spin density n_s. Excitation energy $E_{ex} = 2.41$ eV [81S5].

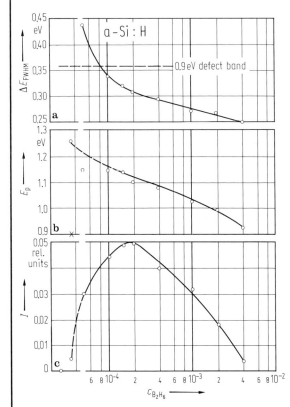

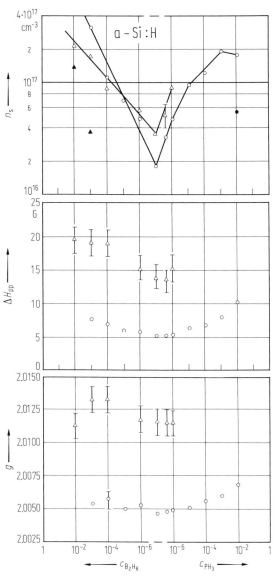

Fig. 98. a-Si:H. Luminescence parameters at 10 K of compensated glow discharge deposited a-Si:H ($T_s = 500$ K, $P = 2$ W, SiH$_4$ with no carrier gas) as a function of the doping level $c_{B_2H_6}$ with B$_2$H$_6$ at a constant doping with PH$_3$, $c_{PH_3} = 10^{-3}$ (concentration of doping gases in SiH$_4$). (a) half-width ΔE_{FWHM}, (b) peak energy E_p and (c) relative luminescence intensity I. $E_{ex} = 2.41$ eV [81S4]. See also Fig. 23.

Fig. 99. a-Si:H. Light induced spin density n_s, peak-to-peak width ΔH_{pp} of the ESR signal and g-values of singly doped glow discharge deposited a-Si:H as a function of the doping level: △ broad line, ○ narrow lines, closed data points in the upper figure are dark ESR data on the only samples with detectable signals. Preparation parameters, see Fig. 97, measurement at 30 K [81S4]. Additional data [77K3, 77B9, 80K3]. For compensated samples, see [81S4]. See also Figs. 40, 41.

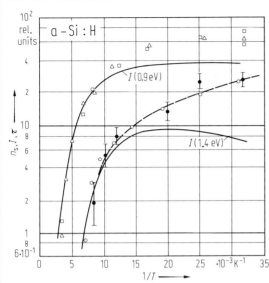

Fig. 100. a-Si:H. Light induced spin density n_s (open symbols), luminescence lifetime τ of undoped samples (full symbols) as a function of reciprocal temperature of glow discharge deposited films ($T_s = 500$ K). ○ undoped, △ P-doped with concentration ratio c_{PH_3} in $SiH_4 = 10^{-3}$, □ B-doped with $c_{B_2H_6}$ in $SiH_4 = 10^{-3}$. The luminescence intensity of the 0.9 eV and 1.4 eV bands are given by the full lines for comparison. All quantities in relative units [80S5].

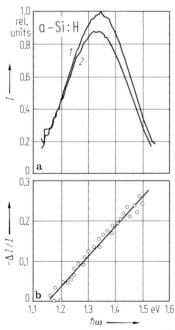

Fig. 102. a-Si:H. (a) Luminescence spectra (intensity vs. photon energy) of glow discharge deposited a-Si:H ($T_s = 520$ K) at 2 K; curve 1: excitation by 1.92 eV photons and 2: in a two beam arrangement excited by 1.92 eV and additional IR light ($\hbar\omega < 0.7$ eV). (b) IR quenching effect $\Delta I/I$ as a function of photon energy [84C1].

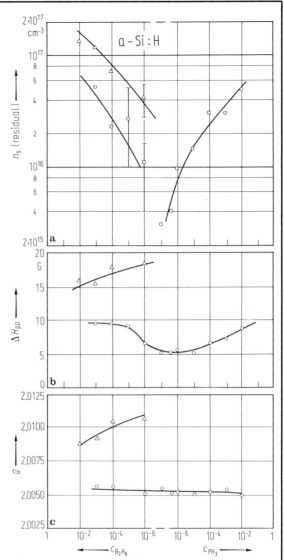

Fig. 101. a-Si:H. Metastable residual LESR (ESR signal after light is turned off) at 30 K of doped glow discharge deposited films ($T_s = 500$ K, same films as in Fig. 99) as a function of the doping level. (a) spin density n_s, (b) half width ΔH_{pp}, (c) g-values. △ broad line, ○ narrow lines [81S4].

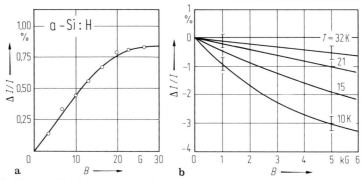

Fig. 103. a-Si:H. Change of luminescence intensity $\Delta I/I$ at 1.35 eV with magnetic field B of glow discharge deposited a-Si:H ($T_s = 500$ K). (a) low field ($B < 30$ G) increase at 10 K (b) high field ($B > 30$ G) decrease at various temperatures [78S3].

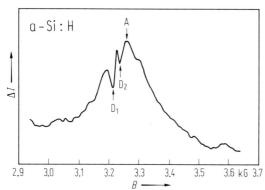

Fig. 104. a-Si:H. Optically detected magnetic resonance spectrum (ODMR, change of total luminescence intensity vs. magnetic field) of glow discharge deposited a-Si:H ($T_s = 570$ K, $c_{SiH_4} = 0.1$ in Ar) at 2 K. The microwave power is chopped at 1 kHz and the coherent change in luminescence is recorded as a function of the magnetic field. Optical excitation energy is 2.41 eV. The spectrum reveals broad enhancing (A) and narrow quenching lines (D_1, D_2) [78M2]. Additional data see [78B].

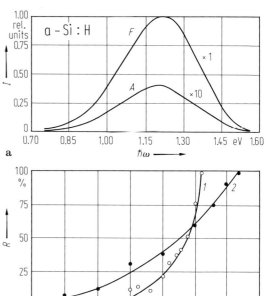

Fig. 106. a-Si:H. (a) Photoluminescence intensity of glow discharge deposited a-Si:H ($T_s = 570$ K, $c_{SiH_4} = 0.1$ in Ar) at 12 K vs. photon energy after excitation with 9 mW (A) and 540 mW (F), $E_{ex} = 2.41$ eV. (b) Recovery R of the fatigue effect by 15 min anneals at the given temperatures. Plotted is $R = \dfrac{I_A - I_F}{I_0 - I_F}$ vs. T_a, where I_0, I_F and I_A denote the intensities at 1.2 eV before fatigue, after fatigue and after annealing at T_a, respectively. Curve 1: as deposited; 2: heat treated at 170 °C before fatiguing [80M5].

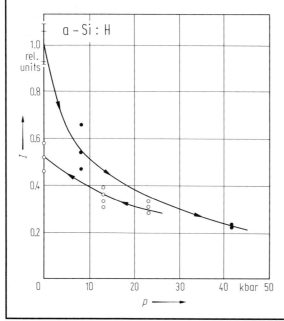

◀

Fig. 105. a-Si:H. Intensity of the photoluminescence (1.4 eV) of glow discharge deposited a-Si:H ($T_s = 520$ K, $c_{SiH_4} = 5 \cdot 10^{-2}$ in argon) as a function of pressure at 10 K. ($E_{ex} = 2.04$ eV, pulse width 10 ns.) Realize pronounced irreversible effects [81W1]. Photoluminescence is quenched above 75 kbar.

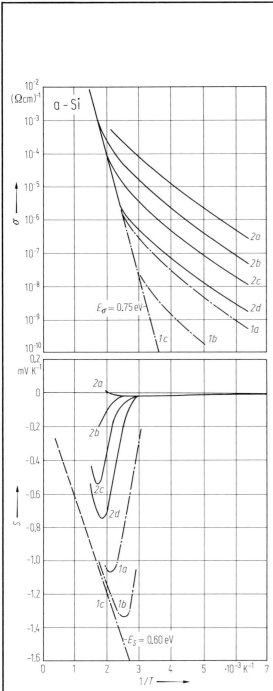

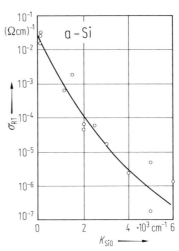

Fig. 108. a-Si. Correlation between the conductivity at 300 K, σ_{RT}, and the oxygen content in O-contaminated film. Plot of σ_{RT} of evaporated amorphous Si films as a function of the infrared absorption coefficient in the SiO peak near 10 μm [74L1].

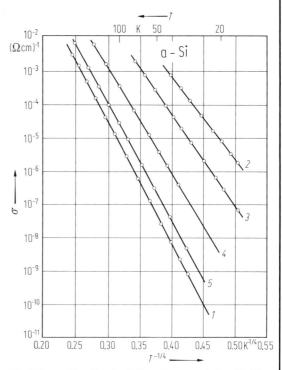

Fig. 107. a-Si. Temperature dependence of conductivity σ and thermoelectric power S of two silicon films deposited at 500 K by evaporation in a vacuum of $2 \cdot 10^{-6}$ Torr. Film 1: deposited with 2 Å/s and annealed at 530 K (1a), 540 K (1b) and 635 K (1c). Film 2 deposited with 16 Å/s and annealed at 500 K (2a), 580 K (2b), 635 K (2c) and 720 K (2d) [74B3, 75B2]. (Annealing time not given.)

Fig. 109. a-Si. Conductivity of evaporated a-Si ($T_s =$ 300 K) vs. $T^{-1/4}$ after bombardment at 14 K with 10^{15} Si$^+$ ions. Curve 1: as deposited; 2: after bombardment and annealing at 45 K; 3: 95 K; 4: 180 K; 5: 300 K. The sample was annealed at each temperature until no further change in conductivity could be observed [78A1].

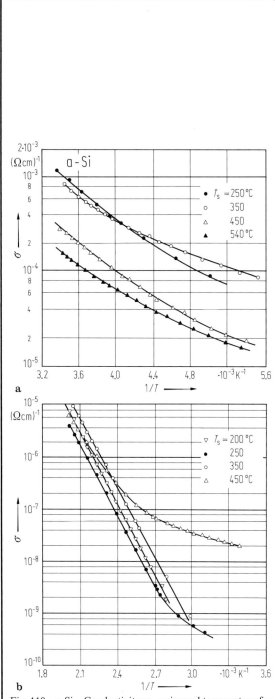

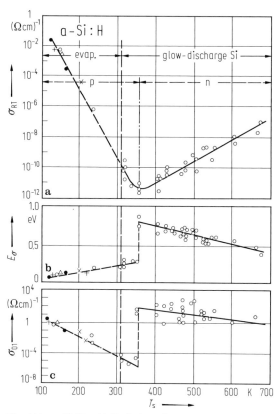

Fig. 111. a-Si:H. (a) Variation of room-temperature conductivity σ_{RT}, (b) activation energy of conductivity E_σ, (c) pre-exponential factor σ_{01} for glow discharge deposited films of a-Si:H with the deposition temperature T_s. Points on the left refer to various evaporated films [74S5]. See for similar results [70C1].

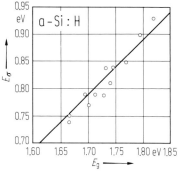

Fig. 110. a-Si. Conductivity vs. reciprocal temperature for a-Si samples deposited by sputtering at various temperatures T_s. (a) unhydrogenated, (b) deposition at a hydrogen partial pressure of $p_H = 9.2 \cdot 10^{-4}$ Torr [77A1]. 1 Torr $\cong$ 133.3 Pa.

Fig. 112. a-Si:H. Activation energy of conductivity E_σ of glow discharge deposited a-Si:H films (wide range of preparation conditions) vs. optical gap E_g [79S2].

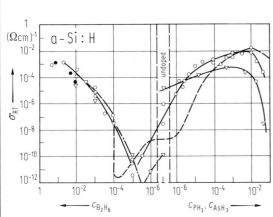

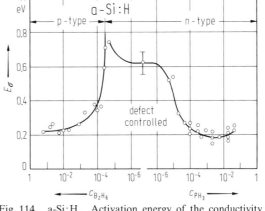

Fig. 113. a-Si:H. Conductivity at 300 K, σ_{RT}, of glow discharge deposited films as a function of the dopant concentration in the gas phase (PH_3 and B_2H_6 in SiH_4): o P and B-doping, $T_s = 525$ K, [77B8, 77B10], ▽ As and B-doping, $T_s = 500$ K [79J4], ● B-doping [77T1], dashed curve: P- and B-doping, $T_s = 550$ K [76S1].

Fig. 114. a-Si:H. Activation energy of the conductivity E_σ of glow discharge deposited a-Si:H ($T_s = 550$ K) as a function of the doping level defined by the concentration of the doping gas in silane, c_{PH_3} and $c_{B_2H_6}$ [76S1].

► Fig. 117. a-Si:F:H. Room-temperature conductivity σ_{RT} and activation energy E_σ of glow discharge deposited a-Si:F:H as a function of the concentration of the doping gas in a 10:1 gas mixture of SiF_4 and H_2; curve 1: AsH_3, 2: PH_3, 3: B_2H_6. Substrate temperature $T_s = 650$ K [80M4]. For comparison curve 4 shows results on a-Si:H films [76S1].

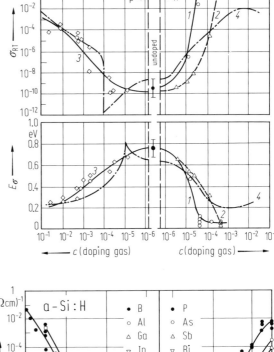

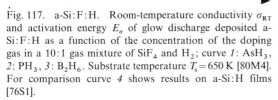

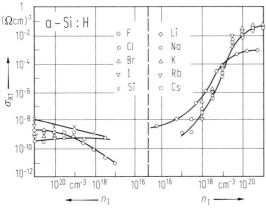

Fig. 115. a-Si:H. Conductivity at 300 K, σ_{RT}, of glow discharge deposited a-Si:H ($T_s \approx 550$ K) as a function of implanted impurity concentration n_I for elements of groups I, IV and VII. Li implantation [79B4], Si implantation [79S3], other data [80K4].

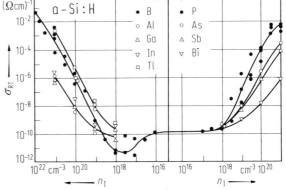

► Fig. 116. a-Si:H. Conductivity at 300 K, σ_{RT}, of glow discharge deposited a-Si:H ($T_s \approx 550$ K) as a function of implanted impurity concentration n_I for elements of groups III and V [80K4].

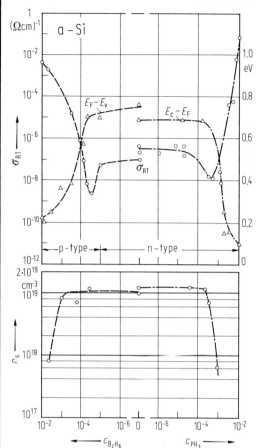

Fig. 118. a-Si. Conductivity at 300 K, σ_{RT}, position of the Fermi level, $E_c - E_F$ and $E_F - E_v$, and spin density n_s ($g = 2.0054$) of CVD deposited a-Si as a function of the nominal doping concentration of PH_3 or B_2H_6 in SiH_4, c_{PH_3} and $c_{B_2H_6}$ [81H2]. P doping at $T_s = 920$ K, B doping at $T_s = 820$ K. See also [79T2, 80S2, 82M2].

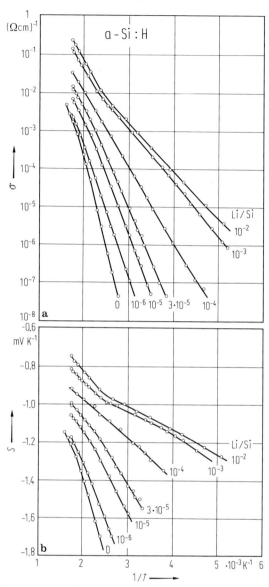

Fig. 120. a-Si:H. Conductivity σ (a) and thermoelectric power S (b) as a function of reciprocal temperature for glow discharge deposited a-Si:H-films ($T_s = 570$ K) implanted stepwise with Li ions of 20 keV energy. Parameter is the concentration ratio of Li to Si atoms in the film [79B3].

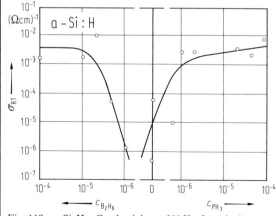

Fig. 119. a-Si:H. Conductivity at 300 K of posthydrogenated CVD deposited films (T_s near 900 K) as a function of the nominal dopant concentration of PH_3 and B_2H_6 in SiH_4, c_{PH_3} and $c_{B_2H_6}$. Posthydrogenation in a rf-hydrogen-plasma at 623 K for 1 h [81H2].

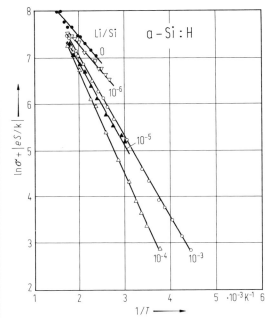

Fig. 121. a-Si:H. $\left(\ln\sigma + \dfrac{eS}{k}\right)$ (σ in $\Omega^{-1}\,cm^{-1}$) as a function of reciprocal temperature of glow discharge deposited a-Si:H ($T_s = 570\,K$) for different doping steps (same data as in Fig. 120). Parameter is the concentration ratio of Li to Si atoms in the films. The slope of these straight lines is identical with the difference of the activation energies of σ and S and points to an activated mobility possibly due to potential fluctuations [79B3, 79B4].

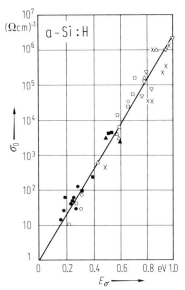

Fig. 122. a-Si:H. Pre-exponential factor σ_0 of glow discharge deposited a-Si:H as a function of the activation energy of the conductivity E_σ. As doping level the concentration of the doping gas in silane is given: PH_3 ($\bullet$ 10^{-1}, $\circ$ 10^{-3}), AsH_3 ($\blacksquare$ 10^{-1}, $\square$ 10^{-3}), $Sb(CH_3)_3$ ($\blacktriangle$ 10^{-1}, $\triangle$ 10^{-3}), $Bi(CH_3)_3$ ($\triangledown$ 10^{-3}), p-type films ($\times$); various deposition conditions [79C].

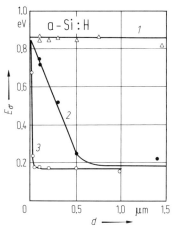

Fig. 124. a-Si:H. Activation energy of the conductivity, E_σ, of glow discharge deposited films ($T_s \approx 570\,K$) as a function of film thickness, d, at different doping levels (PH_3/SiH_4 ratio): curve 1: undoped, 2: $c_{PH_3} = 10^{-6}$, 3: $c_{PH_3} = 10^{-2}$ [83M5].

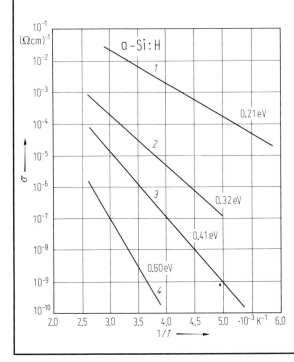

◄

Fig. 123. a-Si:H. Conductivity as a function of reciprocal temperature for various doped and compensated n-type films of glow discharge deposited a-Si:H ($T_s = 500\,K$, $P = 2\,W$, SiH_4 with no carrier gas). Curve 1: $c_{PH_3} = 10^{-3}$; 2: $c_{PH_3} = 10^{-3}$ and $c_{B_2H_6} = 10^{-4}$, 3: $c_{PH_3} = 10^{-3}$ and $c_{B_2H_6} = 2\cdot10^{-4}$, 4: $c_{PH_3} = 10^{-3}$ and $c_{B_2H_6} = 4\cdot10^{-4}$ (concentration of doping gases in SiH_4) [81S4]. Further results [81B2].

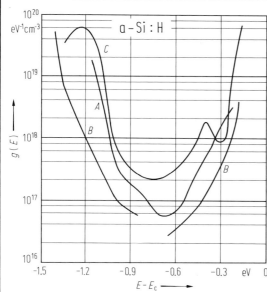

Fig. 125. a-Si:H. Density of states vs. energy of glow discharge deposited a-Si:H films. Curve A: from field effect ($T_s = 550$ K) [80G1], curve B: from capacitance-voltage measurements ($T_s = 525$ K) [79H1], curve C: composite curves from field effect for large number of differently doped films ($T_s = 550$ K) [76M1]. The energy is measured relative to the mobility edge E_c in the conduction band. Further results [80O1, 79J4, 80C3, 80C4, 79B6].

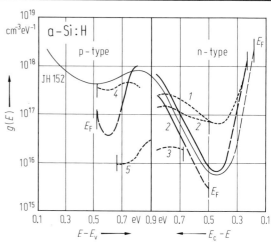

Fig. 126. a-Si:H. Density of states vs. energy of differently doped glow discharge deposited a-Si:H films ($T_s \approx 525$ K) as determined from capacitance transients (full curves) and current transients (dotted curves) [83B5]. Curve 1: 50 ppm PH_3, 2: 30 ppm PH_3, 3: undoped, 4: 200 ppm B_2H_6, 5: 50 ppm B_2H_6. As doping level the concentration of the doping gas in silane is given. Curve JH152 from [82L2]. The energy is measured relative to the mobility edges E_c and E_v in the conduction and valence band, E_F gives the position of the Fermi level. Different data [84T3].

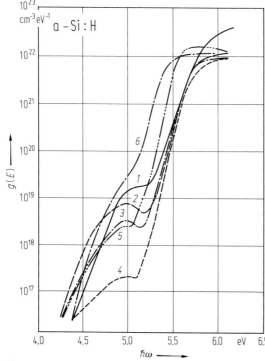

Fig. 127. a-Si:H. Density of occupied states $g(E)$ as a function of energy for glow discharge deposited films at various deposition temperatures T_s as obtained from photoelectric yield spectroscopy. Curve 1: $T_s = 300$ K; 2: 363 K; 3: 443 K; 4: 493 K; 5: 543 K; 6: 653 K. The zero of the energy scale is the vacuum level [83G2].

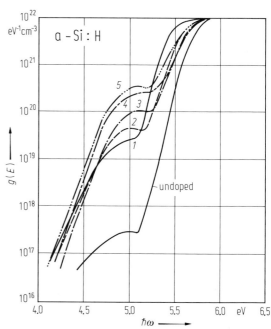

Fig. 128. a-Si:H. Density of occupied states $g(E)$ as a function of energy for P doped glow discharge deposited films ($T_s = 493$ K) at various doping levels (PH_3 concentration in SiH_4) from photoelectric yield spectroscopy [83G2]. Curve 1: $c_{PH_3} < 10^{-4}$, 2: $5 \cdot 10^{-4}$, 3: 10^{-3}, 4: $5 \cdot 10^{-3}$, 5: 10^{-2}. The zero of the energy scale is the vacuum level.

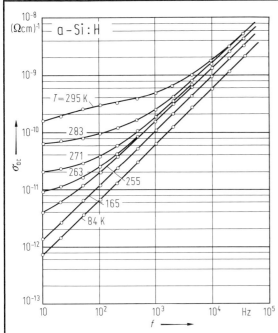

Fig. 129. a-Si:H. ac conductivity vs. frequency for a-Si:H deposited in a rf-glow discharge at 500 K. Parameter: measuring temperature [76A1]. Other results: [77N1] undoped a-Si:H; [80N1] strongly doped a-Si:H; [79R1] evaporated a-Si; [78E1] sputtered a-Si.

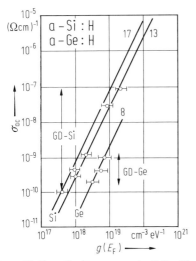

Fig. 130. a-Si:H, a-Ge:H. ac conductivity ($f = 10^4$ Hz, $T = 80$ K) of a-Si:H films deposited by glow discharge (GD) at various T_s as a function of the density of localized states at the Fermi level $g(E_F)$ as determined from field effect. Also shown are data on glow discharge deposited a-Ge:H samples [76A1]. The lines are the theoretical dependences [69A] with the decay length α^{-1} of the wavefunctions given in Å by the number on the curves.

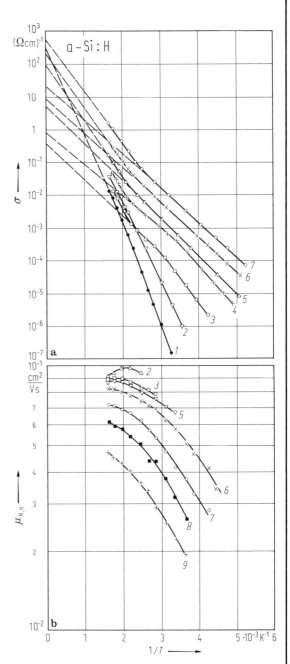

Fig. 131. a-Si:H. (a) Conductivity σ and (b) Hall mobility $\mu_{H,n}$ vs. reciprocal temperature for glow discharge deposited samples ($T_s = 525$ K) doped with PH_3. Concentration c_{PH_3} in silane [vppm] amounted to: 0 (curve 1), 1 (2), 3 (3), 25 (4), 250 (5), 10^3 (6), 10^4 (7), $2 \cdot 10^4$ (8), $4 \cdot 10^4$ (9) [77B8]. Other data, see [77L1, 80D3].

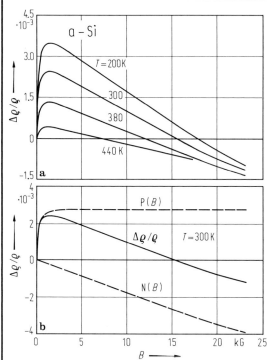

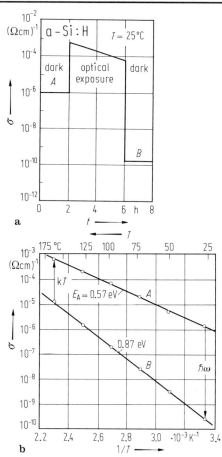

Fig. 132. a-Si. (a) Magnetoresistance $\Delta\varrho/\varrho$ of an evaporated a-Si film as a function of magnetic induction B. Deposition conditions: rate 200 Å/s, base pressure $2\cdot10^{-6}$ Torr, $T_s = 300$ K, annealed to steady state at 550 K. The measurements were taken at the indicated temperatures. (b) Separation of $\Delta\varrho/\varrho$ at 300 K into a positive component $P(B)$ and a negative component $N(B)$ [74M1].

Fig. 134. a-Si:H. (a) Conductivity σ of a glow discharge deposited film ($T_s = 520$ K) as a function of time before, during and after exposure to light of 200 mW/cm² ($\lambda_{ex} = 600\cdots900$ nm). (b) Conductivity vs. reciprocal temperature in the stable (curve A) and metastable (B) state. Above 150 °C annealing reproduces curve A [77S2]. For variation of prefactor σ_0 with activation energy, see [80S3]. Further results on metastable light induced changes, see [84F].

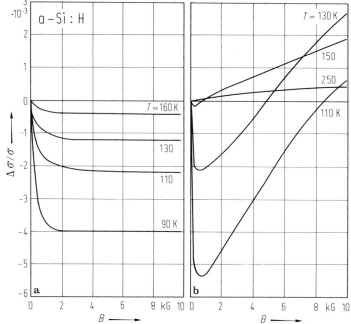

Fig. 133. a-Si:H. Relative change of the conductivity, $\Delta\sigma/\sigma$, of glow discharge deposited films ($T_s = 520$ K) as a function of the magnetic field B for various temperatures. Doping concentration (defined as ratios PH_3/SiH_4 and B_2H_6/SiH_4): (a) $c_{PH_3} = 3\cdot10^{-2}$, (b) $c_{B_2H_6} = 3\cdot10^{-2}$ [81W4].

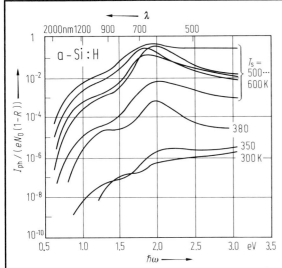

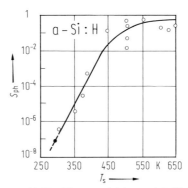

Fig. 135. a-Si:H. Spectral dependence of photoconductivity of undoped glow discharge films deposited at the indicated temperatures. The ordinate is the photocurrent I_{ph} divided by the number of photons per second entering the sample $eN_0(1-R)$ [73L1].

Fig. 136. a-Si:H. Photosensitivity at 1.8 eV defined by $S_{ph} = \eta \dfrac{\tau}{t_{tr}}$ as a function of the deposition temperature T_s of glow discharge deposited films. η denotes the quantum efficiency, τ the lifetime and t_{tr} the transit time of the carriers. ● represents a specimen evaporated at 300 K. Measurements were performed in gap arrangement with an electric field of 4 kV/cm [73L1]. For results on sputtered material, see [79M1, 77M1].

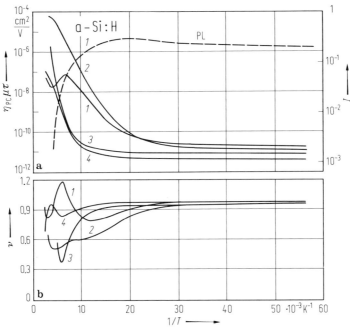

Fig. 137. a-Si:H. Photoconductivity ($\eta_{PC}\mu\tau$-product) (a) and exponent v of the intensity dependence $\sigma_{ph} \propto I^v$ (b) of various glow discharge deposited ($T_s \approx 520$ K) a-Si:H films as a function of reciprocal temperature [83H1]. Curve 1: undoped; 2: $c_{PH_3} = 10^{-4}$; 3: $c_{B_2H_6} = 10^{-3}$, 4: reactively sputtered at 570 K. Doping level defined by concentration of doping gas in silane. Generation rate $3 \cdot 10^{20}$ cm^{-3} s^{-1} and $\lambda = 525$ nm. For comparison the temperature dependence of the luminescence intensity I(PL) at 1.3 eV of sample 1 is shown. Other data [77A3, 79F2, 79W, 80G3, 74S5, 81W2].

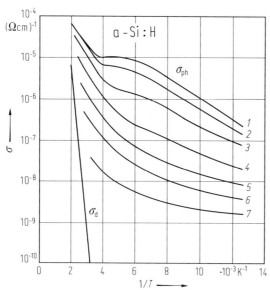

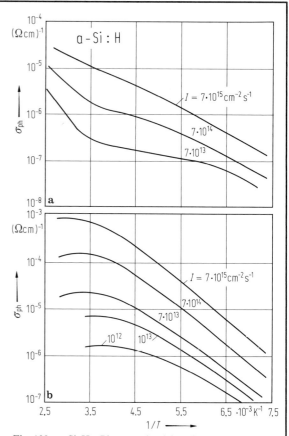

Fig. 138. a-Si:H. Dark conductivity σ_d and photoconductivity σ_{ph}, as a function of reciprocal temperature of an undoped glow discharge deposited film ($T_s \approx 520$ K). A tungsten iodine lamp was used at an incident power of 50 mW/cm². The spin density of dangling bonds is increased by electron bombardment and stepwise decreased by isochronal anneals. Curve 1: $n_s = 5 \cdot 10^{15}$ cm⁻³; 2: $2 \cdot 10^{16}$ cm⁻³; 3: $8.7 \cdot 10^{16}$ cm⁻³; 4: $1.6 \cdot 10^{17}$ cm⁻³; 5: $2.8 \cdot 10^{17}$ cm⁻³; 6: $4.6 \cdot 10^{17}$ cm⁻³; 7: $1.2 \cdot 10^{18}$ cm⁻³ [83D].

Fig. 139. a-Si:H. Photoconductivity of glow discharge deposited a-Si:H ($T_s = 550$ K) as a function of reciprocal temperature. Parameter is the light intensity given in number of photons per s and cm². Excitation by quartz-iodine lamp with suitable filter ($E_{ex} \approx 2$ eV). (a) slightly boron doped (b) slightly phosphorus doped [77A3]. Different results [79F2, 79W, 80G3, 74S5, 81W2, 83D, 83H1].

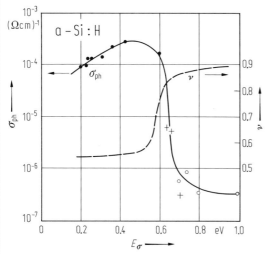

Fig. 140. a-Si:H. Photoconductivity σ_{ph} and exponent ν of the intensity dependence, $\sigma_{ph} \propto I^\nu$, at 300 K as a function of the activation energy of the dark conductivity E_σ for glow discharge deposited films ($T_s \approx 550$ K). Doping by addition of PH_3 or B_2H_6 to SiH_4: ● P doped, + undoped, ○ B doped; $E_{ex} = 2$ eV, unchopped light [77A3]. See also [80F5, 81V3].

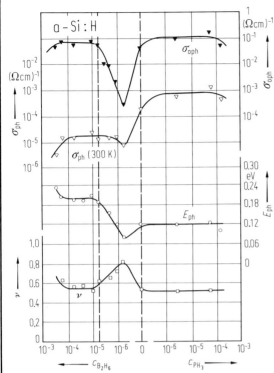

Fig. 141. a-Si:H. Photoconductivity σ_{ph} at 300 K, activation energy E_{ph} of $\sigma_{ph}(T)$, pre-exponential factor $\sigma_{0\,ph}$ and the exponent ν of the intensity dependence ($\sigma_{ph} \propto I^\nu$) of glow discharge films ($T_s = 525$ K) as a function of the doping concentrations $c_{B_2H_6}$ and c_{PH_3}. The doping level is given by the concentration of the doping gas in silane [80F5]. See also [77A3].

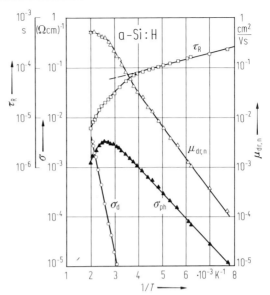

Fig. 142. a-Si:H. Photoconductivity σ_{ph}, dark conductivity σ_d, response time τ_R and drift mobility $\mu_{dr,n}$ of undoped (n-type) glow discharge deposited a-Si:H ($T_s = 530$ K) as a function of reciprocal temperature [78F2]. The response time was determined from the decay of the photocurrent. Activation energy for σ_{ph} is ≈ 0.11 eV. The light intensity was 7.5 mW/cm^2 at $E_{ex} = 1.96$ eV.

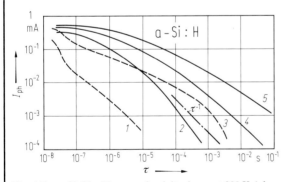

Fig. 143. a-Si:H. Photoconductivity decay at 300 K (photocurrent I_{ph} vs. decay time τ) of P doped a-Si:H deposited by glow discharge at 573 K (full lines) [81H4] and 473 K (dashed lines) [81H5] following a 10 ns excitation flash. Doping concentrations (PH$_3$ in SiH$_4$) c_{PH_3}: curve 1: 10^{-3}; 2: 10^{-3}; 3: 10^{-2}; 4: 10^{-2}; 5: 10^{-1}. Dashed-dotted line shows τ^{-1} dependence.

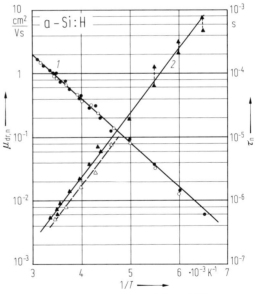

Fig. 144. a-Si:H. (1) Electron drift mobility $\mu_{dr,n}$ as a function of reciprocal temperature of glow discharge deposited a-Si:H ($T_s = 550$ K, $d = 6$ μm) obtained with ● charge collection technique and ○ current transient measurement on pin-junctions. (2) Excess carrier lifetime τ_n as a function of reciprocal temperature obtained from transit time (▲) and delayed field experiments (△) [84S].

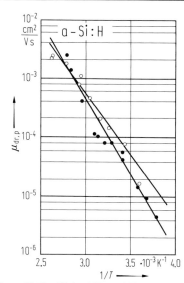

Fig. 145. a-Si:H. Hole drift mobility $\mu_{dr,p}$ of glow discharge deposited a-Si:H films ($T_s = 550$ K) as a function of reciprocal temperature. Both films were compensated with the doping concentrations in silane c_{PH_3} and $c_{B_2H_6}$ being $4.9 \cdot 10^{-5}$ and $5.7 \cdot 10^{-5}$ (●), and $5 \cdot 10^{-4}$ and $5.6 \cdot 10^{-4}$ (○) [78A2].

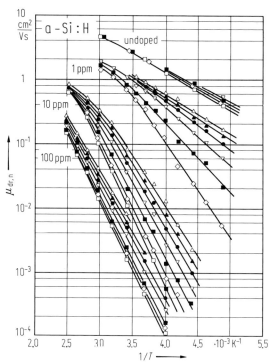

Fig. 146. a-Si:H. Electron drift mobility $\mu_{dr,n}$ in undoped and compensated films of glow discharge deposited a-Si:H ($T_s \approx 500$ K) at various applied electric fields as a function of reciprocal temperature. Doping gases PH_3 and B_2H_6 are added in equal concentrations [ppm] to the SiH_4. Electric field strengths E (in $V\,cm^{-1}$): □ $3 \cdot 10^3$, ◇ $5 \cdot 10^3$, ■ 10^4, ▽ $2 \cdot 10^4$, ● $3 \cdot 10^4$, ▲ $4 \cdot 10^4$, △ $5 \cdot 10^4$ [84M3].

For Fig. 148, see next page.

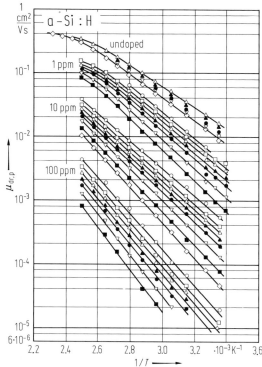

Fig. 147. a-Si:H. Hole drift mobility $\mu_{dr,p}$ in undoped and compensated films of glow discharged deposited a-Si:H ($T_s \approx 500$ K) at various applied fields as a function of reciprocal temperature. Doping gases PH_3 and B_2H_6 are added in equal concentrations [ppm] to the SiH_4. Electric field strengths E (in V/cm): ◇ $5 \cdot 10^3$, ■ 10^4, ▽ $2 \cdot 10^4$, ● $3 \cdot 10^4$, ▲ $4 \cdot 10^4$, △ $5 \cdot 10^4$, □ $6 \cdot 10^4$, ○ $8 \cdot 10^4$ [84M3]. See also [84K2].

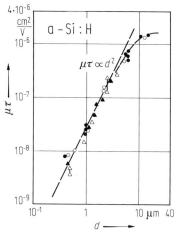

Fig. 149. a-Si:H. Thickness dependence of the $\mu\tau$-product of electrons and holes in junctions and barrier samples (glow discharge samples) as measured by various techniques: ● electrons and ○ holes from transient carrier collection in pin-structures; △ surface photo voltage, collection efficiency and drift length; ▲ primary photocurrent measurements [83S4, 83S5]. Preparation parameters not given.

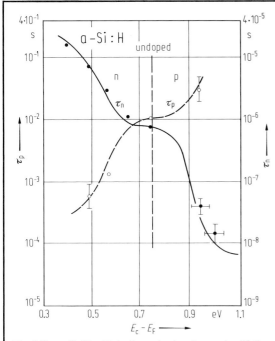

Fig. 148. a-Si:H. Majority and minority carrier lifetimes at 300 K in glow discharge deposited pin structures ($T_s \approx 550$ K) as a function of the position of the Fermi level in the i-layer, τ_n (●) and τ_p (○). The zero of energy is the mobility edge of the conduction band. Delayed field and limiting transit time measurement [84S].

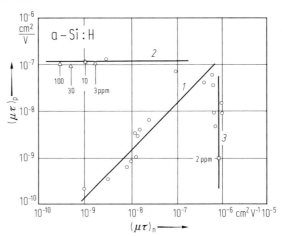

Fig. 150. a-Si:H. $\mu\tau$-product of holes vs. that of the electrons of thick glow discharge deposited films ($T_s \approx 500$ K) from time-of-flight studies at 300 K. Curve 1: undoped films for which $\mu\tau \times$ spin density is constant; 2: B doped; 3: P doped. The doping concentration is indicated unless the doping is unintentional [83S7].

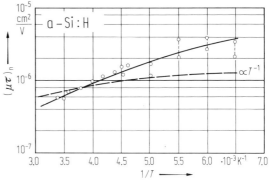

Fig. 152. a-Si:H. $(\mu\tau)_n$ as a function of reciprocal temperature of glow discharge deposited a-Si:H ($T_s = 550$ K, $d = 6$ μm). Same data as in Fig. 144 [84S]. Different results [83S6], see Fig. 151.

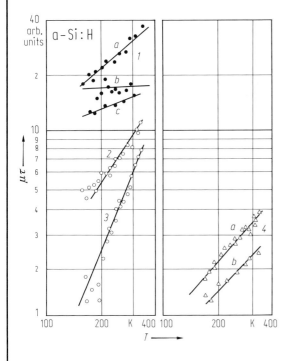

◄

Fig. 151. a-Si:H. $\mu\tau$-products of electrons and holes of glow discharge deposited samples ($T_s \approx 500$ K) as a function of temperature. $(\mu\tau)_n$: curve 1: undoped, spin density of dangling bonds $n_s = 3 \cdot 10^{16}$ cm^{-3} (a), $4.5 \cdot 10^{16}$ cm^{-3} (b) and $5 \cdot 10^{14}$ cm^{-3} (c); curve 2: $c_{B_2H_6} = 10^{-6}$, curve 3: $c_{B_2H_6} = 3 \cdot 10^{-6}$ (B$_2$H$_6$ in SiH$_4$). $(\mu\tau)_p$: curve 4: undoped, $n_s = 2 \cdot 10^{16}$ cm^{-3} (a) and 10^{16} cm^{-3} (b). [83S6]. Different results [84S], see Fig. 152.

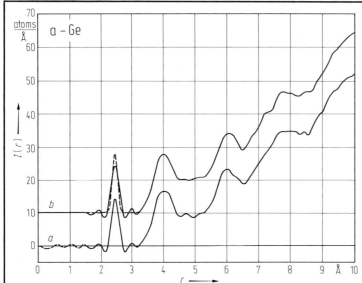

Fig. 153. a-Ge. Radial distribution function $I(r)$ of sputtered films derived from undamped $F(k)$ data to $k_m = 15\,Å^{-1}$ for (curve a) $T_s = 420$ K and (b) $T_s = 620$ K. The dotted line in (a) shows the experimental $I(r)$. The dotted line in (b) shows $I(r)$ after correction for k-space termination effects [73T]. See also [73G1].

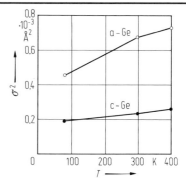

Fig. 154. a-Ge. Mean square relative displacement σ^2 of the first shell vs. temperature in amorphous (o) and crystalline Ge (●) obtained from EXAFS investigations [79R3]. See also [72S3]. The amorphous film was prepared by evaporation at 300 K.

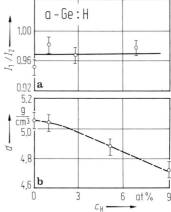

Fig. 155. a-Ge:H. Ratio of the intensities of the first and second peaks of the diffraction interference function (a) and mass density of sputtered a-Ge:H films ($T_s = 300$ K) (b) vs. hydrogen content c_H [76C1].

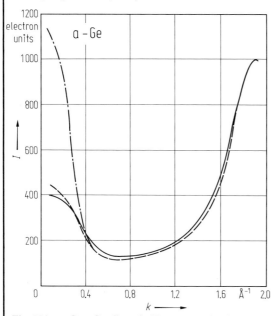

Fig. 156. a-Ge. Small angle X-ray scattering intensity I (electron units) vs. wavevector $k = \dfrac{4\pi \sin\theta}{\lambda}$ for a-Ge prepared by different methods. —— electrolytic deposition, ——— sputtered, —·— evaporated [72S3].

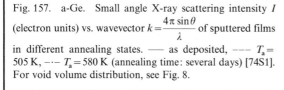

Fig. 157. a-Ge. Small angle X-ray scattering intensity I (electron units) vs. wavevector $k = \dfrac{4\pi \sin\theta}{\lambda}$ of sputtered films in different annealing states. —— as deposited, ——— $T_a = 505$ K, —·— $T_a = 580$ K (annealing time: several days) [74S1]. For void volume distribution, see Fig. 8.

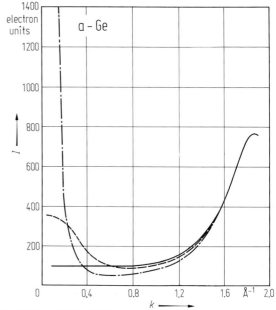

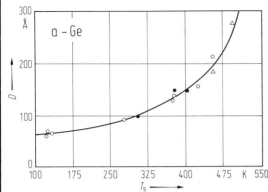

Fig. 158. a-Ge. Diameter of rod-like voids vs. deposition temperature T_s of evaporated Ge-films from electron micrograph studies. ○ in situ, base pressure 10^{-9} Torr, ● base pressure 10^{-7} Torr, △ films deposited at 373 K onto a thin Ge-film evaporated at 450 K and 495 K [74B5].

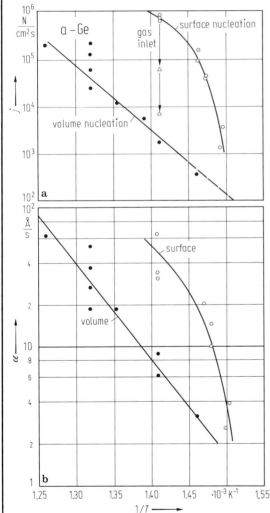

Fig. 159. a-Ge. (a) Nucleation rate j and (b) growth rate α of crystallites in evaporated films of a-Ge vs. reciprocal temperature. The surface nucleation is strongly reduced by contamination. The critical temperature for nucleation is near 380 °C [72B6].

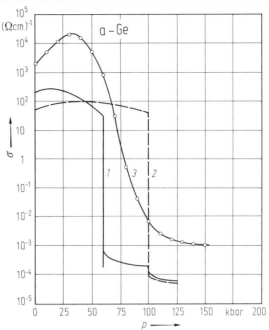

Fig. 160. a-Ge. RT-resistivity vs. pressure. Curve 1: evaporated a-Ge, curve 2: crystalline Ge, curve 3: plasma deposited a-Ge$_{0.98}$O$_{0.02}$ [81M2].

For Fig. 161, see next page.

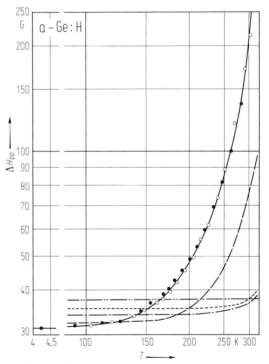

Fig. 162. a-Ge:H. Linewidth of reactively sputtered films ($T_s = 300$ K) vs. temperature. Hydrogen content c_H: —— 0, — — 1 at%, —·—· 2.8 at%, ··—·· 5.1 at%. The short-dashed curve ----- represents results on a deuterated film ($c_D = 3$ at%). The points ○ and ● demonstrate the reproducibility of two pure a-Ge films [76C1].

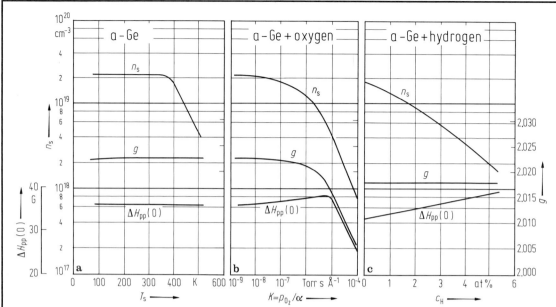

Fig. 161. a-Ge. Spin density n_s, g-value and temperature independent part of the linewidth $\Delta H_{pp}(0)$ vs. (a) substrate temperature T_s, (b) oxygen contamination (ratio of O_2-partial pressure and deposition rate α) and (c) hydrogen content c_H. Films in (a) and (b) were evaporated [76K3], those in (c) were sputtered [76P3].

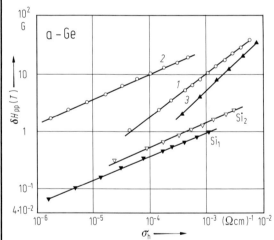

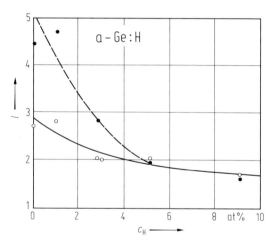

Fig. 163. a-Ge. Temperature dependent part of the linewidth $\delta H_{pp}(T) = \Delta H_{pp} - \Delta H_{pp}(0)$ vs. hopping conductivity for sputtered films [77P1]. Preparation parameters: curve 1: $p_H = 0$, $T_s = 300$ K, undoped, $n_s = 1.6 \cdot 10^{19}$ cm^{-3}; 2: $p_H = 6.6 \cdot 10^{-4}$ Torr, $T_s = 300$ K, p-doped, $n_s = 1.1 \cdot 10^{18}$ cm^{-3}; 3: $p_H = 6.3 \cdot 10^{-5}$ Torr, $T_s = 300$ K, undoped, $n_s = 1.2 \cdot 10^{19}$ cm^{-3}; Si$_1$ and Si$_2$ show the behavior of undoped sputtered a-Si films ($T_s = 300$ K). See Fig. 35 for similar data on a-Si. 1 Torr $\cong 133.3$ Pa.

Fig. 164. a-Ge:H. Lineshape factor l measured at 100 K (○) and 300 K (●) vs. hydrogen content c_H of reactively sputtered films ($T_s = 300$ K) [76C1]. The value for a Lorentzian line is 3.63.

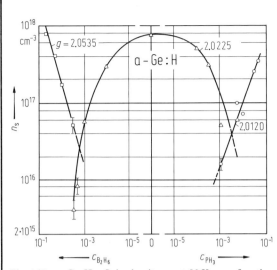

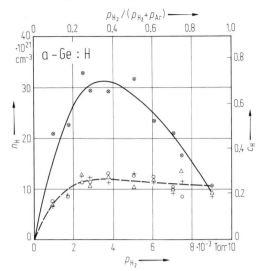

Fig. 165. a-Ge:H. Spin density n_s at 80 K as a function of the doping concentration (PH_3/GeH_4 ratio) c_{PH_3} and (B_2H_6/GeH_4 ratio) $c_{B_2H_6}$. Glow discharge deposited films at 493 K. The g-values of the ESR lines are indicated: $g =$ 2.0225 dangling bonds, $g = 2.0535$ valence band tail holes, $g = 2.0120$ conduction band tail electrons [83S9]; see also [83S10, 83S11].

Fig. 167. a-Ge:H. Hydrogen content n_H and hydrogen concentration $c_H = n_H/n_{Ge}$ of reactively sputtered films as a function of the partial pressure of hydrogen p_{H_2}. The data are determined from gas evolution (o), nuclear reaction (△) and from the integrated intensity of the IR stretching (⊗) and wagging modes (+) [80F1]. $n_{Ge} = 4.5 \cdot 10^{22}$ cm^{-3}.

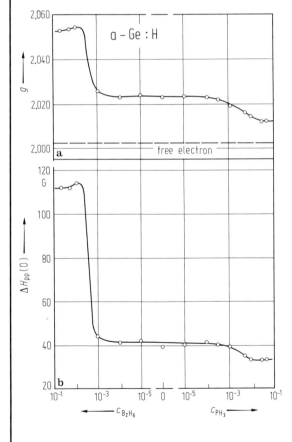

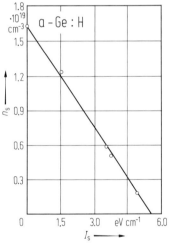

Fig. 168. a-Ge:H. Spin density n_s vs. integrated intensity $I_s = \int K(\omega) d(\hbar\omega)$ of the stretching mode absorption at 0.23 eV for hydrogenated sputtered films ($T_s = 300$ K) [76C1].

◄

Fig. 166. a-Ge:H. g-value (a) and temperature independent linewidth ΔH_{pp} (0) at 80 K (b) as a function of the doping concentrations (PH_3 or B_2H_6/GeH_4 ratio), c_{PH_3} and $c_{B_2H_6}$. Glow discharge deposited films at 493 K [83S9]; see also [83S10, 83S11].

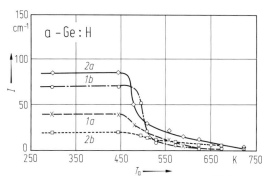

Fig. 169. a-Ge:H. Integrated intensity $I = \int \dfrac{K(\omega)}{\omega}\, d\omega$ of Ge-H stretching modes at 1895 cm^{-1} (curve *1*) and 1970 cm^{-1} (*2*) vs. annealing temperature T_a (annealing time: 1 h) for two hydrogenated films sputtered at 300 K containing (*a*) $9 \cdot 10^{20}$ and (*b*) $9 \cdot 10^{21}$ cm^{-3} hydrogen atoms [79B8]. Behavior of wagging and bending modes is studied in the same reference.

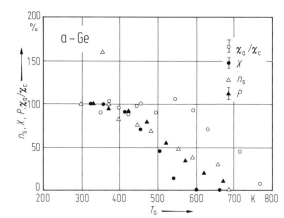

Fig. 170. a-Ge. Spin density n_s, porosity P, stress X and enhancement factor of the diamagnetic susceptibility χ_a/χ_c vs. annealing temperature. Films evaporated at $T_s = 300$ K. The quantities are plotted in % of the 300 K value. The decrease of n_s, X and P indicates annealing out of internal voids, the fact that χ_a/χ_c anneals at higher T_a indicates that the enhancement effect is associated with the bulk of the amorphous film. Values at 300 K: $n_s \approx 10^{20}$ cm^{-3}: $\chi_a/\chi_c = 2.5$ [74P1].

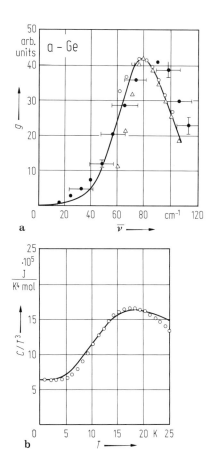

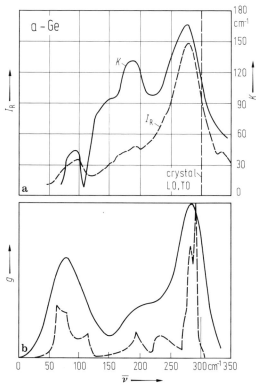

Fig. 172. a-Ge. (a) Vibrational density of states g vs. wavenumber deduced from Raman effect of sputtered a-Ge ($T_s = 300$ K) (o) [72L4], infrared absorption of sputtered a-Ge ($T_s = 300$ K) ($\triangle$) [74S6] and neutron scattering of a-Ge evaporated at 300 K (●) [74A1]. (b) Experimental (o) and calculated (solid line) temperature dependence of the heat capacity of RT evaporated a-Ge [74K1].

Fig. 171. a-Ge. (a) IR spectrum (absorption coefficient vs. wavenumber, solid line) and reduced Raman spectrum (Raman intensity vs. wavenumber, dashed line) of a-Ge sputtered at 300 K [74B2, 75A1]; (b) density of states of crystalline Ge (dashed). The solid line is a broadened density of states [71S2]. See also [82L3, 83M6].

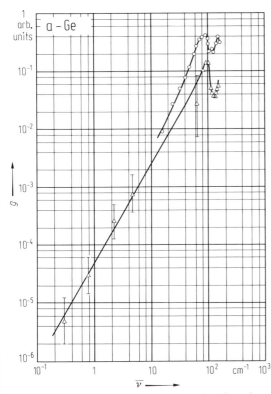

Fig. 173. a-Ge. Comparison of wavenumber dependence of vibrational density of states g of evaporated a-Ge ($T_s = 300\,\text{K}$) determined from neutron scattering (o) [74A1], far infrared and microwave conductivity (△) [76T].

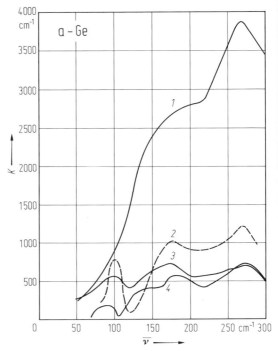

Fig. 174. a-Ge. IR absorption coefficient as a function of the wavenumber. This figure compares results of different authors on evaporated and sputtered material. Curve 1: [73P2], 2: [76T], 3: [74S6], 4: [74B2] (see also Fig. 171). The peaks correspond to the TA, LA and TO modes. Additional data at higher energies in [70T1], on hydrogenated films in [80S8] and [79B8].

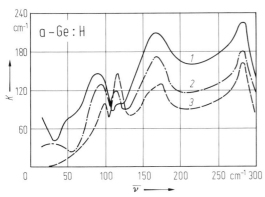

Fig. 175. a-Ge:H. Vibrational absorption spectra of pure and hydrogenated a-Ge prepared by sputtering at 300 K (absorption coefficient vs. wavenumber) [80S8]. Curve 1: pure, 2: $c_H = 6.1$ at%, 3: $c_H = 7.4$ at%.

For Fig. 176, see next page.

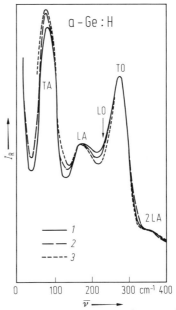

Fig. 177. a-Ge:H. Raman spectra of sputtered films ($T_s = 300\,\text{K}$) between 30 and 400 cm^{-1} (Raman intensity vs. wavenumber). H content calculated from IR absorption: Curve 1: 0, 2: 20 at%, 3: 30 at%. Laser energy $\hbar\omega_L = 2.4$ eV. Assignment of the structure is given in the figure [79B1].

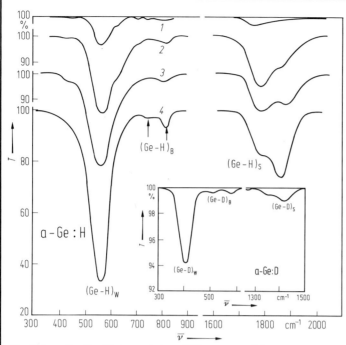

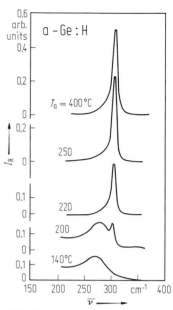

Fig. 176. a-Ge:H. IR transmission of films prepared by sputtering in an Ar/H$_2$ mixture at 300 K vs. wavenumber. H content n_H calculated from the integrated intensity of the IR absorption: Curve 1: $9 \cdot 10^{20}$ cm^{-3}, 2: $9 \cdot 10^{21}$ cm^{-3}, 3: $9 \cdot 10^{21}$ cm^{-3}, 4: $1.3 \cdot 10^{22}$ cm^{-3}. The spectrum in the insert was taken from a deuterated film. The peaks are attributed to the different modes of the Ge—H bonds: wag (Ge—H)$_W$, bend (Ge—H)$_B$, stretch (Ge—H)$_S$. Upon annealing the intensity of the bands decreases because of H effusion [79B8], see Fig. 169.

Fig. 180. a-Ge:H. Raman spectra (Raman intensity vs. wavenumber) of sample sputtered at 300 K for five annealing temperatures (annealing time: 1 h) showing the appearance of the Γ-phonon of crystalline Ge [79B8]. See also [82L3, 83M6].

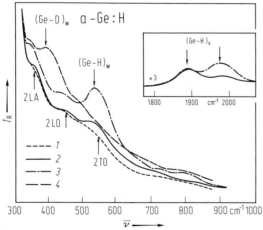

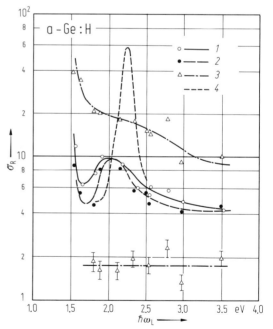

Fig. 178. a-Ge:H. Raman spectra of films, sputtered at 300 K, above 300 cm^{-1} (Raman intensity vs. wavenumber). H content: Curve 1: 0, 2: 20 at%, 3: 30 at%, 4: represents a deuterated sample. Laser energy $\hbar\omega_L = 2.4$ eV. Peaks are attributed to second order structure and wagging (Ge—H)$_W$ and stretching modes (Ge—H)$_S$ [79B1]. Results on electrolytic and sputtered a-Ge [72W].

Fig. 179. a-Ge:H. Raman scattering cross section (relative to CaF$_2$) σ_R as a function of laser energy $\hbar\omega_L$, for films sputtered at 300 K. The lowest curve represents the (Ge—H)$_W$ mode, while the others correspond to the Ge—Ge-vibrations between 130 and 300 cm^{-1}. H content: curve 1: 0, 2: 20 at%, 3: 30 at%. 4: crystalline Ge [79B1].

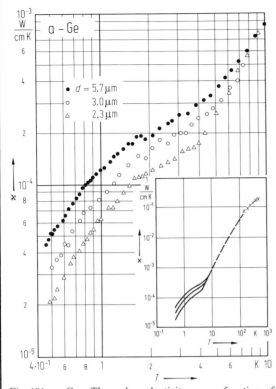

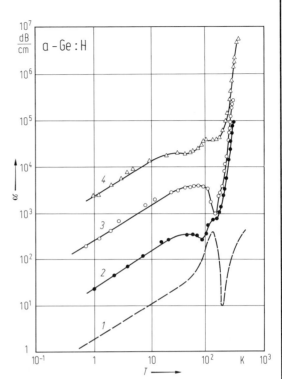

Fig. 181. a-Ge. Thermal conductivity κ as a function of temperature for Ge films evaporated at 300 K of different thickness d [77L2]. The inset shows a comparison of these data with high temperature measurements [74N1]. κ exhibits a plateau near 3 K indicating the existence of localized excitations.

Fig. 182. a-Ge:H. Ultrasonic absorption α (300 MHz) vs. temperature of a-Ge:H films sputtered at 370 K for different hydrogen concentrations c_H: curve 1: 0; 2: 1.5 at%; 3: 9 at%; 4: 19 at% [81B3].

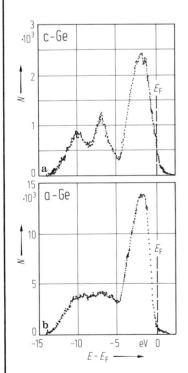

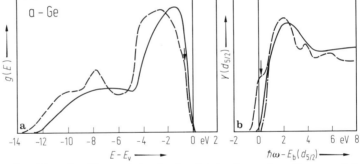

Fig. 184. a-Ge. Photoemission density of states of amorphous Ge evaporated at 300 K (solid line) and crystalline Ge (dashed line) as obtained from XPS studies. The valence band density of states (a) is derived from the electron energy distribution using $\hbar\omega = 25$ eV excitation with a background subtracted. The conduction band density of states (b) is derived from partial yield spectra ($d_{5/2}$ component) at $E - E_v = -8$ eV. The partial yield $Y(d_{5/2})$ is plotted vs. $\hbar\omega - E_b$ ($d_{5/2}$), i.e. relative to the valence band maximum. The conduction band edge in a-Ge has been sharpened (–·–) by 0.5 eV to correct broadening ($d_{5/2}$ core level plus experimental broadening). The pronounced shoulder at $E - E_c = 0$ in the crystalline partial yield spectrum is ascribed to excitonic effects [74E3]. For similar results, see [72L2] and for UPS spectra [71R].

◄ Fig. 183. a-Ge. Corrected XPS-valence band spectra (number of counts vs. binding energy) of crystalline (a) and evaporated ($T_s = 300$ K) amorphous Ge (b) using monochromised Al K_α X-rays. The binding energy is referred to the Fermi level of a thin Au layer [72L2].

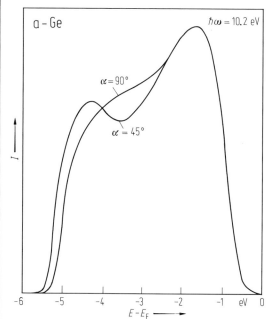

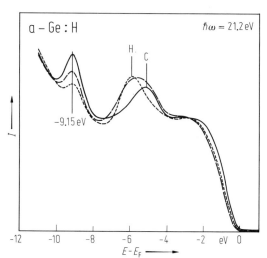

Fig. 185. a-Ge. Energy distribution of electrons excited by $\hbar\omega = 10.2$ eV (intensity vs. binding energy) for films evaporated at different orientations of the substrate. The curves are normalized at -1.8 eV, E_F is the Fermi level. Preparation conditions: 10^{-8} Torr, 1.5 Å/s, thickness 10^3 Å (T_s not given); the angle α of vapor incidence on the substrate was 45° and 90° [74O2]. 1 Torr $\cong 133.3$ Pa.

Fig. 186. a-Ge:H. Valence band XPS spectra excited by $\hbar\omega = 21.2$ eV (intensity vs. binding energy) of glow discharge deposited a-Ge:H films prepared at 300 K with different flow rates: —— 0.02 l/h, — — 0.04 l/h, ---- 0.1 l/h. The peak at 9.15 eV is assigned to the presence of Ar, the structures C and H are hydrogen induced [80G4].

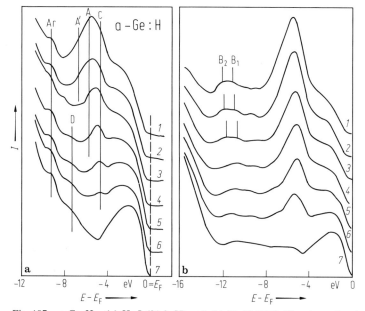

Fig. 187. a-Ge:H. (a) He I (21.2 eV) and (b) He II (40.8 eV) valence band spectra (intensity vs. binding energy) of a-Ge:H deposited in a glow discharge at $T_s = 300$ K from a gas mixture of 10% GeH$_4$ in Ar for a number of annealing steps: curve 1: 300 K; 2: 423 K; 3: 463 K; 4: 493 K; 5: 523 K; 6: 573 K; 7: 673 K. Peaks A and B are assigned to polyhydride configurations and peaks C and D to monohydride configurations. The structure near 9.15 eV is assigned to the 3 p levels of Ar [80G4].

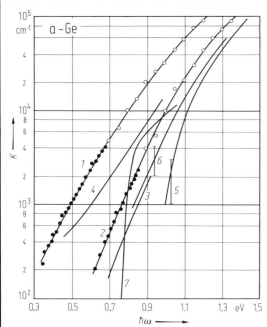

Fig. 188. a-Ge. Absorption coefficient K vs. photon energy. Curves 1 and 2: films sputtered at 300 K and 625 K [73C]; 3: sputtered at $T_s = 300$ K after 100 hours anneal at 425 K [72C1]; 4: evaporated at $T_s = 300$ K [67C]; 5: evaporated at 300 K and annealed to a stable state at 575 K [70T2]; 6: evaporated at 575 K [71T2]; 7: crystalline Ge [55D]. In curves 1 and 2, transmission and ellipsometry data are represented by full circles and open circles.

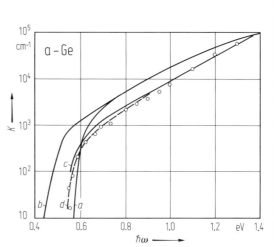

Fig. 189. a-Ge. Absorption coefficient K of evaporated films vs. photon energy. Curves a and b: evaporated at 200 °C and 20 °C under UHV conditions and measured in situ [70D2], c and d: obtained from transmission and photoconductivity on films evaporated in UHV at 450 K. Similar data: [70C2, 73C].

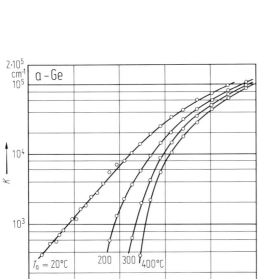

Fig. 190. a-Ge. Absorption edge of evaporated a-Ge deposited at 300 K and annealed step by step with 4°/min to the indicated temperatures (absorption coefficient vs. photon energy). The optical gap E_g thereby increases from 0.87 to 1 eV [74T].

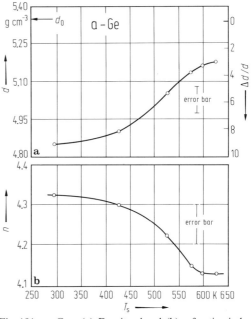

Fig. 191. a-Ge. (a) Density d and (b) refractive index n at 0.15 eV of evaporated films vs. substrate temperature. The scale to the right gives the density deficit $\Delta d/d$ with respect to the crystal [73O]. Similar data, see [71T2, 67W].

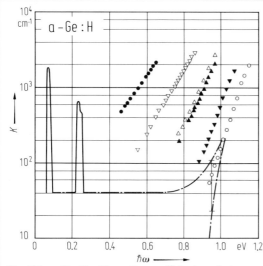

Fig. 192. a-Ge:H. Absorption coefficient vs. photon energy of hydrogenated sputtered films ($T_s = 300$ K). H content: $c_H = 0$ (●), 1 at% (▽), 2.8 at% (△), 3 at% (▲), 5.1 at% (▼), 8 at% (○). Error estimates (dashed line) are given for the 8 at% films [76C1]. See also [84P2].

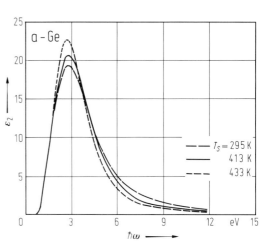

Fig. 195. a-Ge. Imaginary part of the dielectric constant vs. photon energy for three films deposited by evaporation at the indicated temperatures [72B8]. Same data as in Fig. 194.

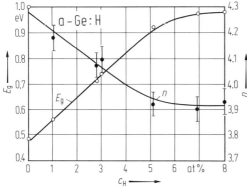

Fig. 193. a-Ge:H. Optical gap E_g defined by the energy at which $K = 10^3$ cm^{-1}, and refractive index at 0.125 eV of hydrogenated sputtered films ($T_s = 300$ K) as a function of hydrogen content [76C1].

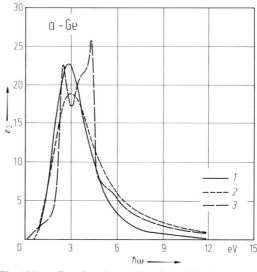

Fig. 196. a-Ge. Imaginary part of the dielectric constant vs. photon energy. Curve 1: near ideal evaporated a-Ge (UHV, $T_s = 430$ K) [72B7], 2: room-temperature evaporated a-Ge [70D3], and 3: crystalline Ge [63P]. Fig. from [74B6]. See also [74T].

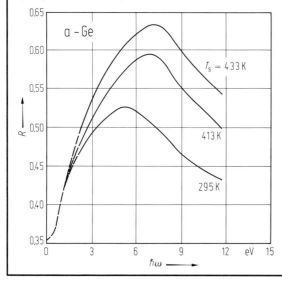

◄

Fig. 194. a-Ge. Reflectance of films deposited by evaporation on substrates held at the indicated temperatures vs. photon energy [72B8].

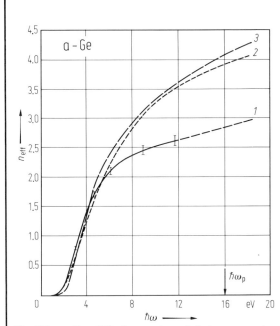

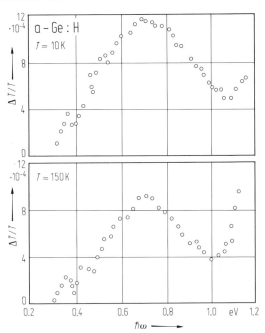

Fig. 197. a-Ge. Effective number of electrons per atom n_{eff} contributing to the optical absorption at photon energy $\hbar\omega$ vs. $\hbar\omega$. The plasma frequency $\hbar\omega_p$ is indicated by the arrow. Curve 1: near ideal (UHV, $T_s = 430$ K) evaporated a-Ge, 2: evaporated at 300 K, 3: crystalline Ge [74B6].

Fig. 198. a-Ge:H. Photoinduced absorption (change of transmission vs. photon energy) spectra of glow discharge deposited a-Ge:H at 10 K and 150 K [84P1]; see also [82O].

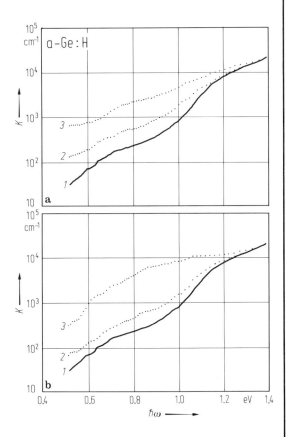

Fig. 199. a-Ge:H. Absorption coefficient vs. photon energy in the subbandgap region (photothermal deflection spectroscopy) of glow discharge deposited films ($T_s \approx 500$ K) doped with P (a) and B (b). Doping concentration (PH$_3$ or B$_2$H$_6$ in GeH$_4$): Fig. (a): curve 1: undoped, 2: $c_{\text{PH}_3} = 10^{-3}$, 3: $c_{\text{PH}_3} = 7 \cdot 10^{-2}$; Fig. (b): curve 1: undoped, 2: $c_{\text{B}_2\text{H}_6} = 10^{-3}$, 3: $c_{\text{B}_2\text{H}_6} = 5 \cdot 10^{-2}$ [83S8].

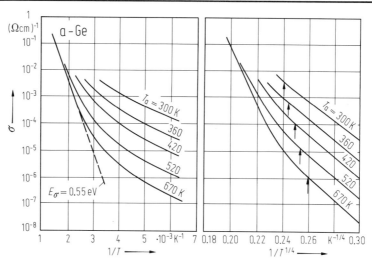

Fig. 200. a-Ge. Electrical conductivity vs. reciprocal temperature for a film evaporated at 300 K onto quartz at a rate of 100 Å/s in a vacuum of $2 \cdot 10^{-6}$ Torr. The film was annealed for 15 min at the indicated temperatures. In a $T^{-1/4}$-plot the curves are linear below the temperatures indicated by the arrows [74M1].

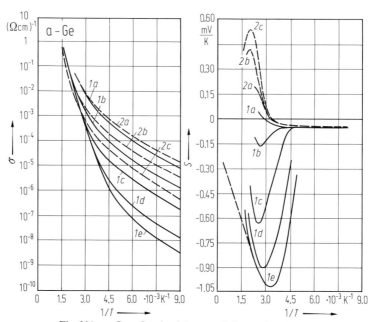

Fig. 201. a-Ge. Conductivity σ and thermoelectric power S vs. reciprocal temperature for two evaporated films in different annealing states. Curve 1: $T_s = 240$ K, $p = 2 \cdot 10^{-6}$ Torr, 50 Å/s and annealed for 15 min at 1a: 370 K, 1b: 445 K, 1c: 495 K, 1d: 540 K and 1e: 585 K; 2: $T_s = 300$ K, $6 \cdot 10^{-8}$ Torr, 100 Å/s and annealed for 15 min at 2a: 445 K, 2b: 575 K, 2c: 660 K [74B3]. Data at lower temperature: [76L2]. 1 Torr $\cong$ 133.3 Pa.

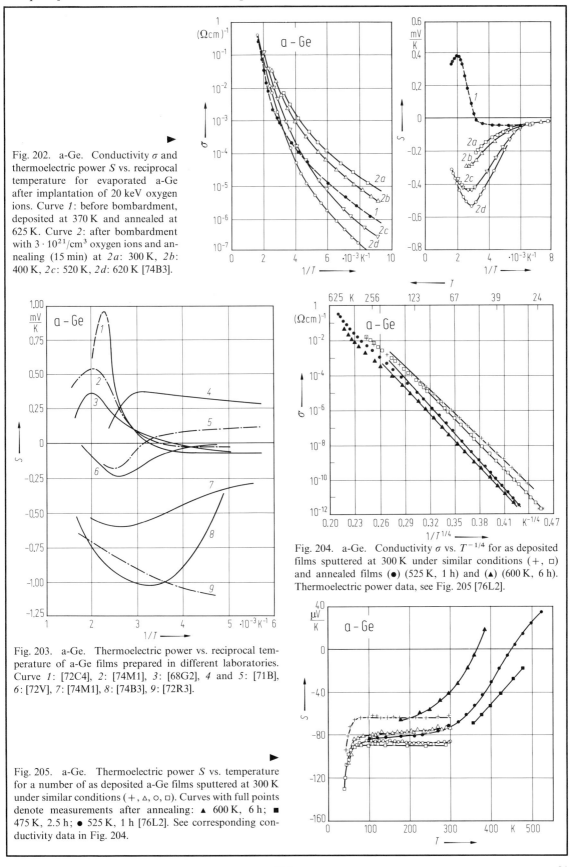

Fig. 202. a-Ge. Conductivity σ and thermoelectric power S vs. reciprocal temperature for evaporated a-Ge after implantation of 20 keV oxygen ions. Curve 1: before bombardment, deposited at 370 K and annealed at 625 K. Curve 2: after bombardment with $3 \cdot 10^{21}/cm^3$ oxygen ions and annealing (15 min) at 2a: 300 K, 2b: 400 K, 2c: 520 K, 2d: 620 K [74B3].

Fig. 203. a-Ge. Thermoelectric power vs. reciprocal temperature of a-Ge films prepared in different laboratories. Curve 1: [72C4], 2: [74M1], 3: [68G2], 4 and 5: [71B], 6: [72V], 7: [74M1], 8: [74B3], 9: [72R3].

Fig. 204. a-Ge. Conductivity σ vs. $T^{-1/4}$ for as deposited films sputtered at 300 K under similar conditions ($+$, $\square$) and annealed films ($\bullet$) (525 K, 1 h) and ($\blacktriangle$) (600 K, 6 h). Thermoelectric power data, see Fig. 205 [76L2].

Fig. 205. a-Ge. Thermoelectric power S vs. temperature for a number of as deposited a-Ge films sputtered at 300 K under similar conditions ($+$, $\triangle$, $\circ$, $\square$). Curves with full points denote measurements after annealing: $\blacktriangle$ 600 K, 6 h; $\blacksquare$ 475 K, 2.5 h; $\bullet$ 525 K, 1 h [76L2]. See corresponding conductivity data in Fig. 204.

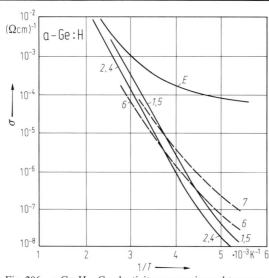

Fig. 206. a-Ge:H. Conductivity σ vs. reciprocal temperature of films deposited in a glow discharge at various substrate temperatures T_s. Curves $1\cdots5$: $T_s = 500$ K, 6: 400 K, 7: 300 K, (E): unannealed evaporated film [76J].

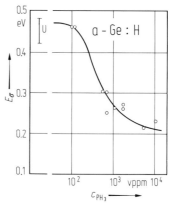

Fig. 208. a-Ge:H. Activation energy of conductivity E_σ vs. doping level given by the concentration c_{PH_3} of the doping gas in GeH$_4$. U refers to undoped films. Films deposited by glow discharge at $T_s = 500$ K [79J5].

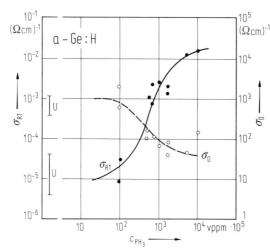

Fig. 207. a-Ge:H. Room-temperature conductivity σ_{RT} and pre-exponential factor σ_0 vs. doping level defined by the concentration of the doping gas in GeH$_4$, c_{PH_3}. U refers to undoped films. Films glow discharge deposited at $T_s = 500$ K [79J5]. For results on hydrogenated sputtered films, see [76P3].

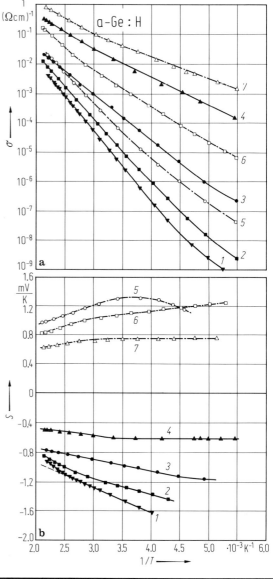

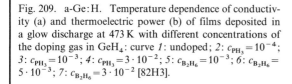

Fig. 209. a-Ge:H. Temperature dependence of conductivity (a) and thermoelectric power (b) of films deposited in a glow discharge at 473 K with different concentrations of the doping gas in GeH$_4$: curve 1: undoped; 2: $c_{PH_3} = 10^{-4}$; 3: $c_{PH_3} = 10^{-3}$; 4: $c_{PH_3} = 3 \cdot 10^{-2}$; 5: $c_{B_2H_6} = 10^{-3}$; 6: $c_{B_2H_6} = 5 \cdot 10^{-3}$; 7: $c_{B_2H_6} = 3 \cdot 10^{-2}$ [82H3].

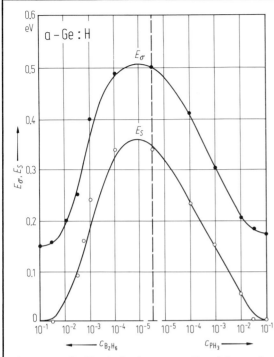

Fig. 210. a-Ge:H. Activation energy, E_σ, of the conductivity and, E_S, of the thermoelectric power near 300 K as a function of the concentrations of the doping gas in GeH_4, c_{PH_3} and $c_{B_2H_6}$. The films were glow discharge deposited at $T_s = 473$ K [82H3].

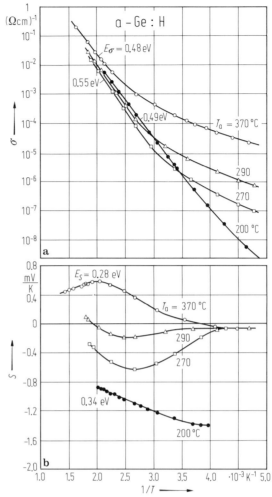

Fig. 211. a-Ge:H. Conductivity (a) and thermoelectric power (b) as a function of reciprocal temperature in various annealing states (5 min anneals). The undoped film was deposited in a glow discharge at $T_s = 473$ K [82H3]. Activation energies are indicated.

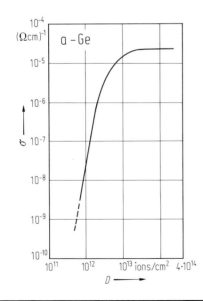

Fig. 212. a-Ge. Conductivity of a-Ge evaporated at 475 K vs. dose D of 100 keV Ar ions at 22 K [73O].

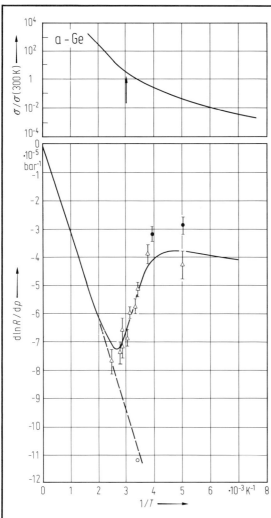

Fig. 213. a-Ge. Normalized conductivity and hydrostatic pressure coefficient of the resistance $\dfrac{\mathrm{d}\ln R}{\mathrm{d}p}$ as a function of reciprocal temperature. △ film deposited by sputtering at $T_s = 623$ K, ● $T_s = 300$ K [74K3]. ○ result on well annealed film [72C5]. Similar temperature dependences are observed in elastoresistance measurements [68G1, 70F, 72F2].

For Fig. 214, see next page.

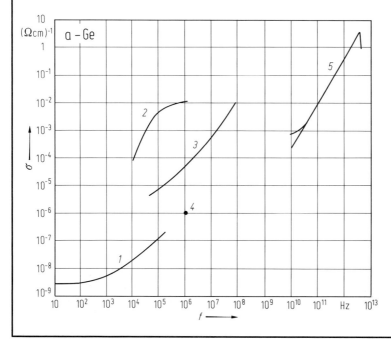

Fig. 215. a-Ge. Conductivity vs. frequency f obtained in different frequency ranges for a-Ge. Curve 1: [74M3] at 80 K; 2: [70C2] at 77 K; 3: [74A2] at 77 K; 4: [71T1] at 80 K; 5: [76T] at 80 K and 10 K. All films evaporated at 300 K.

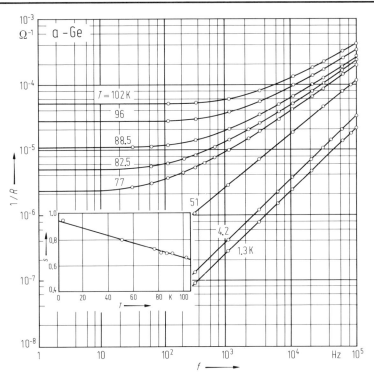

Fig. 214. a-Ge. Conductance of evaporated a-Ge ($T_s =$ 290 K) vs. frequency measured at the indicated temperatures. Inset: Slope s defined by $1/R \propto f^s$ as a function of temperature [80L1]. Additional data: [80L2, 81L4].

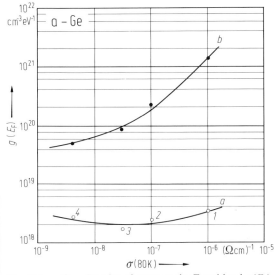

Fig. 216. a-Ge. Density of states at the Fermi level $g(E_F)$ deduced from (curve a) dc conductivity and (b) ac hopping conductivity vs. dc conductivity at 80 K. Curve a: 1: [70C2], 2: [74A2] evaporated at 300 K, 3: [73H1] evaporated, 4: [75A2] evaporated at 300 K, b: [76A1] glow discharge deposited at various T_s.

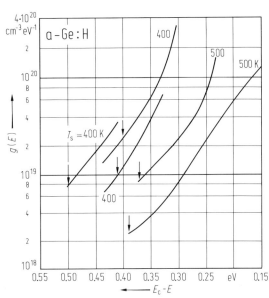

Fig. 217. a-Ge:H. Density of localized states $g(E)$ vs. distance from the mobility edge $E_c - E$ derived from field effect for films deposited by glow discharge at the indicated temperatures. The position of the Fermi level is marked by the arrows [79J5].

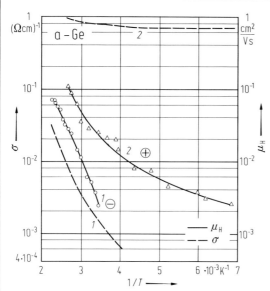

Fig. 218. a-Ge. Hall mobility and conductivity σ of two evaporated films vs. reciprocal temperature. Sample *1*: [74S8] evaporated at 300 K and annealed at 450 K, sample *2*: [73L3] evaporated at 353 K. The Hall effect is positive for sample *2* and negative for sample *1*.

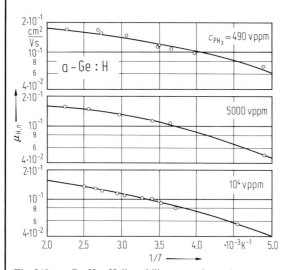

Fig. 219. a-Ge:H. Hall mobility vs. reciprocal temperature for n-type glow discharge films ($T_s = 500$ K) of different doping concentrations c_{PH_3} in GeH$_4$. The sign of the Hall effect is positive [79J5].

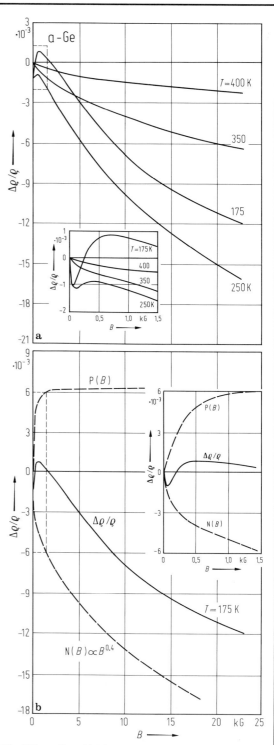

Fig. 220. a-Ge. (a) Magnetoresistance $\Delta\varrho/\varrho$ of an evaporated film vs. magnetic field B at various temperatures. (b) Separation of $\Delta\varrho/\varrho$ at 175 K into a positive and a negative component P (B) and N(B). Sample parameters: 100 Å/s, $T_s = 300$ K, $2\cdot 10^{-6}$ Torr, annealed to a steady state at 670 K [74M1]. Similar results in [73K2, 74C3]. 1 Torr $\hat{=} 133.3$ Pa.

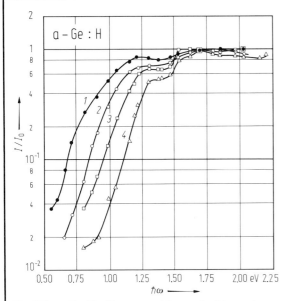

Fig. 221. a-Ge:H. Photocurrent (normalized to maximum of the curve) at 77 K vs. photon energy of sputtered hydrogenated films ($T_s = 300$ K). Curve 1: $c_H = 0$, 2: $c_H = 4.1$ at%, 3: $c_H = 7.4$ at%, 4: $c_H = 12.9$ at% [77M5].

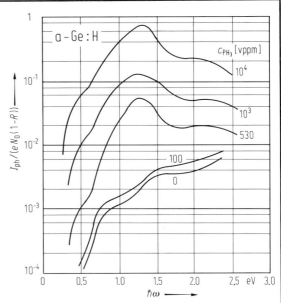

Fig. 222. a-Ge:H. Spectral dependence of the photoconductivity of differently doped n-type glow discharge films ($T_s = 500$ K). The ordinate is the photocurrent divided by the number of photons per second entering the sample. Excitation energy $E_{ex} \approx 1.3$ eV. The doping level is given by the concentration ratio of the dopant gas c_{PH_3} in GeH_4 [79J5].

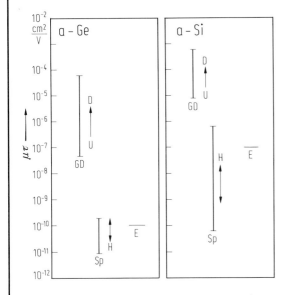

Fig. 223. a-Ge, a-Si. $\mu\tau$-product (mobility times lifetime) derived from photoconductivity experiments for differently prepared films. GD: glow discharge, Sp: sputtered in argon-hydrogen, E: thermal evaporation, U: undoped, D: doped, H: hydrogenated; wide range of deposition conditions [79J5].

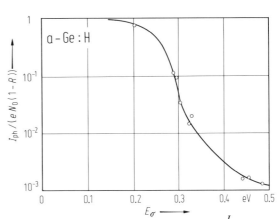

Fig. 224. a-Ge:H. Photoconductive gain $\dfrac{I_{ph}}{e N_0 (1-R)}$ defined by photocurrent divided by the photons per second entering the sample vs. activation energy of the dark conductivity E_σ of differently doped n-type films deposited by glow discharge at 500 K. E_σ is a measure for the distance of the Fermi level from the conducting states. Measurement taken at 300 K in gap geometry at $E_{ex} = 1.3$ eV and a field strength of $2 \cdot 10^3$ V/cm [79J5].

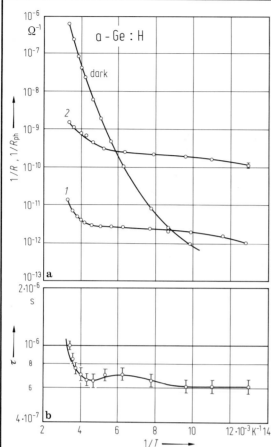

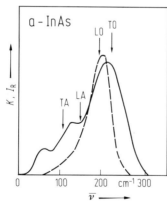

Fig. 225. a-Ge:H. (a) Photoconductance ($1/R$) of hydrogenated sputtered film ($T_s = 300$ K, $c_H = 7.4$ at%) vs. reciprocal temperature. Curve 1: $E_{ex} = 1.96$ eV, intensity: 10^{17} photons/cm²s; 2: 1.48 eV, $3 \cdot 10^{19}$ photons/cm²s. (b) Response time as a function of reciprocal temperature [77M5].

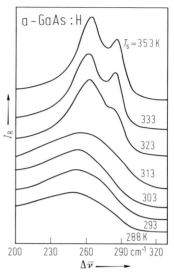

Fig. 226. a-GaAs:H. Raman spectra (Raman intensity vs. Raman shift) of films prepared by sputtering at various substrate temperatures [82W4]. Above $T_s = 323$ K the spectra split into the TO and LO modes of c-GaAs. See also [74L4, 73P2].

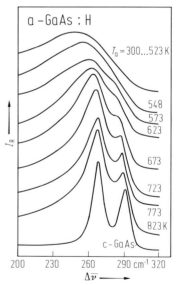

Fig. 227. a-GaAs:H. Raman spectra (Raman intensity vs. Raman shift) of a film prepared by sputtering at 300 K and annealed for 1 h at the indicated temperatures [82W4]. See also [74L4, 73P2].

Fig. 228. a-InAs. Reduced Raman intensity I_R and infrared absorption K vs. wavenumber of sputtered sample ($T_s = 173$ K). Full curve: Raman [74L4], dashed curve: IR absorption [73P2], $K_{max} = 3.7 \cdot 10^3$ cm⁻¹. The position of peaks in the crystalline one phonon density of states is indicated. Similar data on other III-V compounds, see [74L4, 73P2].

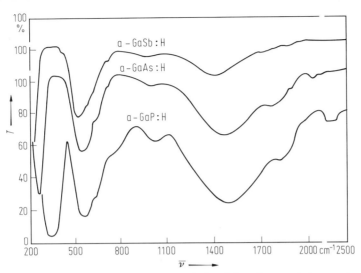

Fig. 229. a-GaP:H, a-GaAs:H, a-GaSb:H. Infrared transmission spectra (transmission vs. wavenumber) of hydrogenated III–V compounds which have been prepared by reactive sputtering at $T_s = 300$ K. The hydrogen content c_H is between 6 and 22 at% [82W4].

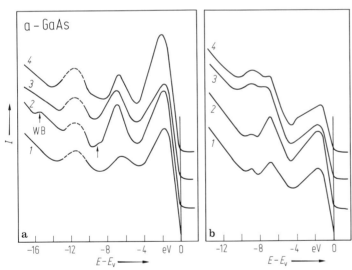

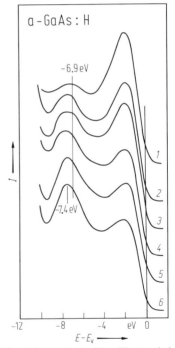

Fig. 230. a-GaAs. Photoemission valence band spectra (photoemission intensity vs. binding energy) excited with $\hbar\omega = 40.8$ eV (a) and 21.2 eV (b). The films are sputter deposited at 300 K and annealed for 1 h at the given temperatures. Curve 1: $T_a = 300$ K, as deposited; 2: 653 K; 3: 753 K; 4: 823 K. In a-GaAs the three peak structure of the valence band is retained. The zero of energy is the top of the valence band [83K2]. The structures near -9 eV and -15 eV (WB) are taken as indication for wrong bonds (As—As). See also [74L4] and for other III–V compounds [74S7].

Fig. 231. a-GaAs:H. Photoemission valence band spectra (photoemission intensity vs. binding energy) excited with $\hbar\omega = 26.9$ eV. The films are sputter deposited at $T_s = 300$ K in an Ar/H$_2$ gas mixture of different hydrogen content: curve 1: 0; 2: 1 vol%; 3: 5 vol%; 4: 10 vol%; 5: 30 vol%; 6: 50 vol%. The zero of the energy scale is the top of the valence band. The structure at -7.4 eV is attributed to the incorporation of hydrogen [83K2].

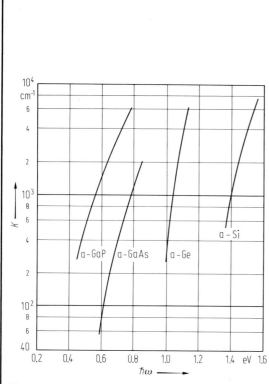

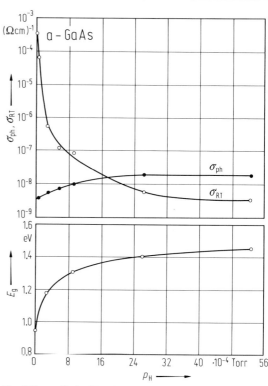

Fig. 232. a-GaP, a-GaAs. Absorption coefficient K vs. photon energy for amorphous films sputtered at various T_s and annealed to a steady state; a-Si, a-Ge for comparison [72C1]. Similar data on evaporated films [81G2], other III–V compounds: a-InSb [71E].

Fig. 233. a-GaAs:H. Optical gap E_g (defined as $\hbar\omega$ at $K=10^4$ cm^{-1}), room-temperature dark conductivity σ_{RT} and photoconductivity σ_{ph} measured with 10^{15} photons/cm^{2}s at $E_{ex}=1.95$ eV vs. hydrogen partial pressure for GaAs:H films sputtered at 277 K [77P2]. Similar data [80H3]. 1 Torr $\cong$ 133.3 Pa.

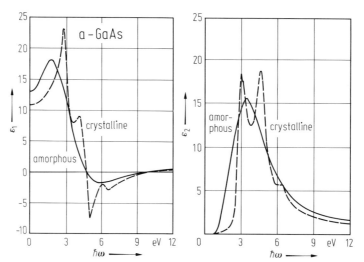

Fig. 234. a-GaAs. Real and imaginary parts of the dielectric constant vs. photon energy of a film prepared by flash-evaporation at 300 K [72S5]. The dashed curves are spectra of crystalline material from [63P]. Similar data on other III–V compounds [72S5, 80G5, 81G2].

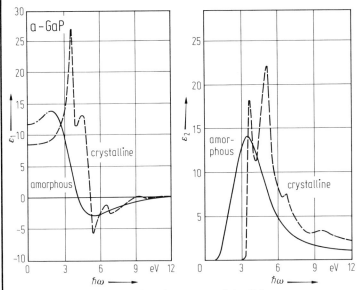

Fig. 235. a-GaP. Real and imaginary parts of the dielectric constant vs. photon energy of a film prepared by flash-evaporation at 300 K [72S5]. The dashed curves show the spectra of crystalline samples [63P].

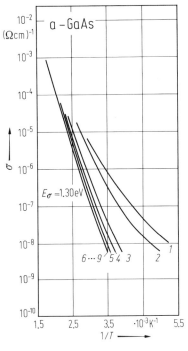

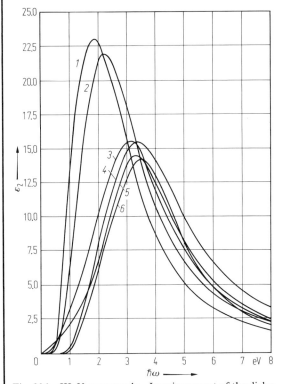

Fig. 236. III–V compounds. Imaginary part of the dielectric constant vs. photon energy of a number of flash-evaporated (240 K $< T_s <$ 300 K) amorphous III–V compounds: Curve 1: InSb, 2: GaSb, 3: InAs, 4: GaAs, 5: InP, 6: GaP [72S5].

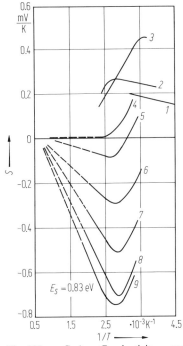

Fig. 237. a-GaAs. Conductivity σ and thermoelectric power S as a function of reciprocal temperature. The film is flash-evaporated at $T_s = 140$ K and annealed for 15 minutes. Curve 1: $T_a = 300$ K, 2: 360 K, 3: 425 K, 4: 460 K, 5: 470 K, 6: 490 K, 7: 520 K, 8: 546 K and 9: 570 K [72B9].

References for 11

51S	Szekeley, G.: J. Electrochem. Soc. **98** (1951) 318.
55D	Dash, W.C., Newman, R.: Phys. Rev. **99** (1955) 1151.
63P	Philipp, H.R., Ehrenreich, H.: Phys. Rev. **129** (1963) 1550.
65S	Sloope, B.W., Tiller, C.O.: J. Appl. Phys. **36** (1965) 3174.
66K1	Krikorian, E., Sneed, R.J.: J. Appl. Phys. **37** (1966) 3665.
66K2	Kane, E.O.: Phys. Rev. **146** (1966) 558.
66Z	Zeppenfeld, K., Raether, H.: Z. Phys. **193** (1966) 471.
67C	Clark, A.H.: Phys. Rev. **154** (1967) 750.
67W	Wales, J., Lowitt, G.J., Hill, R.A.: Thin Solid Films **1** (1967) 137.
68D	Donovan, T.M., Spicer, W.E.: Phys. Rev. Lett. **21** (1968) 1571.
68G1	Grigorovici, R., Vancu, A.: Thin Solid Films **2** (1968) 105.
68G2	Grigorovici, R.: Mater. Res. Bull. **3** (1968) 13.
69A	Austin, I.G., Mott, N.F.: Adv. Phys. **18** (1969) 41.
69B	Brodsky, M.H., Title, R.S.: Phys. Rev. Lett. **23** (1969) 581.
69C	Chittick, R.C., Alexander, J.H., Sterling, H.F.: J. Electrochem. Soc. **116** (1969) 77.
69D	Donovan, T.M., Spicer, W.E., Bennett, J.M.: Phys. Rev. Lett. **22** (1969) 1058.
69G	Grigorovici, R., Manaila, R.: Thin Solid Films **1** (1969) 343.
69L	Light, T.B.: Phys. Rev. Lett. **22** (1969) 999.
69M	Moss, S.C., Graczyk, J.F.: Phys. Rev. Lett. **23** (1969) 1167.
70B1	Beaglehole, D., Zavetova, M.: J. Non-Cryst. Solids **4** (1970) 272.
70B2	Brodsky, M.H., Title, R.S., Weiser, K., Petit, G.D.: Phys. Rev. **B 1** (1970) 2632.
70C1	Chittick, R.C.: J. Non-Cryst. Solids **3** (1970) 255.
70C2	Chopra, K.L., Bahl, S.K.: Phys. Rev. **B 1** (1970) 2545.
70D1	Devenyi, A., Belu, A., Korony, G.: J. Non-Cryst. Solids **4** (1970) 380.
70D2	Donovan, T.M., Ashley, E.J., Spicer, W.E.: Phys. Lett. **A 32** (1970) 85.
70D3	Donovan, T.M., Spicer, W.E., Ashley, E.J., Bennett, J.M.: Phys. Rev. **B 2** (1970) 397.
70F	Fuhs, W., Stuke, J.: Mater. Res. Bull. **5** (1970) 611.
70M	Moss, S.C., Graczyk, J.F.: Proc. 10th Int. Conf. Phys. Semiconductors, Cambridge, Mass. 1970, U.S.A. E.C. Div. Techn. Inform., Oak Ridge, Tenn. **1970**, p. 658.
70O	Osmun, J.W., Fritzsche, H.: Appl. Phys. Lett. **16** (1970) 87.
70S	Shaldervan, A.J., Nakhodkin, N.G.: Sov. Phys. Solid State **11** (1970) 2775 (transl. from Fiz. Tverd. Tela **11** (1969) 3407).
70T1	Tauc, J., Abraham, A., Zallen, R., Slade, M.: J. Non-Cryst. Solids **4** (1970) 279.
70T2	Theye, M.L.: Opt. Commun. **2** (1970) 239.
71B	Buchy, F., Clavaguera, M.T., Germain, Ph.: Proc. Int. Conf. on Low Mobility Materials, Eilat, Israel **1971**, p. 235.
71D	Donovan, T.M., Heinemann, K.: Phys. Rev. Lett. **27** (1971) 1794.
71E	Eckenbach, W., Fuhs, W., Stuke, J.: J. Non-Cryst. Solids **5** (1971) 264.
71J	Jungk, G.: Phys. Status Solidi **(b) 44** (1971) 239.
71R	Ribbing, C.G., Pierce, D.T., Spicer, W.E.: Phys. Rev. **B 4** (1971) 4417.
71S1	Smith, J.E., Brodsky, M.H., Crowder, B.L., Nathan, M.I.: Proc. 2nd Int. Conf. on Light Scattering in Solids, Balkanski, M. (ed.), Paris, Flammarion, **1977**, p. 330.
71S2	Smith Jr., J.E., Brodsky, M.H., Crowder, B.L., Nathan, M.I., Pinczuk, A.: Phys. Rev. Lett. **26** (1971) 642.
71T1	Taft, E.A.: J. Electrochem. Soc. **118** (1971) 1341.
71T2	Theye, M.L.: Mater. Res. Bull. **6** (1971) 103.
71T3	Tiainen, O.J.A.: Phys. Status Solidi **(a) 7** (1971) 583.
72A	Adler, D.: Amorphous Semiconductors, CRC Monoscience Series, London: Butterworth, **1972**.
72B1	Brodsky, M.H., Kaplan, D. Ziegler, J.F.: Appl. Phys. Lett. **21** (1972) 305.
72B2	Brodsky, M.H., Kaplan, D., Ziegler, J.F.: Proc. 11th Int. Conf. Physics of Semicond., Warsaw 1972, Warsaw: PWN-Polish Scientific Publishers, **1972**, p. 529.
72B3	Brown, F.C., Rustgi, O.P.: Phys. Rev. Lett. **28** (1972) 497.
72B4	Brodsky, M.H., Gambino, R.J.: J. Non-Cryst. Solids **8–10** (1972) 739.
72B5	Breitling, G.: J. Non-Cryst. Solids **8–10** (1972) 395.
72B6	Barna, A., Barna, P.B., Pocza, J.F.: J. Non-Cryst. Solids **8–10** (1972) 36.
72B7	Bauer, R.S., Galeener, F.L.: Solid State Commun. **10** (1972) 1171.

72B8	Bauer, R.S., Galeener, F.L., Spicer, W.E.: J. Non-Cryst. Solids **8–10** (1972) 196.
72B9	Beyer, W., Stuke, J.: J. Non-Cryst. Solids **8–10** (1972) 321.
72C1	Connell, G.A.N., Paul, W.: J. Non-Cryst. Solids **8–10** (1972) 215.
72C2	Cargill III, G.S.: Phys. Rev. Lett. **28** (1972) 1372.
72C3	Connell, G.A.N.: Phys. Status Solidi **(b) 53** (1972) 213.
72C4	Chopra, K.L., Bahl, S.K.: Thin Solid Films **12** (1972) 211.
72C5	Camphausen, D.L., Connell, G.A.N., Paul, W.: J. Non-Cryst. Solids **8–10** (1972) 223.
72F1	Fischer, J.E., Donovan, T.M.: J. Non-Cryst. Solids **8–10** (1972) 202.
72F2	Fuhs, W.: Phys. Status Solidi **(a) 10** (1972) 201.
72G	Gudat, W., Koch, E.E., Yu, P.Y., Cardona, M., Penchina, C.M.: Phys. Status Solidi **(b) 52** (1972) 505.
72H	Herd, S.R., Chaudhari, P., Brodsky, M.H.: J. Non-Cryst. Solids **7** (1972) 309.
72L1	Lewis, A.: Phys. Rev. Lett. **29** (1972) 1555.
72L2	Ley, L., Kowalczyk, S., Pollak, R., Shirley, D.A.: Phys. Rev. Lett. **29** (1972) 1088.
72L3	LeComber, P.G., Madan, A., Spear, W.E.: J. Non-Cryst. Solids **11** (1972) 219.
72L4	Lannin, J.S.: Solid State Commun. **11** (1972) 1523.
72P1	Pierce, D.T., Spicer, W.E.: Phys. Rev. **B 5** (1972) 3017.
72P2	Pierce, D.T., Ribbing, G., Spicer, W.E.: J. Non-Cryst. Solids **8–10** (1972) 959.
72R1	Rudee, M.L.: Thin Solid Films **12** (1972) 207.
72R2	Renner, O.: Thin Solid Films **12** (1972) 13.
72R3	Rockstadt, H.K., DeNeufville, J.P.: Proc. 11th Int. Conf. on Physics of Semicond., Warsaw 1972, Warsaw: Polish Scientific Publishers **1972**, p. 542.
72S1	Smith Jr., J.E., Brodsky, M.H., Crowder, B.L., Nathan, M.I.: J. Non-Cryst. Solids **8–10** (1972) 179.
72S2	Spear, W.E., LeComber, P.G.: J. Non-Cryst. Solids **8–10** (1972) 727.
72S3	Shevchik, N.J., Paul, W.: J. Non-Cryst. Solids **8–10** (1972) 381.
72S4	Sauvage, J.A., Mogab, C.J., Adler, D.: Philos. Mag. **25** (1972) 1305.
72S5	Stuke, J., Zimmerer, G.: Phys. Status Solidi **(b) 49** (1972) 513.
72T	Tan, S.I., Berry, B.S., Crowder, B.L.: Appl. Phys. Lett. **20** (1972) 88.
72V	Vescan, L., Telnic, M., Popescu, C.: Phys. Status Solidi **(b) 54** (1972) 733.
72W	Wihl, M., Cardona, M., Tauc, J.: J. Non-Cryst. Solids **8–10** (1972) 172.
73A	Aggarwal, S.C.: Phys. Rev. **B 7** (1973) 685.
73B	Bauer, R.S.: 5th Int. Conf. Amorphous and Liquid Semiconductors, Garmisch-Partenkirchen 1973, London: Taylor & Francis **1974**, p. 595.
73C	Connell, G.A.N., Temkin, R.J., Paul, W.: Adv. Phys. **22** (1973) 643.
73G1	Graczyk, J.F., Chaudhari, P.: Phys. Status Solidi **(b) 58** (1973) 163.
73G2	Goebel, H., Dettmer, K., Kessler, F.R.: Phys. Status Solidi **(a) 16** (1973) 61.
73H1	Hauser, J.J., Staudinger, A.: Phys. Rev. **B 8** (1973) 607.
73H2	Hudgens, S.J.: Phys. Rev. **B 7** (1973) 2481.
73H3	Hauser, J.J.: Phys. Rev. **B 8** (1973) 2678.
73J	Jungk, G.: Phys. Status Solidi **(b) 55** (1973) 579.
73K1	Knotek, M.L., Donovan, T.M.: Phys. Rev. Lett. **30** (1973) 652.
73K2	Kubelik, J., Triska, A.: Czech. J. Phys. **B 23** (1973) 123.
73L1	Loveland, R.J., Spear, W.E., Al-Sharbaty, A.: J. Non-Cryst. Solids **13** (1973/74) 55.
73L2	Laude, L.D., Willis, R.P., Fitton, B.: Solid State Commun. **12** (1973) 1007.
73L3	Lomas, R.A., Hamshire, M.J., Tanlinson, R.D., Knott, K.F.: Phys. Status Solidi **(a) 16** (1973) 385.
73O	Olley, J.A.: Solid State Commun. **13** (1973) 1441.
73P1	Paul, W., Connell, G.A.N., Temkin, R.J.: Adv. Phys. **22** (1973) 529.
73P2	Prettl, W., Shevchik, N.J., Cardona, M.: Phys. Status Solidi **(b) 59** (1973) 241.
73P3	Paul, D.K., Mitra, S.S.: Phys. Rev. Lett. **31** (1973) 1000.
73R	Rudee, M.L.: Philos. Mag. **25** (1973) 1149.
73S1	Stimets, R.W., Waldman, J., Lin, J., Chang, T.S., Temkin, R.J., Connell, G.A.N.: Solid State Commun. **13** (1973) 1485.
73S2	Shevchik, N.J., Paul, W.: J. Non-Cryst. Solids **13** (1973) 1.
73S3	Shevchik, N.J.: Phys. Rev. Lett. **31** (1973) 1245.
73T	Temkin, R.J., Paul, W., Connell, G.A.N.: Adv. Phys. **22** (1973) 581.
74A1	Axe, J.D., Keating, D.T., Cargill, G.S., Alben, R.: AIP Conf. Proc. **20** (1974) 279.

74A2	Arizumi, T., Yoshida, A., Saji, K.: Proc. 5th Int. Conf. on Amorphous and Liquid Semicond., Garmisch-Partenkirchen 1973, Stuke, J., Brenig, W., (eds.). London: Taylor & Francis, **1974**, p. 1065.
74A3	Arizumi, T., Yoshida, A., Baba, T., Shinakawa, K., Nitta, S.: AIP Conf. Proc. **20** (1974) 363.
74A4	Anderson, G.W., Davey, J.E., Comas, J., Saks, N.S.: J. Appl. Phys. **45** (1974) 4528.
74B1	Bauer, R.S.: AIP Conf. Proc. **20** (1974) 126.
74B2	Brodsky, M.H., Lurio, A.: Phys. Rev. **B 9** (1974) 1646.
74B3	Beyer, W., Stuke, J.: Proc. 5th Int. Conf. on Amorphous and Liquid Semiconductors, Garmisch-Partenkirchen 1973, Stuke, J., Brenig, W., (eds.). London: Taylor & Francis, **1974**, p. 251.
74B4	Bahl, S.K., Bhagat, S.M., Glosser, R.: Proc. 5th Int. Conf. on Amorphous and Liquid Semiconductors, Garmisch-Partenkirchen 1973, Stuke, J., Brenig, W., (eds.). London: Taylor & Francis, **1974**, p. 69.
74B5	Barna, A., Barna, P.B., Bodó, Z., Pócza, J.F., Pozsgai, I., Radnóczi, G.: Proc. 5th Int. Conf. on Amorphous and Liquid Semiconductors, Garmisch-Partenkirchen 1973, Stuke, J., Brenig, W., (eds) London: Taylor & Francis, **1974**, p. 109.
74B6	Bauer, R.S.: Proc. 5th Int. Conf. on Amorphous and Liquid Semiconductors, Garmisch-Partenkirchen 1973, Stuke, J., Brenig, W., (eds.) London: Taylor & Francis, **1974**, p. 595.
74B7	Botila, T., Croitoriu, N., Ioanid, G., Stoica, T., Vescan, L.: AIP Conf. Proc. **20** (1974) 33.
74C1	Czapla, A., Szczyrbowski, J.: Acta Phys. Pol. **A 45** (1974) 193.
74C2	Connell, G.A.N., Temkin, R.J.: Phys. Rev. **B 9** (1974) 5323.
74C3	Clark, A.H., Cohen, M., Campi, M., Lanyon, H.P.D.: J. Non-Cryst. Solids **16** (1974) 117.
74D	Donovan, T.M., Knotek, M.L., Fischer, J.E.: Proc. 5th Int. Conf. on Amorphous and Liquid Semiconductors, Garmisch-Partenkirchen 1973, Stuke, J., Brenig, W., (eds.) London: Taylor & Francis, **1974**, p. 549.
74E1	Engemann, D., Fischer R.: Proc. 12th Int. Conf. Physics of Semiconductors, Pilkuhn, M.H., (ed.) Stuttgart: Teubner, **1974**, p. 1042.
74E2	Engemann, D., Fischer, R.: Proc. 5th Int. Conf. on Amorphous and Liquid Semiconductors, Garmisch-Partenkirchen 1973, Stuke, J., Brenig, W., (eds.) London: Taylor & Francis, **1974**, p. 947.
74E3	Eastman, D.E., Freeouf, J.L., Erbudak, M.: AIP Conf. Proc. **20** (1974) 95.
74E4	Elliott, P.C., Yoffe, A.D., Davis, E.A.: AIP Conf. Proc. **20** (1974) 311.
74H1	Hesse, H.J., Fuhs, W., Langer, K.H.: Proc. 5th Int. Conf. on Amorphous and Liquid Semiconductors, Garmisch-Partenkirchen 1973, Stuke, J., Brenig, W., (eds.) London: Taylor & Francis. **1974**, p. 79.
74H2	Hirose, M., Hayama, M., Osaka, Y.: Jpn. J. Appl. Phys. **13** (1974) 1399.
74H3	Hirose, M., Suzuki, T., Yoshifuji, S., Osaka, Y.: Jpn. J. Appl. Phys. **13** (1974) 40.
74J	Johannessen, J.S.: Phys. Status Solidi **(a) 26** (1974) 571.
74K1	King, C.N., Phillips, W.A., de Neufville, J.P.: Phys. Rev. Lett. **32** (1974) 538.
74K2	Knotek, M.L.: AIP Conf. Proc. **20** (1974) 297.
74K3	Kastner, M., Connell, G.A.N., Lewis, A., Paul, W., Temkin, R.J.: Proc. 5th Int. Conf. on Amorphous and Liquid Semiconductors, Garmisch-Partenkirchen 1973, Stuke, J., Brenig, W., (eds.) London: Taylor & Francis, **1974**, p. 237.
74L1	LeComber, P.G., Loveland, R.J., Spear, W.E., Vaughan, R.A.: Proc. 5th Int. Conf. on Amorphous and Liquid Semiconductors, Garmisch-Partenkirchen 1973, Stuke, J., Brenig, W., (eds.) London: Taylor & Francis, **1974**, p. 245.
74L2	Lucovsky, G.: Proc. 5th Int. Conf. on Amorphous and Liquid Semiconductors, Garmisch-Partenkirchen 1973, Stuke, J., Brenig, W., (eds.) London: Taylor & Francis, **1974**, p. 1099.
74L3	Lytle, F.W., Sayers, D.E., Eikum, A.K.: J. Non-Cryst. Solids **13** (1974) 69.
74L4	Lewis, A.J., Connell, G.A.N., Paul, W., Pawlik, J.R., Temkin, R.J.: AIP Conf. Proc. **20** (1974) 27.
74L5	Lannin, J.S.: AIP Conf. Proc. **20** (1974) 260.
74M1	Mell, H.: Proc. 5th Int. Conf. on Amorphous and Liquid Semiconductors, Garmisch-Partenkirchen 1973, Stuke, J., Brenig, W., (eds.) London: Taylor & Francis, **1974**, p. 203.
74M2	Malhotra, A.K., Neudeck, G.W.: Appl. Phys. Lett. **24** (1974) 557.
74M3	Mell, H.: AIP Conf. Proc. **20** (1974) 357.
74N1	Nath, P., Chopra, K.L.: Phys. Rev. **B 10** (1974) 3412.
74N2	Narasimhan, K.L., Guha, S.: J. Non-Cryst. Solids **16** (1974) 143.
74O1	Ottaviani, G., Marrello, V., Sigurd, D., Mayer, J.W., McCaldin, J.O.: J. Appl. Phys. **45** (1974) 1730.

74O2	Orlowski, B., Spicer, W.E., Baer, A.D.: AIP Conf. Proc. **20** (1974) 241.
74O3	Olley, J.A., Yoffee, A.D.: Proc. 5th Int. Conf. on Amorphous and Liquid Semicond., Garmisch-Partenkirchen 1973, Stuke, J., Brenig, W., (eds.) London: Taylor & Francis, **1974**, p. 73.
74P1	Paesler, M.A., Agarwal, S.C., Hudgens, S.J., Fritzsche, H.: AIP Conf. Proc. **20** (1974) 37.
74P2	Pollak, R.A.: AIP Conf. Proc. **20** (1974) 90.
74S1	Shevchik, N.J., Paul, W.: J. Non-Cryst. Solids **16** (1974) 55.
74S2	Spicer, W.E.: Proc. 5th Int. Conf. on Amorphous and Liquid Semiconductors, Garmisch-Partenkirchen 1973, Stuke, J., Brenig, W., (eds.) London: Taylor & Francis, **1974**, p. 499.
74S3	Shevchik, N.J.: AIP Conf. Proc. **20** (1974) 72.
74S4	Spear, W.E., Loveland, R.J., Al-Sharbaty, A.: J. Non-Cryst. Solids **15** (1974) 410.
74S5	Spear, W.E.: Proc. 5th Int. Conf. on Amorphous and Liquid Semiconductors, Garmisch-Partenkirchen 1973, Stuke, J., Brenig, W., (eds.) London: Taylor & Francis, **1974**, p. 1.
74S6	Stimets, R.W., Waldman, J., Lin, J., Chang, T.S., Temkin, R.J., Connell, G.A.N.: Proc. 5th Int. Conf. on Amorphous and Liquid Semiconductors, Garmisch-Partenkirchen 1973, Stuke, J., Brenig, W., (eds.) London: Taylor & Francis, **1974**, p. 1239.
74S7	Shevchik, N.J., Tejeda, J., Cardona, M.: Phys. Rev. **B 9** (1974) 2627.
74S8	Seager, C.H., Knotek, M.L., Clark, A.H.: Proc. 5th Int. Conf. on Amorphous and Liquid Semiconductors, Garmisch-Partenkirchen 1973, Stuke, J., Brenig, W., (eds.) London: Taylor & Francis, **1974**, p. 1133.
74S9	Shimomura, O., Minomura, S., Sakai, N., Asaumi, K., Tamura, K., Fukushima, J., Endo, H.: Philos. Mag. **29** (1974) 547.
74T	Theye, M.L.: Proc. 5th Int. Conf. on Amorphous and Liquid Semiconductors, Garmisch-Partenkirchen 1973, Stuke, J., Brenig, W., (eds.) London: Taylor & Francis, **1974**, p. 479.
74Y	Yates, D.A., Penchina, C.M.: Proc. 5th Int. Conf. on Amorphous and Liquid Semiconductors, Garmisch-Partenkirchen 1973, Stuke, J., Brenig, W., (eds.) London: Taylor & Francis, **1974**, p. 617.
75A1	Alben, R., Weaire, D., Smith Jr., J.E., Brodsky, M.H.: Phys. Rev. **B 11** (1975) 2271.
75A2	Agarwal, S.C., Guha, S., Narasimhan, K.L.: J. Non-Cryst. Solids **18** (1975) 429.
75B1	Bahl, S.K., Bhagat, S.M.: J. Non-Cryst. Solids **17** (1975) 409.
75B2	Beyer, W., Stuke, J.: Phys. Status Solidi **(a) 30** (1975) 511.
75B3	Beyer, W., Stuke, J., Wagner, H.: Phys. Status Solidi **(a) 30** (1975) 231.
75B4	Barthwal, S.K., Nath, P., Chopra, K.L.: Solid State Commun. **16** (1975) 723.
75C	Connell, G.A.N.: Phys. Status Solidi **(b) 69** (1975) 9.
75E	Engemann, D., Fischer, R., Richter, F.W., Wagner, H.: Proc, 6th Int. Conf. Amorphous and Liquid Semiconductors, Kolomiets, B.T., (ed.) Leningrad: Nauka **1975**, p. 217.
75F1	Fritzsche, H., Hudgens, S.J.: Proc. 6th Int. Conf. on Amorphous and Liquid Semiconductors, Kolomiets, B.T., (ed.) Leningrad: Nauka **1975**, p. 6.
75F2	Fischer, R., Vornholz, D.: Phys. Status Solidi **(b) 68** (1975) 561.
75K	Knotek, M.L.: Solid State Commun. **17** (1975) 1431.
75M	Malhotra, A.K., Neudeck, G.W.: J. Appl. Phys. **46** (1975) 2690.
75N	Neudeck, G.W., Malhotra, A.K.: J. Appl. Phys. **46** (1975) 2662.
75P	Pandya, D.K., Barthwal, S.K., Chopra, K.L.: Phys. Status Solidi **(a) 32** (1975) 489.
75R	Renucci, J.B., Tyte, R.N., Cardona, M.: Phys. Rev. **B 11** (1975) 3885.
75S	Spear, W.E., LeComber, P.G.: Solid State Commun. **17** (1975) 1193.
76A1	Abkowitz, M., LeComber, P.G., Spear, W.E.: Comm. on Physics **1** (1976) 175.
76A2	Araki, M., Ozaki, H.: Solid State Commun. **18** (1976) 1603.
76B1	Bahl, S.K., Bhagat, S.M.: J. Non-Cryst. Solids **21** (1976) 279.
76B2	Brodsky, M.H., Title, R.S.: AIP Conf. Proc. **31** (1976) 97.
76B3	Blum, N.A., Feldman, C.: J. Non-Cryst. Solids **22** (1976) 29.
76B4	Barthwal, S.K., Chopra, K.L.: Phys. Status Solidi **(a) 36** (1976) 345.
76C1	Connell, G.A.N., Pawlik, J.R.: Phys. Rev. **B 13** (1976) 787.
76C2	Chik, K.P., Pui-Kong, Lim: Thin Solid Films **35** (1976) 45.
76C3	Chopra, K.L., Pandya, D.K.: Phys. Status Solidi **(a) 36** (1976) 89.
76D	DiSalvo, F.J., Bagley, B.G., Hutton, R.S., Clark, A.H.: Solid State Commun. **19** (1976) 97.
76E1	Engemann, D., Fischer, R.: AIP Conf. Proc. **31** (1976) 37.
76E2	Enck, R.C., Pfister, G.: in: Photoconductivity and Related Phenomena, Mort, J., Pai, D.M., (eds.) New York: Elsevier **1976**, p. 297.
76G	Greene, J.E., Mei, L.: Thin Solid Films **37** (1976) 429.

76H1	Hauser, J.J.: Solid State Commun. **19** (1976) 1049.
76H2	Hirose, M., Che, I., Osaka, Y.: in: Electronic Phenomena in Non-Cryst. Solids, Kolomiets, B.T., (ed.) Acad. Science USSR, **1976**, p. 255.
76J	Jones, D.I., Spear, W.E., LeComber, P.G.: J. Non-Cryst. Solids **20** (1976) 259.
76K1	Knights, J.C.: AIP Conf. Proc. **31** (1976) 296.
76K2	Kinney, W.I.: J. Non-Cryst. Solids **21** (1976) 275.
76K3	Kubler, L., Gewinner, G., Koulmann, J.J., Jaegle, A.: Phys. Status Solidi **(b) 78** (1976) 149.
76K4	Knights, J.C.: Jpn. J. Appl. Phys. **18**, Suppl. 1 (1976) 101.
76L1	Lewis, A.J.: Phys. Rev. **B 14** (1976) 658.
76L2	Lewis, A.J.: Phys. Rev. **B 13** (1976) 2565.
76M1	Madan, A., LeComber, P.G., Spear, W.E.: J. Non-Cryst. Solids **20** (1976) 239.
76M2	Müller, G., Kalbitzer, S.: Proc. 4th Int. Conf. on the Physics of Non-Crystalline Solids, Clausthal-Zellerfeld 1976, Frischat, G.H., (ed.) Trans. Tech. S.A. **1977**, p. 278.
76M3	Moustakas, T.D., Connell, G.A.N., Paul, W.: in: Electronic Phenomena in Non-Cryst. Solids, Kolomiets, B.T., (ed.) Acad. Science USSR, **1976**, p. 310.
76P1	Pankove, J.I., Carlson, D.E.: Appl. Phys. Lett. **29** (1976) 620.
76P2	Paul, W., Lewis, A.J., Connell, G.A.N., Moustakas, T.D.: Solid State Commun. **20** (1976) 969.
76P3	Pawlik, I.R., Connell, G.A.N., Prober, D.: in: Electronic Phenomena in Non-Cryst. Solids, Kolomiets, B.T., (ed.) Acad. Science USSR, **1976**, p. 304.
76R	Rehm, W., Engemann, D., Fischer, R., Stuke, J.: Proc. 13th Int. Conf. on Physics of Semiconductors, Rome **1976**, p. 525.
76S1	Spear, W.E., LeComber, P.G.: Philos. Mag. **33** (1976) 935.
76S2	Spear, W.E., LeComber, P.G.: in: Photoconductivity and Related Phenomena, Mort, J., Pai, D.M., (eds.) New York: Elsevier, **1976**, p. 213.
76T	Taylor, P.C., Strom, U., Hendrickson, J.R., Bahl, S.K.: Phys. Rev. **B 13** (1976) 1711.
76V1	Voget-Grote, U., Stuke, J., Wagner, H.: AIP Conf. Proc. **31** (1976) 91.
76V2	Vescan, L., Croitoriu, N.: in: Electronic Phenomena in Non-Cryst. Solids, Kolomiets, B.T., (ed.) Acad. Science USSR, **1976**, p. 244.
77A1	Anderson, D.A., Moustakas, T., Paul, W.: Proc. 7th Int. Conf. on Amourphous and Liquid Semiconductors, Edinburgh 1977, Spear, W.E., (ed.), Univ. Edinburgh, CICL **1977**, p. 334.
77A2	Allan, D., LeComber, P.G., Spear, W.E.: Proc. 7th Int. Conf. on Amorphous and Liquid Semiconductors, Edinburgh 1977, Spear, W.E., (ed.) Univ. Edinburgh, CICL **1977**, p. 323.
77A3	Anderson, D.A., Spear, W.E.: Philos. Mag. **36** (1977) 695.
77B1	Brodsky, M.H., Frisch, M.A., Ziegler, J.F., Lanford, W.A.: Appl. Phys. Lett. **30** (1977) 561.
77B2	Brodsky, M.H.: Thin Solid Films **40** (1977) L 23.
77B3	Barna, A., Barna, P.B., Radnoczi, G., Toth, L., Thomas, P.: Phys. Status Solidi **(a) 41** (1977) 81.
77B4	Brodsky, M.H., Cardona, M., Cuomo, J.J.: Phys. Rev. **B 16** (1977) 3556.
77B5	Bermejo, D., Cardona, M., Brodsky, M.H.: Proc. 7th Int. Conf. on Amorphous and Liquid Semiconductors, Edinburgh 1977, Spear, W.E., (ed.) Univ. Edinburgh, CICL **1977**, p. 343.
77B6	Beeman, D., Alben, R.: Adv. Phys. **26** (1977) 339.
77B7	Brodsky, M.H., Cuomo, J.J., Evangelisti, F.: Proc. 7th Int. Conf. on Amorphous and Liquid Semiconductors, Edinburgh 1977, Spear, W.E., (ed.) Univ. Edinburgh, CICL **1977**, p. 397.
77B8	Beyer, W., Mell, H., Overhof, H.: Proc. 7th Int. Conf. on Amorphous and Liquid Semiconductors, Edinburgh 1977, Spear, W.E., (ed.) Univ. Edinburgh, CICL **1977**, p. 328.
77B9	Biegelsen, D.K., Knights, J.C.: Proc. 7th Int. Conf. on Amorphous and Liquid Semiconductors, Edinburgh 1977, Spear, W.E., (ed.) Univ. Edinburgh, CICL **1977**, p. 429.
77B10	Beyer, W., Mell, H., Overhof, H.: in: Proc. 7th Int. Conf. on Amorphous and Liquid Semiconductors, Edinburgh 1977, Spear, W.E., (ed.) Univ. Edinburgh, CICL **1977**, p. 333.
77C1	Clark, G.J., White, C.W., Allred, D.D., Appleton, B.R., Magee, C.W., Carlson, D.E.: Appl. Phys. Lett. **31** (1977) 582.
77C2	Chevallier, J., Wieder, H., Onton, A., Guarnieri, C.R.: Solid State Commun. **24** (1977) 867.
77C3	Chopra, K.L., Randhawa, H.S., Malhotra, L.K.: Thin Solid Films **47** (1977) 203.
77D	DeChelle, F., Raisin, C., Robin-Kandare, S.: Thin Solid Films **46** (1977) 187.
77E1	Emin, D.: Philos. Mag. **35** (1977) 1189.
77E2	Engemann, D., Fischer, R., Mell, H.: Proc. 7th Int. Conf. on Amorphous and Liquid Semiconductors, Edinburgh 1977, Spear, W.E., (ed.) Univ. Edinburgh, CICL **1977**, p. 387.
77E3	Engemann, D., Fischer, R.: Phys. Status Solidi **(b) 79** (1977) 195.

77F	Friedman, L.: Philos. Mag, **36** (1977) 553.
77H	Hauser, J.J., DiSalvo, F.J., Hutton, R.S.: Philos. Mag. **35** (1977) 1557.
77J	Jones, D.I., LeComber, P.G., Spear, W.E.: Philos. Mag. **B 36** (1977) 541.
77K1	Knights, J.C., Biegelsen, D.K., Solomon, I.: Solid State Commun. **22** (1977) 133.
77K2	Knights, J.C., Hayes, T.M., Mikkelson Jr., J.C.: Phys. Rev. Lett. **39** (1977) 712.
77K3	Knights, J.C.: Proc. 7th Int. Conf. on Amorphous and Liquid Semiconductors, Edinburgh 1977, Spear, W.E., (ed.) Univ. Edinburgh, CICL **1977**, p. 433.
77L1	LeComber, P.G., Jones, D.I., Spear, W.E.: Philos. Mag. **35** (1977) 1173.
77L2	Lohneysen, H.v., Steglich, F.: Phys. Rev. Lett. **39** (1977) 1420.
77M1	Moustakas, T.D., Anderson, D.A., Paul, W.: Solid State Commun. **23** (1977) 155.
77M2	Müller, G., Kalbitzer, S., Spear, W.E., LeComber, P.G.: Proc. 7th Int. Conf. on Amorphous and Liquid Semiconductors, Edinburgh 1977, Spear, W.E., (ed.) Univ. Edinburgh, CICL **1977**, p. 442.
77M3	Moore, A.R.: Appl. Phys. Lett. **31** (1977) 762.
77M4	Movaghar, B., Schweitzer, L.: Phys. Status Solidi **(b) 80** (1977) 491.
77M5	Moustakas, T.D., Paul, W.: Phys. Rev. **B 16** (1977) 1564.
77N1	Nitta, S., Shimakawa, K., Sakaguchi, K.: J. Non-Cryst. Solids **24** (1977) 137.
77N2	Nashashibi, T.S., Austin, I.G., Searle, T.M.: Proc. 7th Int. Conf. on Amorphous and Liquid Semiconductors, Edinburgh 1977, Spear, W.E., (ed.) Univ. Edinburgh, CICL **1977**, p. 392.
77N3	Nashashibi, T.S., Austin, I.G., Searle, T.M.: Philos. Mag. **35** (1977) 831.
77P1	Pawlik, J.R., Paul, W.: Proc. 7th Int. Conf. on Amorphous and Liquid Semiconductors, Edinburgh 1977, Spear W.E., (ed.) Univ. Edinburgh, CICL **1977**, p. 437.
77P2	Paul, W., Moustakas, T.D., Anderson, D.A., Freeman, E.: Proc. 7th Int. Conf. on Amorphous and Liquid Semiconductors, Edinburgh 1977, Spear, W.E., (ed.) Univ. Edinburgh, CICL **1977**, p. 467.
77R1	v. Roedern, B., Ley, L., Cardona, M.: Phys. Rev. Lett. **39** (1977) 1576.
77R2	Rehm, W., Fischer, R., Stuke, J., Wagner, H.: Phys. Status Solidi **(b) 79** (1977) 529.
77S1	Spear, W.E.: Adv. Phys. **26** (1977) 811.
77S2	Staebler, D.L., Wronski, C.R.: Appl. Phys. Lett. **31** (1977) 292.
77S3	Solomon, I., Biegelsen, D., Knights, J.C.: Solid State Commun. **22** (1977) 505.
77S4	Stuke, J.: Proc. 7th Int. Conf. on Amorphous and Liquid Semiconductors, Edinburgh 1977, Spear, W.E., (ed.) Univ. Edinburgh, CICL **1977**, p. 406.
77T1	Tsai, C.C., Fritzsche, H., Tanielian, M., Gaczi, P.J., Persans, P.D., Vesaghi, M.A.: Proc. 7th Int. Conf. on Amorphous and Liquid Semiconductors, Edinburgh 1977, Spear, W.E., (ed.) Univ. Edinburgh, CICL **1977**, p. 339.
77T2	Testardi, L.R., Hauser, J.J.: Solid State Commun. **21** (1977) 1039.
77T3	Thomas, P.A., Kaplan, D.: AIP Conf. Proc. **31** (1977) 85.
77W1	Welber, B., Brodsky, M.H.: Phys. Rev. **B 16** (1977) 3660.
77W2	Wronski, C.R.: IEEE Trans. **ED-24** (1977) 351.
77Z	Zanzucchi, P.J., Wronski, C.R., Carlson, D.E.: J. Appl. Phys. **48** (1977) 5227.
78A1	Apsley, N., Davis, E.A., Yoffe, A.D., Troup, A.P.: J. Phys. **C 11** (1978) 4983.
78A2	Allen, D.: Philos. Mag. **B 38** (1978) 381.
78B	Biegelsen, D.K., Knights, J.C., Street, R.A., Tsang, C., White, R.M.: Philos. Mag. **B 37** (1978) 477.
78C1	Carlson, D.E., Magee, C.W.: Appl. Phys. Lett. **33** (1978) 81.
78C2	Carlson, D.E., Crandall, R.S., Goldstein, B., Hanak, J.J., Moore, A.R., Pankove, J.I., Staebler, D.L.: Final Report (SAN-1286-8), U.S. Department of Energy **1978**, p. 44.
78C3	Candea, R.M., Hudgens, S.J., Kastner, M.: Philos. Mag. **B 37** (1978) 119.
78D	DelCueto, J.A., Shevchik, N.J.: J. Phys. **C 11** (1978) L 829.
78E1	Elliot, R.S.: Philos. Mag. **B 38** (1978) 325.
78E2	Engemann, D., Fischer, R., Knecht, J.: Appl. Phys. Lett. **32** (1978) 567.
78F1	Freeman, E.C., Paul, W.: Phys. Rev. **B 18** (1978) 4288.
78F2	Fuhs, W., Milleville, M., Stuke, J.: Phys. Status Solidi **(b) 89** (1978) 495.
78F3	Fritzsche, H., Tsai, C.C., Persans, P.: Solid State Technol. **21** (1978) 55.
78G1	Grimsditch, M., Senn, W., Winterling, G.: Solid State Commun. **26** (1978) 229.
78G2	Guha, S., Narasimhan, K.L.: Thin Solid Films **50** (1978) 151.
78K1	Kaplan, D., Sol, N., Velasco, G., Thomas, P.A.: Appl. Phys. Lett. **35** (1978) 440.
78K2	Knights, J.C., Lucovsky, G., Nemanich, R.J.: Philos. Mag. **B 37** (1978) 467.

78K3	Kaplan, D., Sol, N., Velasco, G.: Appl. Phys. Lett. **33** (1978) 440.

78K3   Kaplan, D., Sol, N., Velasco, G.: Appl. Phys. Lett. **33** (1978) 440.

78K4   Knights, J.C., Lujan, R.A.: J. Appl. Phys. **49** (1978) 1291.

78M1   Movaghar, B., Schweitzer, L.: J. Phys. **C 11** (1978) 125.

78M2   Morigaki, K., Dunstan, D.J., Cavenett, R.C., Dawson, P., Nicholls, J.E., Nitta, S., Shimakawa, K.: Solid State Commun. **26** (1978) 981.

78M3   Mell, H., Movaghar, B., Schweitzer, L.: Phys. Status Solidi **(b) 88** (1978) 531.

78M4   Miller, D.J., Haneman, D.: Solid State Commun. **27** (1978) 91.

78P   Paesler, M.A., Anderson, D.A., Freeman, E.C., Moddel, G., Paul, W.: Phys. Rev. Lett. **41** (1978) 1492.

78R   Reilly, O.J., Spear, W.E.: Philos. Mag, **B 38** (1978) 295.

78S1   Senn, W., Winterling, G., Grimsditch, M., Brodsky, M.H.: Physics of Semiconductors, Edinburgh 1978, London: Inst. of Physics **1979**, p. 709.

78S2   Solomon, I., Dietl, T., Kaplan, D.: J. Phys. (Paris) **39** (1978) 1241.

78S3   Street, R.A., Biegelsen, D.K., Knights, J.C., Tsang, C., White, R.M.: Solid State Electron. **21** (1978) 1461.

78S4   Street, R.A., Knights, J.C., Biegelsen, D.K.: Phys. Rev. **B 18** (1978) 1880.

78T1   Tanielian, M., Fritzsche, H., Tsai, C.C., Symbalisty, E.: Appl. Phys. Lett. **33** (1978) 353.

78T2   Tsang, C., Street, R.A.: Philos. Mag. **B 37** (1978) 601.

78T3   Thomas, P.A., Brodsky, M.H., Kaplan, D., Lepine, D.: Phys. Rev. **B 18** (1978) 3059.

78T4   Taniguchi, M., Hirose, M., Osaka, Y.: J. Cryst. Growth **45** (1978) 126.

79A1   Ast, D.G., Brodsky, M.H.: Inst. Phys. Conf. Ser. **43** (1979) 1159.

79A2   Austin, I.G., Nashashibi, T.S., Searle, T.M., LeComber, P.G., Spear, W.E.: J. Non-Cryst. Solids **32** (1979) 373.

79A3   Anderson, D.A., Moddel, G., Collins, R.W., Paul, W.: Solid State Commun. **31** (1979) 677.

79A4   Austin, I.G., Richards, K., Searle, T.M., Thompson, M.J., Alkais, M.M., Thomas, J.P., Allison, J.: Phys. of Semiconductors, Edinburgh 1978, London: Inst. of Physics **1979**, p. 1155.

79A5   Ackley, D.E., Tauc, J., Paul, W.: Phys. Rev. Lett. **43** (1979) 715.

79B1   Bermejo, D., Cardona, M.: J. Non-Cryst. Solids **32** (1979) 405.

79B2   Biegelsen, D.K., Street, R.A., Tsai, C.C., Knights, J.C.: Phys. Rev. **B 20** (1979) 4839.

79B3   Beyer, W., Overhof, H.: Solid State Commun. **31** (1979) 1.

79B4   Beyer, W., Fischer, R., Overhof, H.: Philos. Mag. **B 39** (1979) 205.

79B5   Beyer, W., Barna, A., Wagner, H.: Appl. Phys. Lett. **35** (1979) 539.

79B6   Balberg, I., Carlson, D.E.: Phys. Rev. Lett. **43** (1979) 58.

79B7   Brodsky, M.H. (ed.): Amorphous Semiconductors, Topics in Applied Physics **36**, New York: Springer **1979**.

79B8   Bermejo, D., Cardona, M.: J. Non-Cryst. Solids **32** (1979) 421.

79C   Carlson, D.E., Wronski, C.R.: in: Amorphous Semiconductors, Topics in Applied Physics **36**, Brodsky, M.H., (ed.) New York: Springer **1979**, p. 287.

79D1   D'Antonio, P., Konnert, J.H.: Phys. Rev. Lett. **43** (1979) 1161.

79D2   Döhler, G.: Phys. Rev. **B 19** (1979) 2083.

79E   Ewald, D., Milleville, M., Weiser, G.: Philos. Mag. **B 40** (1979) 291.

79F1   Fritzsche, H., Tsai, C.C.: Sol. Energy Mater. **1** (1979) 471.

79F2   Fritzsche, H., Tanielian, M., Tsai, C.C., Gaczi, P.J.: J. Appl. Phys. (USA) **50** (1979) 3366.

79F3   Freeman, E.C., Paul, W.: Phys. Rev. **B 20** (1979) 716.

79F4   Fischer, R.: in: Amorphous Semiconductors, Topics in Applied Physics **36**, Brodsky, M.H., (ed.) New York: Springer **1979**, p. 159.

79G   Graczyk, J.F.: Phys. Status Solidi **(a) 55** (1979) 231.

79H1   Hirose, M., Suzuki, T., Döhler, G.H.: Appl. Phys. Lett. **34** (1979) 234.

79H2   Holzenkämpfer, E., Richter, F.W., Stuke, J., Voget-Grote, U.: J. Non-Cryst. Solids **32** (1979) 327.

79J1   Janai, M., Allred, D.D., Booth, D.C., Seraphin, B.O.: Sol. Energy Mater. **1** (1979) 11.

79J2   Jones, D.I., Gibson, R.A., LeComber, P.G., Spear, W.E.: Sol. Energy Mater. **2** (1979) 93.

79J3   Janai, M., Karlsson, B.: Sol. Energy Mater. **1** (1979) 387.

79J4   Jan, Z.S., Bube, R.H., Knights, J.C.: J. Electron. Mater. **8** (1979) 47.

79J5   Jones, D.I., Spear, W.E., LeComber, P.G., Li, S., Martins, R.: Philos. Mag. **B 39** (1979) 147.

79K1   Knights, J.C., Lujan, R.A.: Appl. Phys. Lett. **35** (1979) 244.

79K2   Knights, J.C.: Jpn. J. Appl. Phys. **18** (1979) 101.

79K3   Knights, J.C., Lucovsky, G., Nemanich, R.J.: J. Non-Cryst. Solids **32** (1979) 393.

79K4   Kurita, S., Czaja, W., Kinmond, S.: Solid State Commun. **32** (1979) 879.

79K5   Kumate, K., Matsumoto, N.: Jpn. J. Appl. Phys. **18** (1979) 1789.

79L1   Leadbetter, A.J., Rashid, A.A.M., Richardson, R.M., Wright, A.F., Knights, J.C.: Solid State Commun. **33** (1979) 1161.

79L2   LeComber, P.G., Spear, W.E.: in: Amorphous Semiconductors, Topics of Applied Physics **36**, Brodsky, M.H., (ed.) New York: Springer **1979**, p. 251.

79L3   Lucovsky, G., Nemanich, R.J., Knights, J.C.: Phys. Rev. **B 19** (1979) 2064.

79M1   Moustakas, T.D.: J. Electron. Mater. (USA) **8** (1979) 391.

79M2   McMillan, J.A., Petersen, E.M.: J. Appl. Phys. **50** (1979) 5238.

79M3   Madan, A., Ovshinsky, S.R., Benn, E.: Philos. Mag. **40** (1979) 259.

79M4   Milleville, M., Fuhs, W., Demond, F.J., Mannsperger, H., Müller, G., Kalbitzer, S.: Appl. Phys. Lett. **34** (1979) 173.

79M5   Mott, N.F., Davis, E.A.: Electronic Processes in Non-Crystalline Materials, Oxford: Clarendon Press, **1979**.

79M6   Matsumoto, N., Kumabe, K.: Jpn. J. Appl. Phys. **18** (1979) 1011.

79M7   Malina, V., Kohout, J., Misek, J., Zelinka, J.: Thin Solid Films **58** (1979) 43.

79N   Naidu, B.S., Reddy, P.J.: Thin Solid Films **61** (1979) 379.

79P1   Perrin, J., Solomon, I., Bourdon, B., Fontenille, J., Ligeon, E.: Thin Solid Films **62** (1979) 327.

79P2   Papa, T., Sette, D., Stagni, L.: J. Phys. **D 7** (1979) 2024.

79R1   Rieder, G.: Phys. Rev. **B 20** (1979) 607.

79R2   Rehm, W., Fischer, R.: Phys. Status Solidi **(b) 94** (1979) 595.

79R3   Rabe, P., Tolkiehn, G., Werner, A.: J. Phys. **C 12** (1979) L 545.

79R4   v. Roedern, B., Ley, L., Cardona, M.: Solid State Commun. **29** (1979) 415.

79R5   v. Roedern, B., Ley, L., Cardona, M., Smith, F.W.: Philos. Mag. **B 40** (1979) 433.

79S1   Senemaud, C., Costa Lima, M.T.: J. Non-Cryst. Solids **33** (1979) 141.

79S2   Solomon, I., Perrin, J., Broudon, B.: Inst. Phys. Conf. Ser. **43** (1979) 689.

79S3   Spear, W.E., LeComber, P.G., Kalbitzer, S., Müller, G.: Philos. Mag. **B 39** (1979) 159.

79S4   Snell, A.J., Mackenzie, K.D., LeComber, P.G., Spear, W.E.: Philos. Mag. **B 40** (1979) 1.

79S5   Searle, T.M., Nashashibi, T.S., Austin, I.G., Devonshire, R., Lockwood, G.: Philos. Mag. **B 39** (1979) 389.

79S6   Street, R.A., Biegelsen, D.K., Stuke, J.: Philos. Mag. **B 40** (1979) 451.

79S7   Starosta, K., Zelinka, J., Berkova, D., Kohout, J.: Thin Solid Films **61** (1979) 241.

79S8   Saito, N., Fujiyasu, H., Yamada, S.: Phys. Status Solidi **(a) 51** (1979) 235.

79T1   Tsai, C.C., Fritzsche, H.: Sol. Energy Mater. **1** (1979) 29.

79T2   Taniguchi, M., Osaka, Y., Hirose, M.: J. Electron. Mater. **8** (1979) 689.

79T3   Tsang, C., Street, R.A.: Phys. Rev. **B 19** (1979) 3027.

79T4   Tsai, C.C.: Phys. Rev. **B 19** (1979) 2041.

79V   Viscor, P., Allen, D.: Thin. Solid Films **62** (1979) 269.

79W   Williams, R.H., Varma, R.R., Spear, W.E., LeComber, P.G.: J. Phys. **C 12** (1979) L 209.

80A1   Al-Jalali, S., Weiser, G.: J. Non-Cryst. Solids **41** (1980) 1.

80A2   Ast, D.G., Brodsky, M.H.: J. Non-Cryst. Solids **35/36** (1980) 611.

80A3   Ast, D.G., Brodsky, M.H.: Philos. Mag. **B 41** (1980) 273.

80B1   Booth, D.C., Allred, D.D., Seraphin, B.O.: J. Non-Cryst. Solids **35/36** (1980) 213.

80B2   Bruyere, J.C., Deneuville, A., Mini, A., Fontenille, J., Danielou, R.: J. Appl. Phys. **51** (1980) 2199.

80B3   Brodsky, M.H., Leary, P.A.: J. Non-Cryst. Solids **35/36** (1980) 487.

80B4   Beyer, W., Stritzker, B., Wagner, H.: J. Non-Cryst. Solids **35/36** (1980) 321.

80B5   Beichler, J., Fuhs, W., Mell, H., Welsch, H.M.: J. Non-Cryst. Solids **35/36** (1980) 587.

80B6   Biegelsen, D.K., Street, R.A., Tsai, C.C., Knights. J.C.: J. Non-Cryst. Solids **35/36** (1980) 285.

80C1   Cody, G.D., Abeles, B., Wronski, C.R., Lanford, W.A.: J. Non-Cryst. Solids **35/36** (1980) 463.

80C2   Crandall, R.S.: Phys. Rev. Lett. **44** (1980) 749.

80C3   Cohen, J.D., Lang, D.V., Bean, J.C., Harbison, J.P.: J. Non-Cryst. Solids **35/36** (1980) 581.

80C4   Cohen, J.D., Lang, D.V., Harbison, J.P.: Phys. Rev. Lett. **45** (1980) 197.

80C5   Collins, R.W., Paesler, M.A., Paul, W.: Solid State Commun. **34** (1980) 833.

80C6   Collins, R.W., Paesler, M.A., Moddel, G., Paul, W.: J. Non-Cryst. Solids **35/36** (1980) 681.

80C7   Chik, K.P., Feng, S.Y., Poon, S.K.: Solid State Commun. **33** (1980) 1019.

80C8   Cody, G.D., Abeles, B., Wronski, C.R., Stephens, R.B., Brooks, B.: Solar Cells **2** (1980) 227.

80D1   De Neufville, J.P., Moustakas, T.D., Ruppert, A.F., Lanford, W.A.: J. Non-Cryst. Solids **35/36** (1980) 481.

80D2	Deneuville, A., Bruyère, J.C., Mini, A., Kahil, H., Danielou, R., Ligeon, E.: J. Non-Cryst. Solids **35/36** (1980) 469.
80D3	Dresner, J.: Appl. Phys. Lett. **37** (1980) 742.
80F1	Fang, C.J., Gruntz, K.J., Ley, L., Cardona, M., Demond, F.J., Müller, G., Kalbitzer, S.: J. Non-Cryst. Solids **35/36** (1980) 255.
80F2	Fang, C.J., Ley, L., Shanks, H.R., Gruntz, K.J., Cardona, M.: Phys. Rev. **B 22** (1980) 6140.
80F3	Fritzsche, H.: Sol. Energy Mater. **3** (1980) 447.
80F4	Fuhs, W., Milleville, M.: Phys. Status Solidi **(b) 98** (1980) K 29.
80F5	Fischer, R., Rehm, W., Stuke, J., Voget-Grote, U.: J. Non-Cryst. Solids **35/36** (1980) 687.
80F6	Friederich, A., Kaplan, D.: J. Phys. Soc. Jap. **49** (Suppl. A) (1980) 1233.
80F7	Fan, J.C.C., Zeiger, H.J., Gale, R.P., Chapman, R.L.: Appl. Phys. Lett. **36** (1980) 158.
80G1	Goodman, N.B., Fritzsche, H.: Philos. Mag. **B 42** (1980) 149.
80G2	Goodman, N.B., Fritzsche, H., Ozaki, H.: J. Non-Cryst. Solids **35/36** (1980) 599.
80G3	Griffith, R.W., Kampas, F.J., Vanier, P.E., Hirsch, M.D.: J. Non-Cryst. Solids **35/36** (1980) 391.
80G4	Gruntz, K.J., Ley, L., Cardona, M., Johnson, R., Harbeke, G., v. Roedern, B.: J. Non-Cryst. Solids **35/36** (1980) 453.
80G5	Gheorghiu, A., Theye, M.L.: J. Non-Cryst. Solids **35/36** (1980) 397.
80H1	Haumeder, M.v., Strom, U., Hunklinger, S.: Phys. Rev. Lett. **44** (1980) 84.
80H2	Hauschild, D., Fischer, R., Fuhs, W.: Phys. Status Solidi **(b) 102** (1980) 563.
80H3	Hargreaves, M., Thompson, M.J., Turner, D.: J. Non-Cryst. Solids **35/36** (1980) 403.
80J	Jan, Z.S., Bube, R.H., Knights, J.C.: J. Appl. Phys. **51** (1980) 3378.
80K1	Knights, J.C.: J. Non-Cryst. Solids **35/36** (1980) 159.
80K2	Knights, J.C., Street, R.A., Lucovsky, G.: J. Non-Cryst. Solids **35/36** (1980) 279.
80K3	Knights, J.C., Lucovsky, G.: CRC Critical Reviews in Solid State and Mater Sciences, **1980**, 211.
80K4	Kalbitzer, S., Müller, G., LeComber, P.G., Spear, W.E.: Philos. Mag. **B 41** (1980) 439.
80K5	Konagai, M., Takahashi, K.: Appl. Phys. Lett. **36** (1980) 599.
80K6	Kruehler, W.W., Plaettner, R.D., Moeller, M., Rauscher, B., Stetter, W.: J. Non-Cryst. Solids **35/36** (1980) 333.
80K7	Kumeda, M., Shimizu, T.: Jpn. J. Appl. Phys. **19** (1980) L 197.
80K8	Krishnaswamy, S.V., Messier, R., Tsong, T.T., Ng, Y.S., McLane, S.B.: J. Non-Cryst. Solids **35/36** (1980) 531.
80L1	Long, A.R., Balkan, N.: J. Non-Cryst. Solids **35/36** (1980) 415.
80L2	Long, A.R., Balkan, N.: Philos. Mag. **B 41** (1980) 287.
80L3	Leadbetter, A.J., Rashid, A.A.M., Richardson, R.M., Wright, A.F., Knights, J.C.: Solid State Commun. **33** (1980) 973.
80M1	Müller, G., Demond, F.J., Kalbitzer, S., Damjantschitsch, H., Mannsberger, H., Spear, W.E., LeComber, P.G., Gibson, R.A.: Philos. Mag. **B 41** (1980) 571.
80M2	Müller, G., Kalbitzer, S.: Philos. Mag. **B 41** (1980) 307.
80M3	Minomura, S., Tsuji, K., Oyanagi, H., Fujii, Y.: J. Non-Cryst. Solids **35/36** (1980) 513.
80M4	Madan, A., Ovshinsky, S.R.: J. Non-Cryst. Solids **35/36** (1980) 731.
80M5	Morigaki, K., Hirabayashi, I., Nakayama, M., Nitta, S., Shimakawa, K.: Solid State Commun. **33** (1980) 851.
80M6	Morigaki, K., Cavenett, B.C., Dawson, P., Nitta, S., Shimakawa, K.: J. Non-Cryst. Solids **35/36** (1980) 633.
80M7	Mell, H., Schweitzer, L., Voget-Grote, U.: J. Non-Cryst. Solids **35/36** (1980) 639.
80M8	Moore, A.R.: Appl. Phys. Lett. **37** (1980) 327.
80M9	Matsuda, A., Yamasaki, S., Nakagawa, K., Okushi, H., Tamaka, K., Itzima, S., Matsumura, M., Yamamoto, H.: Jpn. J. Appl. Phys. **19** (1980) L 305.
80M10	Moddel, G., Anderson, D.A., Paul, W.: Phys. Rev. **B 22** (1980) 1918.
80N1	Nitta, S., Shimakawa, K., Nonumura, S.: J. Non-Cryst. Solids **35/36** (1980) 339.
80N2	Nashashibi, T.S., Searle, T.M., Austin, I.G., Richards, K., Thompson, M.J., Allison, J.: J. Non-Cryst. Solids **35/36** (1980) 675.
80N3	Nemanich, R.J., Knights, J.C.: J. Non-Cryst. Solids **35/36** (1980) 243.
80N4	Nakhodkin, N.G., Bardamid, A.F., Shaldervan, A.I., Chenakui, S.P.: Thin Solid Films **65** (1980) 209.
80N5	Noolandi, J., Hong, K.M., Street, R.A.: Solid State Commun. **34** (1980) 45.
80O1	Overhof, H., Beyer, W.: J. Non-Cryst. Solids **35/36** (1980) 375.

80O2	Oguz, S., Collins, R.W., Paesler, M.A., Paul, W.: J. Non-Cryst. Solids **35/36** (1980) 231.
80O3	O'Connor, P., Tauc, J.: Solid State Commun. **36** (1980) 947.
80P1	Peercy, P.S., Stein, H.J., Ginley, D.S.: Appl. Phys. Lett. **36** (1980) 678.
80P2	Pajasova, L., Abraham, A., Gregora, I., Zavetova, M.: Sol. Energy Mater. **4** (1980) 1.
80P3	Paesler, M.A., Paul, W.: Phil. Mag. **B 41** (1980) 393.
80P4	Paul, D.K., Blake, J., Oguz, S., Paul, W.: J. Non-Cryst. Solids **35/36** (1980) 501.
80P5	Paul, D.K., v. Roedern, B., Oguz, S., Blake, J., Paul, W.: J. Phys. Soc. Jpn. **49** (Suppl. A) (1980) 1261.
80P6	Paul, W.: Solid State Commun. **34** (1980) 283.
80R1	Reimer, J.A., Vaughan, R.W., Knights, J.C.: Phys. Rev. Lett. **44** (1980) 193.
80R2	Rehm, W., Fischer, R., Beichler, J.: Appl. Phys. Lett. **37** (1980) 445.
80R3	Roedern, B.v., Moddel, G.: Solid State Commun. **35** (1980) 467.
80S1	Sol, N., Kaplan, D., Dienmegard, D., Dubreuil, D.: J. Non-Cryst. Solids **35/36** (1980) 291.
80S2	Shanks, H., Fang, C.J., Ley, L., Cardona, M., Demond, F.J., Kalbitzer, S.: Phys. Status Solidi **(b) 100** (1980) 43.
80S3	Staebler, D.L., Wronski, C.R.: Appl. Phys. **51** (1980) 3262.
80S4	Spear, W.E., Allan, D., LeComber, P.G., Ghaith, A.: Philos. Mag. **B 41** (1980) 419.
80S5	Street, R.A., Biegelsen, D.K.: J. Non-Cryst. Solids **35/36** (1980) 651.
80S6	Street, R.A., Knights, J.C.: Philos. Mag. **B 42** (1980) 551.
80S7	Street, R.A., Biegelsen, D.K.: Solid State Commun. **33** (1980) 1159.
80S8	Shen, S.C., Fang, C.J., Cardona, M., Genzel, L.: Phys. Rev. **B 22** (1980) 2913.
80S9	Shah, J., Baigley, B.G., Alexander, F.B.: Solid State Commun. **36** (1980) 199.
80T1	Taniguchi, M., Hirose, M., Osaka, Y.: J. Non-Cryst. Solids **35/36** (1980) 189.
80T2	Tanielian, M., Chatani, M., Fritzsche, H., Smid, V., Persans, P.D.: J. Non-Cryst. Solids **35/36** (1980) 575.
80T3	Tiedje, T., Abeles, B., Morel, D.L., Moustakas, T.D., Wronski, C.R.: Appl. Phys. Lett. **36** (1980) 695.
80T4	Theye, M.L., Gheorghiu, A., Launois, H.: J. Phys. **C 13** (1980) 6569.
80T5	Tiedje, T., Wronski, C.R., Cebulka, J.M.: J. Non-Cryst. Solids **35/36** (1980) 447.
80V1	Voget-Grote, U., Kümmerle, W., Fischer, R., Stuke, J.: Philos. Mag. **B 41** (1980) 127.
80V2	Vardeny, Z., O'Connor, P., Ray, S., Tauc, J.: Phys. Rev. Lett. **44** (1980) 1267.
80V3	Victorovitch, P., Moddel, G.: J. Appl. Phys. **51** (1980) 4847.
80W	Weiser, G., Ewald, D., Milleville, M.: J. Non-Cryst. Solids **35/36** (1980) 447.
80Z1	Zemek, J., Zavetova, M., Koc, S.: J. Non-Cryst. Solids **37** (1980) 15.
80Z2	Zellama, K., Germain, P., Squelard, S., Monge, J., Ligeon, E.: J. Non-Cryst. Solids **35/36** (1980) 225.
80Z3	Zesch, J.C., Lujan, R.A., Deline, V.R.: J. Non-Cryst. Solids **35/36** (1980) 273.
81A1	Agrawal, B.K.: Solid State Commun. **37** (1981) 271.
81A2	Alimoussa, L., Carchano, H., Thomas, J.P.: J. Phys. (Paris) **42**, C4 (1981) 683.
81B1	Bhatia, K.L., Haumeder, M.v., Hunklinger, S.: Solid State Commun. **37** (1981) 943.
81B2	Beyer, W., Mell, H., Overhof, H.: J. Phys. (Paris) **42**, C4 (1981) 103.
81B3	Bhatia, K.L., Haumeder, M.v., Hunklinger, S.: J. Phys. (Paris) **42**, C4 (1981) 365.
81C	Cody, G.D., Tiedje, T., Abeles, B., Brooks, B., Goldstein, Y.: Phys. Rev. Lett. **47** (1981) 1480.
81D1	Dersch, H., Stuke, J., Beichler, J.: Phys. Status Solidi **(b) 105** (1981) 265.
81D2	Delahoy, D., Griffith, R.W.: J. Appl. Phys. **52** (1981) 6337.
81D3	Dersch, H., Stuke, J., Beichler, J.: Phys. Status Solidi **(b) 107** (1981) 307.
81F	Fan, J.C.C., Anderson, C.: J. Appl. Phys. **52** (1981) 4003.
81G1	Gruntz, K.J., Ley, L., Johnson, R.L.: Phys. Rev. **B 24** (1981) 2069.
81G2	Gheorghiu, A., Theye, M.L.: Philos. Mag. **B 44** (1981) 285.
81G3	Gheorghiu, A., Rappenau, T., Dupin, J.P., Theye, M.L.: J. Phys. (Paris) **42**, C4 (1981) 881.
81H1	Hasegawa, S., Kasajima, T., Shimizu, T.: Philos. Mag. **B 43** (1981) 149.
81H2	Hirose, M.: J. Phys. (Paris) **42**, C4 (1981) 705.
81H3	Higashi, G.S., Kastner, M.: Phys. Rev. **B 24** (1981) 2295.
81H4	Hvam, J.M., Brodsky, M.H.: Phys. Rev. Lett. **46** (1981) 371.
81H5	Hvam, J.M., Brodsky, M.H.: J. Phys. (Paris) **42**, C4 (1981) 551.
81J	Jackson, W.B., Amer, N.H.: AIP Conf. Proc. **73** (1981) 263.
81K1	Knights, J.C., Lujan, R.A., Rosenblum, M.P., Street, R.A., Biegelsen, D.K., Reimer, J.A.: Appl. Phys. Lett. **38** (1981) 331.

81K2	Kniffler, N., Müller, W.W., Pirrung, J.M., Hänisch, N., Schröder, B., Geiger, J.: J. Phys. (Paris) **42**, C4 (1981) 811.
81L1	Leadbetter, A.J., Rashid, A.A.M., Colenutt, N., Wright, A.F., Knights, J.C.: Solid State Commun. **38** (1981) 957.
81L2	Ley, L., Gruntz, K.J., Johnson, R.L.: AIP Conf. Proc. **73** (1981) 161.
81L3	Ley, L., Richter, H., Kärcher, R., Johnson, R.L., Reichardt, J.: J. Phys. (Paris) **42**, C4 (1981) 753.
81L4	Long, A.R., Hogg, W.R., Balkan, N., Ferrier, R.P.: J. Phys. (Paris) **42**, C4 (1981) 107.
81M1	Miller, J.N., Lindau, I., Spicer, W.E.: Philos. Mag. **B 43** (1981) 273.
81M2	Minomura, S.: J. Phys. (Paris) **42**, C4 (1981) 181.
81M3	Morigaki, K., Sano, Y., Hirabayashi, I.: Solid State Commun. **39** (1981) 947.
81N	Nonomura, S., Okamoto, H., Nishino, T., Hamakawa, Y.: J. Phys. (Paris) **42**, C4 (1981) 761.
81P1	Paul, W., Anderson, D.A.: Sol. Energy Mater. **5** (1981) 229.
81P2	Powell, M.J.: Philos. Mag. **B 43** (1981) 93.
81R1	Richter, H., Schröder, B., Geiger, J.: J. Non-Cryst. Solids **43** (1981) 153.
81R2	Reimer, J.A.: J. Phys. (Paris) **42**, C4 (1981) 715.
81R3	Reimer, J.A., Vaughan, R.W., Knights, J.C.: Phys. Rev. **B 24** (1981) 3360.
81R4	Reinelt, M., Kalbitzer, S.: J. Phys. (Paris) **42**, C4 (1981) 843.
81S1	Shimizu, T., Kumeda, M., Kiriyama, Y.: Solid State Commun. **37** (1981) 699.
81S2	Shen, S.C., Cardona, M.: Phys. Rev. **B 23** (1981) 5322.
81S3	Spear, W.E., Al-Ani, H., LeComber, P.G.: Philos. Mag. **B 43** (1981) 781.
81S4	Street, R.A., Biegelsen, D.K., Knights, J.C.: Phys. Rev. **B 24** (1981) 969.
81S5	Street, R.A.: Adv. Phys. **30** (1981) 593.
81S6	Sussmann, R.S., Ogden, R.: Philos. Mag. **B 44** (1981) 137.
81S7	Street, R.A.: Phys. Rev. **B 23** (1981) 861.
81S8	Spear, W.E., Gibson, R.A., Young, D., LeComber, P.C., Müller, G., Kalbitzer, S.: J. Phys. (Paris) **42**, C4 (1981) 1143.
81T1	Tawada, Y., Okamoto, H., Hamakawa, Y.: Appl. Phys. Lett. **39** (1981) L237.
81T2	Thompson, C.R., Johnson, N.M., Street, R.A.: J. Phys. (Paris) **42**, C4 (1981) 617.
81T3	Tiedje, T., Cebulka, J.M., Morel, D.M., Abeles, B.: Phys. Rev. Lett. **46** (1981) 1425.
81T4	Tiedje, T., Moustakas, T.D., Cebulka, J.M.: J. Phys. (Paris) **42**, C4 (1981) 155.
81V1	Veprek, S., Iqbal, Z., Oswald, H.R., Sarott, F.A., Wagner, J.J.: J. Phys. (Paris) **42**, C4 (1981) 251.
81V2	Vardeny, Z., O'Connor, P., Ray, S., Tauc, J.: Phys. Rev. Lett. **46** (1981) 1223.
81V3	Vanier, P.E., Delahoy, A.E., Griffith, R.W.: AIP Conf. Proc. **73** (1981) 227.
81W1	Weinstein, B.A.: Phys. Rev. **B 23** (1981) 787.
81W2	Wronski, C.R., Daniel, R.E.: Phys. Rev. **B 23** (1981) 794.
81W3	Welsch, H.M., Fuhs, W., Greeb, K.H., Mell, H.: J. Phys. (Paris) **42**, C4 (1981) 567.
81W4	Weller, D., Mell, H., Schweitzer, L., Stuke, J.: J. Phys. (Paris) **42**, C4 (1981) 143.
81Y1	Yamasaki, S., Hata, N., Yoshida, T., Oheda, H., Matsuda, A., Okushi, H., Tanaka, K.: J. Phys. (Paris) **42**, C4 (1981) 297.
81Y2	Yamasaki, S., Nakagawa, K., Yamamoto, H.: AIP Conf. Proc. **73** (1981) 258.
82B	Beyer, W., Wagner, H.: J. Appl. Phys. **53** (1982) 8145.
82C1	Carlos, W.E., Taylor, P.C.: Phys. Rev. **B 26** (1982) 3605.
82C2	Collins, R.W., Paul, W.: Phys. Rev. **B 25** (1982) 5257.
82C3	Collins, R.W., Viktorovitch, P., Weisfield, R.L., Paul, W.: Phys. Rev. **B 26** (1982) 6642.
82C4	Carlos, W.E., Taylor, P.C.: Phys. Rev. **B 25** (1982) 1435.
82C5	Cohen, J.D., Harbison, J.P., Wecht, K.W.: Phys. Rev. Lett. **48** (1982) 109.
82H1	Hamakawa, Y. (ed.): Amorphous Semiconductor Technologies & Devices, Amsterdam: North-Holland, **1982**.
82H2	Hasegawa, S., Imai, Y.: Philos. Mag. **B 46** (1982) 239.
82H3	Hauschildt, D., Stutzmann, M., Stuke, J., Dersch, H.: Sol. Energy Mater. **8** (1982) 319.
82J	Jackson, W.B., Amer, N.M.: Phys. Rev. **B 25** (1982) 5559.
82K1	Kärcher, R., Ley, L.: Solid State Commun. **43** (1982) 415.
82K2	Klazes, R.H., v.d.Broek, M.H.L.M., Bezemer, J., Radelaar, S.: Philos. Mag. **B 25** (1982) 377.
82K3	Kirby, P.B., Paul, W., Ray, S., Tauc, J.: Solid State Commun. **42** (1982) 533.
82K4	Kinvalainen, P., Heleskivi, J., Leppihalme, M., Gyllenberg-Gastrin, U., Isotalo, H.: Phys. Rev. **B 26** (1982) 2041.

82K5	Kirby, P.B., Paul, W.: Phys. Rev. **B 25** (1982) 5773.
82L1	Lim, K.S., Konagai, M., Takahashi, K.: Jpn. J. Appl. Phys. **21** (1982) 1473.
82L2	Lang, V.D., Cohen, J.D., Harbison, J.P.: Phys. Rev. **B 25** (1982) 5285.
82L3	Lannin, J.S., Pilione, L.J., Kshirsagar, S.T., Messier, R., Ross, R.C.: Phys. Rev. **B 26** (1982) 3506.
82L4	Löhneysen, H.v., Schink, H.J.: Phys. Rev. Lett. **48** (1982) 1121.
82M1	Matsumura, M., Furukawa, S.: in: Amorphous Semiconductor Technologies & Devices, Hamakawa, Y., (ed.), Amsterdam: North-Holland **1982**, p. 88.
82M2	Margarino, J., Kaplan, D., Friedrich, A., Deneuville, A.: Philos. Mag. **B 45** (1982) 285.
82N1	Noguchi, T., Usui, S., Sawada, A., Kanoh, Y., Kikuchi, M.: Jpn. J. Appl. Phys. **21** (1982) L485.
82N2	Nashashibi, T.S., Austin, I.G., Searle, T.M., Gibson, R.A., Spear, W.E., LeComber, P.G.: Philos. Mag. **B 45** (1982) 553.
82O	O'Connor, P., Tauc, J.: Phys. Rev. **B 25** (1982) 2748.
82P	Persans, J.P.: Philos. Mag. **B 46** (1982) 435.
82S1	Street, R.A., Biegelsen, D.K.: Solid State Commun. **44** (1982) 501.
82S2	Street, R.A.: Phys. Rev. **B 26** (1982) 3588.
82S3	Street, R.A.: Phys. Rev. Lett **49** (1982) 1187.
82S4	Street, R.A.: Philos. Mag. **B 46** (1982) 273.
82S5	Segul, Y., Carrere, F., Bui, A.: Thin Solid Films **92** (1982) 303.
82T1	Tauc, J.: in: Festkörperprobleme (Advances in Solid State Physics) **XXII**, Grosse, P., (ed.) Braunschweig: Vieweg **1982**, p. 1.
82T2	Tanielian, M.: Philos. Mag. **B 45** (1982) 435.
82T3	Tanaka, K., Yamasaki, S.: Sol. Energy Mater. **8** (1982) 277.
82T4	Theye, M.L., Gheorghiu, A.: Sol. Energy Mater. **8** (1982) 331.
82V	Vanier, P.E., Griffith, R.W.: J. Appl. Phys. **53** (1982) 3098.
82W1	Wilson, B.A., Kerwin, T.P.: Phys. Rev. **B 25** (1982) 5276.
82W2	Wronski, C.R., Abeles, B., Tiedje, T., Cody, G.D.: Solid State Commun. **44** (1982) 1423.
82W3	Weber, K., Grünewald, M., Fuhs, W., Thomas, P.: Phys. Status Solidi **(b) 110** (1982) 133.
82W4	Wang, Z.P., Ley, L., Cardona, M.: Phys. Rev. **B 26** (1982) 3249.
82Y	Yamasaki, S., Oheda, H., Matsuda, A., Okushi, H., Tanaka, K.: J. Appl. Phys. **21** (1982) L539.
83A	Aker, B., Peng, S.Q., Cai, S., Fritzsche, H.: J. Non-Cryst. Solids **59/60** (1983) 509.
83B1	Beyer, W., Wagner, H.: J. Non-Cryst. Solids **59/60** (1983) 161.
83B2	Biegelsen, D.K., Street, R.A., Jackson, W.B.: Physica **117B/118B** (1983) 899.
83B3	Biegelsen, D.K., Stutzmann, M.: J. Non-Cryst. Solids **59/60** (1983) 137.
83B4	Bellissent, R., Chevenas-Paule, A., Roth, M.: J. Non-Cryst. Solids **59/60** (1983) 229.
83B5	Beichler, J., Mell, H., Weber, K.: J. Non-Cryst. Solids **59/60** (1983) 257.
83C	Chevenas-Paule, A., Bourret, A.: J. Non-Cryst. Solids **59/60** (1983) 233.
83D	Dersch, H., Schweitzer, L., Stuke, J.: Phys. Rev. **B 28** (1983) 4678.
83E	Ellis, F.B., Gordon, R.G., Paul, W., Yacobi, B.G.: J. Non-Cryst. Solids **59/60** (1983) 719.
83F	Fuhs, W., Welsch, H.M., Booth, D.C.: Phys. Status Solidi **(b) 120** (1983) 197.
83G1	Glade, A., Fuhs, W., Mell, H.: J. Non-Cryst. Solids **59/60** (1983) 269.
83G2	Griep, S., Ley, L.: J. Non-Cryst. Solids **59/60** (1983) 253.
83G3	Gheorghiu, A., Ouchene, M., Rappeneau, T., Theye, M.L.: J. Non-Cryst. Solids **59/60** (1983) 621.
83H1	Hoheisel, M., Carius, R., Fuhs, W.: J. Non-Cryst. Solids **59/60** (1983) 457.
83H2	Hirabayashi, I., Morigaki, K.: J. Non-Cryst. Solids **59/60** (1983) 133.
83I1	Ito, H., Kawakyu, Y., Higuchi, T., Ide, K.: J. Non-Cryst. Solids **59/60** (1983) 585.
83I2	Itozaki, H., Fujita, N., Igarashi, T., Hitotsuyanagi, H.: J. Non-Cryst. Solids **59/60** (1983) 589.
83J1	Janai, M., Weil, R., Pratt, B.: J. Non-Cryst. Solids **59/60** (1983) 743.
83J2	Jackson, W.B., Nemanich, R.J.: J. Non-Cryst. Solids **59/60** (1983) 353.
83J3	Jackson, W.B., Thompson, M.J.: Physica **117B/118B** (1983) 883.
83J4	Jackson, W.B., Street, R.A., Thompson, M.J.: Solid State Commun. **47** (1983) 435.
83K1	Kärcher, R., Johnson, R.L., Ley, L.: J. Non-Cryst. Solids **59/60** (1983) 593.
83K2	Kärcher, R., Wang, Z.P., Ley, L.: J. Non-Cryst. Solids **59/60** (1983) 629.
83L	Löhneysen, H.v.: J. Non-Cryst. Solids **59/60** (1983) 1087.
83M1	Matsumura, M., Furukawa, S.: J. Non-Cryst. Solids **59/60** (1983) 739.
83M2	Morimoto, A., Kumeda, M., Shimizu, T.: J. Non-Cryst. Solids **59/60** (1983) 537.
83M3	Minomura, S., Tsuji, K., Wakagi, M., Ishidate, T., Inoue, K., Shibuya, M.: J. Non-Cryst. Solids **59/60** (1983) 541.

83M4	Matsumura, H., Maeda, M., Furukawa, S.: J. Non-Cryst. Solids **59/60** (1983) 517.
83M5	Müller, G., Winterling, G., Kalbitzer, S., Reinelt, M.: J. Non-Cryst. Solids **59/60** (1983) 469.
83M6	Maley, N., Pilione, L.J., Kshirsagar, S.T., Lannin, J.S.: Physica **117B/118B** (1983) 880.
83N1	Nozawa, K., Yamaguchi, Y., Hanna, J., Shimizu, I.: J. Non-Cryst. Solids **59/60** (1983) 533.
83N2	Nitta, S., Hatano, A., Yamada, M., Watanabe, M., Kawai, M.: J. Non-Cryst. Solids **59/60** (1983) 553.
83O1	Overhof, H., Beyer, W.: Philos. Mag. **B 47** (1983) 377.
83O2	Okushi, H., Takahama, T., Tokumaru, Y., Yamasaki, S., Oheda, A., Tanaka, K.: Phys. Rev. **B 27** (1983) 5184.
83R1	Reinelt, M., Kalbitzer, S., Müller, G.: J. Non-Cryst. Solids **59/60** (1983) 169.
83R2	Rudder, R.A., Cook, J.W., Lucovsky, G.: Appl. Phys. Lett. **43** (1983) 871.
83R3	Reichardt, J., Ley, L., Johnson, R.L.: J. Non-Cryst. Solids **59/60** (1983) 329.
83R4	Rhodes, A.J., Bhat, P.K., Austin, I.G., Searle, T.M., Gibson, R.A.: J. Non-Cryst. Solids **59/60** (1983) 365.
83S1	Scott, B.A., Olbricht, W.L., Reimer, J.A. Meyerson, B.S., Wolford, D.J.: J. Non-Cryst. Solids **59/60** (1983) 659.
83S2	Schmitt, J.P.M.: J. Non-Cryst. Solids **59/60** (1983) 649.
83S3	Shanks, H.R., Kamitakahara, W.A., McClelland, J.F., Carlone, C.: J. Non-Cryst. Solids **59/60** (1983) 197.
83S4	Steemers, H., Spear, W.E., LeComber, P.G.: Philos. Mag. **B 47** (1983) L83.
83S5	Spear, W.E.: J. Non-Cryst. Solids **59/60** (1983) 1.
83S6	Street, R.A., Zesch, J.: J. Non-Cryst. Solids **59/60** (1983) 449.
83S7	Street, R.A., Zesch, J., Thompson, M.J.: Appl. Phys. Lett. **43** (1983) 672.
83S8	Skumanich, A., Amer, N.M.: J. Non-Cryst. Solids **59/60** (1983) 249.
83S9	Stutzmann, M., Stuke, J., Dersch, H.: Phys. Status Solidi **(b) 115** (1983) 141.
83S10	Stutzmann, M., Stuke, J.: Solid State Commun. **47** (1983) 635.
83S11	Stutzmann, M., Stuke, J.: Phys. Status Solidi **(b) 120** (1983) 225.
83T	Tsai, C.C., Knights, J.C., Lujan, R.A., Wacker, B., Stafford, B.L., Thompson, M.J.: J. Non-Cryst. Solids **59/60** (1983) 731.
83V1	Vardeny, Z.: J. Non-Cryst. Solids **59/60** (1983) 317.
83V2	Vanecek, M., Kocka, J., Stuchlik, J., Kozisek, Z., Stika, O., Triska, A.: Sol. Energy Mater. **8** (1983) 411.
83W1	Wagner, H., Beyer, W.: Solid State Commun. **48** (1983) 585.
83W2	Wilson, B.A., Hu, P., Harbison, J.P., Jedju, T.M.: Phys. Rev. Lett. **50** (1983) 1490.
83W3	Wilson, B.A., Hu, P., Harbison, J.P., Jedju, T.M.: J. Non-Cryst. Solids **59/60** (1983) 341.
83Y	Yoshida, M., Morigaki, K.: J. Non-Cryst. Solids **59/60** (1983) 357.
84A	Aker, B.: J. Non-Cryst. Solids **66** (1984) 19.
84B1	Biegelsen, D.K., Street, R.A., Jackson, W.A., Weisfield, R.L.: J. Non-Cryst. Solids **66** (1984) 139.
84B2	Beyer, W., Overhof, H.: in: Hydrogenated Amorphous Silicon, Semiconductor and Semimetals, Vol. 21 B, Willardson, R.K., Beer, A.C., (general eds.), New York: Academic Press **1984**.
84B3	Boyce, J.B., Thompson, M.J.: J. Non-Cryst. Solids **66** (1984) 127.
84C1	Carius, R., Fuhs, W., Hoheisel, M.: J. Non-Cryst. Solids **66** (1984) 151.
84C2	Crandall, R.S.: in: Hydrogenated Amorphous Silicon, Semiconductor and Semimetal, Vol. 21 B, Willardson, R.K., Beer, A.C., (general eds.) New York: Academic Press, **1984**.
84C3	Carius, R., Jahn, K., Siebert, W., Fuhs, W.: J. of Lumin. **1984**, in press.
84F	Fuhs, W.: Festkörperprobleme (Advances in Solid State Physics) **XXIV**, Grosse, P., (ed.) Braunschweig: Vieweg, **1984**, p. 133.
84H1	Harbison, J.P.: J. Non-Cryst. Solids **66** (1984) 87.
84H2	Heiden, E.D.v.d., Ohlson, W.D., Taylor, P.C.: J. Non-Cryst. Solids **66** (1984) 115.
84I	Ibaraki, N., Fritzsche, H.: J. Non-Cryst. Solids **66** (1984) 231.
84J	Joannopoulos, J.D., Lucovsky, G. (ed.): The Physics of Hydrogenated Amorphous Silicon I, II, Topics in Applied Physics, Vol. 55–56, Berlin, Heidelberg, New York: Springer **1984**.
84K1	Knights, J.C.: in: The Physics of Hydrogenated Amorphous Silicon I, Topics in Applied Physics, Vol. 55, Berlin, Heidelberg, New York: Springer **1984**.
84K2	Kirby, P.B., Eggert, J.R., Mackenzie, K.D., Paul, W.: J. Non-Cryst. Solids **66** (1984) 181.
84L1	Ley, L.: in: The Physics of Hydrogenated Amorphous Silicon II, Topics in Applied Physics **56**, Berlin, Heidelberg, New York: Springer **1984**, p. 61.

84L2	Lang, D.V., Cohen, J.D., Harbison, J.P., Chen. M.C., Sergent, A.M.: J. Non-Cryst. Solids **66** (1984) 217.
84M1	Mahan, A.H., Williamson, D.L., Madan, A.: Appl. Phys. Lett. **44** (1984) 220.
84M2	Mescheder, U., Weiser, G.: AIP Conf. Proc. **1984**, in press.
84M3	Marshall, J.M., Street, R.A., Thompson, M.J.: J. Non-Cryst. Solids **66** (1984) 175.
84O	Orton, J.W., Powell, M.J.: Philos. Mag. **B 50** (1984) 10.
84P1	Pfost, D., Hsiang-na, L., Vardeny, Z., Tauc, J.: Phys. Rev. **B 30** (1984) 1083.
84P2	Persans, P.D., Ruppert, A.F., Chan, S.S., Cody, G.D.: Solid State Commun. **51** (1984) 203.
84R1	Roedern, B.v., Mahan, A.H., Könenkamp, R., Williamson, D.L., Sanchez, A., Madan, A.: J. Non-Cryst. Solids **66** (1984) 13.
84R2	Ross, R.C., Johncock, A.G., Chan, A.R.: J. Non-Cryst. Solids **66** (1984) 81.
84S	Spear, W.E., Steemers, H.L.: J. Non-Cryst. Solids **66** (1984) 163.
84T1	Tsai, C.C., Knights, J.C., Thompson, M.J.: J. Non-Cryst. Solids **66** (1984) 45.
84T2	Taylor, P.C.: in: Hydrogenated Amorphous Silicon, Semiconductor and Semimetals Vol. 21C, Williardson, R.K., Beer, A.C., (eds.) New York: Academic Press, **1984**.
84T3	Tanaka, K., Okushi, H.: J. Non-Cryst. Solids **66** (1984) 205.

12 Organic semiconductors

12.0 Introduction

12.0.1 General remarks

Organic chemistry allows a wide variety of molecular structures to be formed, containing carbon as a key atom. Bonding forces in the crystalline state can be purely van der Waals; static dipolar bonding, charge transfer interaction, ionic and/or hydrogen bonding may be superimposed.

The great majority of these crystals are good insulators in the dark, but many of them (if sufficiently pure) become photoconductive under illumination. Special members exhibit only a small gap between valence and conduction states, leading to considerable dark conductivity, which can approach those orders of magnitude which are common for well conducting inorganic semiconductors; there are also some metallic-like conducting organic crystals (i.e. with vanishing thermal activation and hence zero energy gap) exhibiting conductivities which are close to those typical for metals. Conduction in most better conducting representatives is quasi one-dimensional; cf. [74S1, 75K2, 75S1, 76S1, 77K1, 77P1, 78M1, 78T1, 79B1, 79D, 79H, 80A1, 80S6, 81B1, 81C1, 81E1, 82K3, 82M1, 83S9, 83S10] and [84H] (bibliography). Most of these systems undergo a metal to semiconductor transition at lower temperatures, some remain metallic under normal or enhanced pressure. A few of them even become superconducting around 1K [80J, 80R1, 80R2, 81B2, 81D1; cf. 80A2, 80G; 83P2, 83Y]; see also the reviews [82J1, 82J2, 84B2, 84G2, 84J].

The aim of this chapter on organic semiconductors is to give an illustrative survey of the rapidly expanding field, based on a choice of experimental results which were obtained with *single crystals* (where appropriate), and which are believed to be both, characteristic and sufficiently reliable, rather than to collect all available data. Such selection necessarily bears a considerable degree of arbitrariness and the future will well elucidate the importance of material which has been ommitted here since it neither fit into the present understanding nor was independent experimental confirmation available or simply in order not to blow up this contribution.

To those readers who are not familiar with the field and who wish to consult introductory literature, the review [80B1] is strongly recommended. A more complete literature survey may be obtained from a bibliography [69L1, 83S1, 84S2] and from several monographs and review articles on organic semiconductors [62B, 63L, 64K1, 64O, 65K, 66T, 67B1, 67G, 67H1, 67P, 68K, 69F1, 69L1, 70B1, 70C1, 70S1, 71C, 71P1, 72S1, 73H1, 74K1, 74M1, 74M2, 75G1, 75K1, 75S2, 76G1, 76G2, 76K1, 76K2, 76M1, 77K2, 77L, 77S1, 78H, 79B2, 79G1, 80B1, 80M1, 80R3, 80S4, 81H, 81K1, 81P1, 81W1, 82H, 82P1, 83G3].

The molecules of most organic semiconductors studied so far contain conjugated carbon $2p_z$-electrons, forming a system of comparatively weakly bound and hence strongly polarizable delocalized π-electrons, superimposed on a more rigid framework of σ-bonds, established by more strongly bound and more localized carbon $2sp_x p_y$ or $2sp_x$-hybridized electrons.

Gas and solid state electron affinities, A_g and A_c respectively, of these conjugated π-electron systems are positive for most members (i.e. an additional electron is bound under gain of energy). Solid state ionization energies, I_c, and band gap energies E_g as well as optical excitation energies of the lowest neutral or charge transfer states of π-electrons lie typically in an experimentally easily accessible energy range of less than $6\cdots 8$ eV, whereas for σ-electrons the respective energies are beyond these values ($I_c > 8$ eV).

The additional fact that molecules with rigid bond systems (especially those with conjugated π-electrons) tend to form good single crystals more easily than those with a mobile molecular backbone, has led to a strong preference and hence dominance of the former material class.

In section 12.1 on wide gap photoconductive organic semiconductors, or more precisely "semiinsulators" (which comprise mainly one-component van der Waals crystals and "weak" donor-acceptor complexes), charge carrier mobilities will be considered of prime importance. Since charge carrier mobilities are very sensitively influenced by even small concentrations of impurities (on the order of ≤ 1 ppm, cf. Figs. 32/33), the compilation will concentrate on compounds for which achievement of high purity has been possible, and for which trap-uninfluenced "microscopic" mobilities have been obtained. Historically, unsubstituted aromatic molecules, containing a few benzene rings, have been favored, since they melt without decomposition and hence can be zone-refined, cf. [80K2].

By far the most thoroughly studied organic photoconductor has been *anthracene*. By comparison with a few other organic photoconductors the conclusion can be drawn that the basic properties of the anthracene crystal can be considered to be somehow representative for the entire material class of purely van der Waals-bonded, small molecular weight, conjugated π-electron systems.

For anthracene and several other organic crystals temperature dependent "microscopic" i.e. trap-unperturbed electron and hole drift mobility data are available which will be tabulated or represented by figures. A few published Hall effect experiments [69K, 69S1, 70S2, 70D2, 71P2, 71S] have led to contradictory results. Therefore these data will be omitted. The only cyclotron resonance experiment [74B2, 77B1] has not yet been reproduced independently.

The sequence of the tabulated data is as follows: In alphabetical order of the compounds the molecular structure is given first, followed by crystal structure data and mechanical properties. Next, data on *neutral* (excitonic, vibronic, and phononic) and on *ionic* (charge carrier) levels, including charge carrier band positions, band gap energies and polarization energies are collected; these are followed by mobility data.

Charge carrier generation and recombination processes and dielectric and optical tensor data are listed in another subparagraph.

Large dielectric relaxation times as a consequence of small conductivity and high rate constants for bimolecular recombination of charge carriers (as a consequence of small dielectric constants) are typical for most organic photoconductors. Since recombination often leads to excitons which decay radiatively, charge carrier double injection and electroluminescence are also important topics.

Finally, it will be essential to consider charge carrier trap states, both of chemical and physical nature. A few experiments done with well purified and subsequently intentionally doped crystals are available [75P]. Whereas magnetic field effects on *transport* have not yet been reliably established, a magnetophotoconductivity effect, based on magnetic field dependent *detrapping* by the interaction of mobile triplet state excitons with trapped (doublet state) charge carriers may well deserve mentioning, cf. [79Z1].

In section 12.2 on systems with intermediate or high dark conductivity, most representatives belong to the classes of strong *donor: acceptor complexes and radical ion salts* which are distinguished by pairs of complementary molecules with extreme differences in their redox properties, (which arise from exotic chemical structures) leading to partial or full charge transfer in the ground state.

In a first approximation charge transfer can be considered to occur if the residual difference between the ionization energy of the donor D, and the electron affinity of the acceptor A, (both taken for the solid state) can be compensated by the gain of lattice energy due to the ionic contribution (Madelung energy) associated with charge transfer.

Further, a charge transfer ground state is often favored by chemical stabilization of at least one of the partners in the cation/anion pair formed. Chemical stabilization can occur by aromatization of non-aromatic π-electron systems, by charge delocalization over large aromatic π-electron systems, and/or by formation of stable, in most cases inorganic, coordination complex counter-ions.

These systems can be classified (cf. [74S1]) as:

I) Radical cation salts, where an organic radical cation (such as perylene$^{\cdot+}$) is combined with an (in most cases) inorganic counter-ion (such as PF_6^-).

II) Radical anion salts, where an organic radical anion (such as TCNQ$^{\cdot-}$) is combined with an organic or inorganic cation, (such as N-methylphenazinium$^+$, or K$^+$, respectively).

III) Charge transfer complexes with at least partially ionic ground state, due to *fractional* charge transfer from the donor to the acceptor molecules (i.e. charge transfer which is not dictated by chemical stoichiometry).

It is frequently encountered that the molecular radical ion combines with its neutral species to form a *complex* donor or acceptor, such as $\{(perylene)_2\}^{\cdot+}$ or $\{(TCNQ)_2\}^{\cdot-}$, instead of a *simple* one.

The organic molecules under consideration are usually planar and form sandwich like stacks in most cases. Depending on the crystallographic packing architecture, *mixed* (i.e. $\cdots D^{\delta+} A^{\delta-} D^{\delta+} A^{\delta-} D^{\delta+} \cdots$) and *segregated* (i.e. $\cdots D^{\delta+} D^{\delta+} D^{\delta+} D^{\delta+} \cdots$ and $\cdots A^{\delta-} A^{\delta-} A^{\delta-} A^{\delta-} \cdots$) stacking are distinguished; $(0 < \delta \leq 1$; $\delta = 1$: full charge transfer). These stacks can be *regular*, i.e. with equal spacing between intrastack neighbors, or *alternating* ("dimerized").

For dark conductivity to occur, the gap between valence and conduction band, E_g, must be sufficiently narrow. To achieve a narrow band gap, a donor with particularly low ionization energy and an acceptor with high electron affinity must be combined. This necessarily means using components near the borderline to chemical instability. That is why most dark-conductive complexes and their constituent molecules are not stable at their melting point, thus preventing efficient purification by the very effective and universal zone refining method. Chemical purity of organic dark conductors is therefore often low, and the reproducibility of experimental data is frequently rather poor.

Fortunately, in the better (semi-)conducting systems with high concentrations of charge carriers, at least *trapping* by impurities should be of minor importance. However, thermal ionization of impurity donor and acceptor levels (giving rise to "extrinsic" contributions to the conductivity), as well as impurity scattering and charge or spin density wave pinning by impurities, and their influence on phase transitions, such as metal to semiconductor (Peierls) transitions, can play an important role. The whole field is presently in a phase of very rapid development.

Direct charge carrier transport investigations are scarce for dark-conductive organic materials. Information on transport has been drawn basically from magnitude, temperature dependence and anisotropy of conductivity, location and shape of phase transitions seen in conductivity, and their structural correlatives. These data have been complemented sometimes by thermo-EMF, IR reflectivity, ESR and spin susceptibility measurements. An interesting problem, unsolved for many members of this material class, is the frequently observed non-correlation between the temperature dependence of conductivity and spin $S = 1/2$ concentration. This may, among other, more fundamental reasons, also be a consequence of ill-defined, impure systems.

The material situation in terms of clearly defined (microscopic) physical and chemical properties is even worse with *photo- and semiconducting polymers*, where, besides chemical purity, the parameters molecular structure (chain length and conformation) and solid state aggregation can also vary strongly from sample to sample, and thus often lead to very poor reproducibility. As a consequence of disorder several microscopic parameters, such as site energies and transfer integrals, exhibit a distribution rather than sharp values. This very important material class is therefore omitted here. The reader is referred to the literature (see e.g. [68K, 70B2, 75G1, 76G1, 76G2, 79B2, 80M1, 81B5, 82E2, 82M3, 84B1]).

In section 12.3 a comparative representation of some general properties of selected compounds is given. The main intention of this chapter is to serve as a guide for selecting compounds with the required energy level positions. The tables and figures represent information on correlations between chemical structure and physical properties.

12.0.2 Symbols, definitions and abbreviations

a) Symbols, definitions

Symbol	Unit	Property, definition
$a, b, c; \alpha, \beta, \gamma$	Å; deg	lattice parameters
a_g, a_u, b_g, b_u etc.		symmetry assignment of vibrons (intramolecular vibrations)
A_c	eV	electron affinity of the crystal
A_g	eV	electron affinity of the free (gas phase) molecule
c_{ij}	dyn cm^{-2}	elastic moduli
C_p	J mol^{-1} K^{-1}	heat capacity
d	g cm^{-3}	density
e	C	elementary charge ($|e| = 1{,}6021892(46) \cdot 10^{-19}$ C)
E	eV	energy
E	V m^{-1}	electric field strength
$E_{a(d)}$	eV	energy of acceptor (donor) state
E_A	eV	activation energy, mostly of conductivity, defined by $\sigma = \sigma_0 \exp(-E_A/kT)$
E_{CT}	eV	energy for the formation of electron–hole pairs which are still within their Coulombic interaction range, (charge transfer energy), classified according to the relative crystallographic coordinates of the ion pair, e.g. E_{CT} $[\frac{1}{2}\frac{1}{2}0]$
E_g	eV	threshold energy for the production of free–free electron–hole pairs at infinite distance (commonly interpreted as "band gap" energies) (i.e. including thermal contributions to the final separation of the ("geminate") electron–hole pair against its residual Coulomb binding energy)
E_{S_1}	eV	lowest excited singlet exciton energy level
E_{T_1}	eV	lowest triplet exciton energy level
E_t	eV	carrier trapping levels, characterized by the activation energy for detrapping transitions to the band. In the only host material (anthracene) where traps have been incorporated intentionally and studied, electron traps, $E_{t,n}$, can be considered alternatively as deep acceptors of energy $E_a (E_a \gg E_{t,n}, E_a + E_{t,n} \approx E_g)$, hole traps, $E_{t,p}$, as deep donors of energy $E_d (E_d \gg E_{t,p}, E_d + E_{t,p} \approx E_g)$

Symbol	Unit	Property, definition
E_{vac}	eV	vacuum level
f	Hz	frequency
$g(\omega)$	Hz^{-1}	density of states
h	Js	Planck constant ($h = 6.626176 \cdot 10^{-34}$ Js)
$\boldsymbol{H}$	Oe	magnetic field strength
ΔH_m	J mol^{-1}	enthalpy of fusion
ΔH_s	J mol^{-1}	enthalpy of sublimation
$\boldsymbol{i}_{(n,p)}$	A cm^{-2}	current density (of electrons, holes)
I	W cm^{-2}	light intensity
$I_{h\nu}$	cm^{-2} s^{-1}	photon flux density
I_c	eV	photoemission threshold energy defined as threshold of photoelectron yield (η) as a function of photon energy in a plot $\eta^{1/3}$ vs. $\hbar\omega$ (cf. original literature or [82P1])
I_c^{th}, I_c^p	eV	threshold and (first) peak ionization energy of the (poly) crystalline solid, obtained from photoelectron spectroscopy
I_g^a, I_g^v	eV	adiabatic (threshold) and vertical (peak) ionization energy of the highest occupied molecular orbital of the free (gas phase) molecule, experimental value
I_{ph}	A	photocurrent
k	J K^{-1}	Boltzmann constant (k $= 1.380662(44) \cdot 10^{-23}$ J K^{-1})
$\boldsymbol{k}$	cm^{-1}	wavevector of electrons
$\boldsymbol{k}_F$	cm^{-1}	Fermi wavevector
K	cm^{-1}	absorption coefficient
m_0	g	electron mass
m_p	m_0	effective mass of holes
n	cm^{-3}	electron concentration
n_{ij}	–	temperature exponent of the tensor component μ_{ij} (see under μ_{ij})
$n_{\alpha, \beta, \gamma}$	–	principal refractive indices
p	bar	pressure
p	cm^{-3}	hole concentration
$P^{(\pm)}$	eV	lattice polarization energy associated with the hole (electron) state
R	cm	radius
R	–	reflectance
R_H	cm^3 C^{-1}	Hall coefficient
S	V K^{-1}	thermopower, Seebeck coefficient
S	–	spin quantum number
$4t$	eV	bandwidth (t stands for the transfer integral to the next neighbor)
T	K, °C	temperature
T_c	K	superconductor transition temperature
T_m	K	melting point
$T_{MS(MI)}$	K	metal-semiconductor (metal-insulator) transition temperature
T_{tr}	K	phase transition temperature
U	V	voltage
V	cm^{-3}	volume
$2V$	deg	angle between the optical axes
Z	–	number of formula units in the unit cell
α	K^{-1}	thermal expansion coefficient
γ_{np}	cm^3 s^{-1}	electron-hole recombination constant, defined by the reaction equation $-dn/dt = -dp/dt = \gamma_{np} \cdot n \cdot p$
γ_{SS}	cm^3 s^{-1}	singlet exciton bimolecular annihilation constant, leading to free charge carrier pairs; defined by the reaction equation $dn/dt = dp/dt = \gamma_{SS} S_1^2$ where S_1 is the singlet exciton concentration
γ_{ST}	cm^3 s^{-1}	annihilation constant for singlet – triplet collisions, leading to free charge carrier pairs; defined by the reaction equation $dn/dt = dp/dt = \gamma_{ST} S_1 T_1$, where T_1 is the triplet exciton concentration
δ^+, δ^-	e/molecule	fractional positive or negative charge

Symbol	Unit	Property, definition
$\varepsilon_{ij}(\theta)$	– (deg)	components (orientation angle) of the dielectric tensor, defined in the same notation as for the mobility tensor
ε_1	–	real part of dielectric constant
ζ	–	reduced wavevector coordinate
η	–	quantum yield
λ	cm	wavelength
μ_B	$J\,T^{-1}$	Bohr magneton ($\mu_B = 9.274078\,(36) \cdot 10^{-24}\,J\,T^{-1}$)
μ_{ij}	$cm^2\,V^{-1}\,s^{-1}$	charge carrier drift mobility tensor components, relating the drift velocity $\boldsymbol{v}$ (with components v_i) of holes and electrons, respectively, to the applied electric field $\boldsymbol{E}$, with components E_j according to the definition equation (see e.g. [57N])

$$v_i = \mu_{ij}\,E_j.$$

The experimental technique used for the determination of the drift mobility data given in this compilation has always been based on the time-of-flight method, ([60K], cf. [60L1]; see also [74K1]), if not otherwise stated. The quantity which is measured in a time-of-flight experiment for a general crystal orientation is the drift velocity component parallel to the applied electric field, $v_{\parallel}$. The ratio of cause, E, and response, $v_{\parallel}$, the "mobility in the given direction", $\mu_{\parallel}$, is in general a certain linear combination of all 6 mobility tensor components. Therefore, in a general situation, at least 6 measurements in 6 independent directions are required for an evaluation of the 6 independent tensor elements of the mobility tensor, cf. [57N].

The tensor components are given (where appropriate) in the respective orthogonalized crystallographic axes system $\boldsymbol{a}/a$, $\boldsymbol{b}/b$, $\boldsymbol{c}/c$ (for orthorhombic), $\boldsymbol{a}/a$, $\boldsymbol{b}/b$, $\boldsymbol{c}^*/c^*$ (for monoclinic) and $\boldsymbol{a}'/a'$, $\boldsymbol{b}^*/b^*$, $\boldsymbol{c}/c$ (for triclinic crystal symmetry), where $\boldsymbol{a}'/a' = (\boldsymbol{b}^*/b^*) \times (\boldsymbol{c}/c)$, $\boldsymbol{b}^*/b^* = (\boldsymbol{c}/c) \times (\boldsymbol{a}/a)$. In matrix notation

$$\mu_{ij} = \begin{pmatrix} \mu_{aa} & \mu_{ab} & \mu_{ac} \\ \mu_{ab} & \mu_{bb} & \mu_{bc} \\ \mu_{ac} & \mu_{bc} & \mu_{cc} \end{pmatrix}$$

where the star has to be added to the subscript c in monoclinic symmetry and the subscripts a and b have to be replaced by a' and b* in triclinic symmetry.

From symmetry requirements $\mu_{ab} = \mu_{bc} = 0$ in monoclinic crystals, and all mixed indices components $\mu_{ab} = \mu_{bc} = \mu_{ac} = 0$ for orthorhombic and higher symmetry crystals.

The principal axes mobility components are designated μ_{11}, μ_{22} and μ_{33}; in matrix notation

$$\mu'_{ij} = \begin{pmatrix} \mu_{11} & 0 & 0 \\ 0 & \mu_{22} & 0 \\ 0 & 0 & \mu_{33} \end{pmatrix}'$$

with the directions 1, 2, 3 chosen as close as possible to the orthogonalized crystallographic directions $\boldsymbol{a}^{(')}$, $\boldsymbol{b}^{(*)}$, $\boldsymbol{c}^{(*)}$, respectively. For orthorhombic (and higher) symmetry $\mu_{11} = \mu_{aa}$, $\mu_{22} = \mu_{bb}$, $\mu_{33} = \mu_{cc}$. For monoclinic symmetry $\mu_{22} = \mu_{bb}$; the other two principal directions are generally oblique with respect to the crystallographic axes $\boldsymbol{a}$ and $\boldsymbol{c}^*$, forming an angle θ between $\boldsymbol{a}$ and the principal direction 1. This angle θ is counted in the same direction as the monoclinic angle β ($\beta > 90°$), defined as $\beta = (360/2\pi) \cdot \arcsin \dfrac{\boldsymbol{a} \times \boldsymbol{c}}{\boldsymbol{a} \cdot \boldsymbol{c}}$. Since the angle θ is not fixed by symmetry, the tensor is allowed to rotate (e.g. with temperature) about the crystallographic $\boldsymbol{b}$ axis in monoclinic crystals, whereas it is completely free to rotate about any axis in triclinic crystals.

Symbol	Unit	Property, definition
		Some of the numerical tensor data are also represented by approximate analytical expressions
		$$\mu_{ij}(T) = \mu_{ij}(300\text{ K}) \left\{ \frac{T[\text{K}]}{300} \right\}^{n_{ij}}$$
		$$\theta(T) = \theta(300\text{ K}) + m_\theta(T[\text{K}] - 300)$$
		where n_{ij} designates the temperature exponent of the tensor component μ_{ij}
$\mu_{n(p)}$	$\text{cm}^2\,\text{V}^{-1}\,\text{s}^{-1}$	electron (hole) mobility
$\bar{v}$	cm^{-1}	wavenumber
$\bar{v}_M^*\,(h v_M^*)$	cm^{-1} (eV)	wavenumber (energy) of excited states of the (radical) ion of the molecule under consideration; M^{*+}: cation, M^{*-}: anion
ϱ	$\Omega\,\text{cm}$	electrical resistivity
$\Delta\varrho/\varrho$	–	magnetoresistance
σ	$\Omega^{-1}\,\text{cm}^{-1}$	electrical conductivity
σ_{dark}	$\Omega^{-1}\,\text{cm}^{-1}$	dark conductivity
σ_{opt}	$\Omega^{-1}\,\text{cm}^{-1}$	optical conductivity, from Drude – Lorentz fit of optical reflection spectrum
σ_{S_1}	cm^2	cross section for the production of free charge carrier pairs by photoionization of singlet excitons, defined by the reaction equation $dn/dt = dp/dt = \sigma_{S_1} \cdot I_{hv} \cdot S_1$
σ_{T_1}	cm^2	cross section for the production of free charge carrier pairs by photoionization of triplet excitons, defined by the reaction equation $dn/dt = dp/dt = \sigma_{T_1} \cdot I_{hv} \cdot T_1$
τ	s	electron relaxation time
$\tau_{S_1}\,(\tau_{T_1})$	–	fluorescence lifetime of the S_1 state (T_1 state)
ω	rad s^{-1}	circular frequency
$\hbar\omega$	eV	photon energy
ω_p	s^{-1}	plasma resonance frequency

b) Abbreviations

A, a	acceptor		lum	luminescence
c	crystal(line)		M	molecule, metal
CB	conduction band		opt	optical
CDW	charge density wave		R	Raman active
CT	charge transfer		rec	recombination
D, d	donor		RT	room temperature
dc	direct current		SCL	space-charge-limited (current)
e	electron		SDW	spin density wave
EMF	electromotive force		TDR	triplet-doublet resonance
ESR	electron spin resonance		TSC	thermally stimulated current
exc	excitation		UV	ultraviolet
I	insulator		vac	vacuum
IR	infrared active		VB	valence band
inj	injection			

Physical property	Numerical value	Experimental conditions	Experimental method, remarks	Ref.

12.1 Photoconductive wide band gap organic semiconductors

12.1.1 Anthracene, $C_{14}H_{10}$

molecular structure:

crystal structure: monoclinic, space group C_{2h}^5-$P2_1/a$, $Z=2$

lattice parameters (lengths in Å):

		T [K]		
a	8.562 (6)	290	X-ray diffraction	
b	6.038 (8)		the a, b, c lattice parameters of	64M;
c	11.184 (8)		perdeuterated anthracene are	cf. 50M,
β	124°42 (6)′		slightly smaller (298 K, neutron	83P1
			diffraction [72L])	
a	8.696	353	X-ray diffraction	56K
b	6.150			
c	11.310			
β	124°28′			
a	8.443 (6)	95	X-ray diffraction	64M;
b	6.002 (7)			cf. 83P1
c	11.124 (8)			
β	125°36 (8)′			
a	8.37 (3)	16	neutron diffraction, $C_{14}D_{10}$	82C1
b	6.00 (2)			
c	11.12 (4)			
β	125.4 (1)°			

molecular packing diagram: Fig. 1
thermal expansion: Fig. 2; cf. [68R, 79J]
change of lattice parameters with pressure: Figs. 3 and 4
cleavage plane: (001)

elastic moduli (in 10^{10} dyn/cm²):

c_{22}	11.6	RT	for sound velocities see [67W2,	67A;
			78H]; for the other coefficients,	cf. 71K2
	11.7		see [78E; cf. 78H, p. 29, 80S4]	69H1
	13.8			68D
	11.48			78H

density:

d	1.28 g/cm³			64M

melting point:

T_m	489 K			81W3

enthalpy of fusion:

ΔH_m	28.86 kJ/mol			81W3

enthalpy of sublimation:

ΔH_s	100.5 (8) kJ/mol	$T=119.3$ °C		73M2

vapor pressure:

p	≈0.1 Torr	$T=115$ °C		73M2
	≈1 Torr	145 °C		
	40 Torr	216 °C		

Physical property	Numerical value	Experimental conditions	Experimental method, remarks	Ref.
energy levels and lifetimes of excitons:				
		T [K]		
E_{S_1}	3.1117 eV	4.2; $E \parallel b$	optical absorption,	75W1
	3.1616 eV	4.2; $E \parallel a$	fluorescence	
τ_{S_1}	5.0 (5) ns	20	fluorescence	82B1
E_{T_1}	1.83 eV	300	excitation spectrum, phosphorescence	68A
	1.826 eV	83	photocurrent excitation spectroscopy, Fig. 24 (for details, see figure caption)	75K5
	1,8274 eV	4; $E \parallel a$	photocurrent excitation spectroscopy, Fig. 25	77F
	1.8273 eV	4.2; $E \parallel a$	delayed fluorescence	81P2,
	1.82996 eV	4.2; $E \parallel b$	excitation spectroscopy	79P; cf. 69D3
τ_{T_1}	25 ms	295	delayed fluorescence	80K2
	45 ms	295	delayed fluorescence	80K2
		perdeuterated anthracene		

absorption coefficients as a function of photon energy: Fig. 162
reflection spectra: see [70C2, 78S6, 83T]

vibron wavenumbers (in cm^{-1}):

		T [K]		
$\bar{\nu}(b_{1g}^*)$	247/244.5	4.2	single crystal; from vibrational	84K3;
$\bar{\nu}(b_{2g}^*)$	290	293	structure of electronic transi-	cf. 64K3,
$\bar{\nu}(b_{3g})$	390	293	tions, Raman scattering, and	67W2,
$\bar{\nu}(a_g)$	395	293	infrared absorption. Symmetry	68C2,
$\bar{\nu}(b_{3g})$	478.5/481	4.2	assignments for the point group	73A,
$\bar{\nu}(b_{1g}^*)$	477	293	D_{2h} of the free molecule, (with	73R,
$\bar{\nu}(b_{2g}^*)$	577	293	the subscripts 1, 2, 3 chosen in	74G1,
$\bar{\nu}(a_g)$	622	293	the same way as for naphtha-	74K3,
$\bar{\nu}(b_{1g})$	747	293	lene, 12.1.12). The symmetry	77F
$\bar{\nu}(a_g)$	753	293	point group D_{2h} allows for the	
$\bar{\nu}(b_{2g}^*)$	762.5/764	4.2	$N = 24$ atoms of the molecule	
$\bar{\nu}(b_{2g}^*)$	896	293	$3N - 6$ normal "internal" vi-	
$\bar{\nu}(b_{1g}^*)$	904	293	brational modes which are dis-	
$\bar{\nu}(b_{2g}^*)$	918/916	4.2	tributed among the symmetry	
$\bar{\nu}(b_{3g})$	954/959	4.2	types in the following way:	
$\bar{\nu}(b_{2g}^*)$	980/978.5	4.2		
$\bar{\nu}(a_g)$	1008	293	$12a_g + 5a_u^* + 11b_{3g} + 6b_{3u}^*$	
$\bar{\nu}(b_{3g})$	1103	293	$+ 4b_{1g}^* + 11b_{1u} + 6b_{2g}^* + 11b_{2u}$.	
$\bar{\nu}(a_g)$	1163	293		
$\bar{\nu}(b_{3g})$	1187	293	In addition, many combinations	
$\bar{\nu}(a_g)$	1261	293	and overtones are possible	
$\bar{\nu}(b_{3g})$	1274	293	which are not listed. All gerade	
$\bar{\nu}(b_{3g})$	1376	293	vibrations are Raman-active, cf.	
$\bar{\nu}(a_g)$	1403	293	[73A, 73R]; a_u vibrations are	
$\bar{\nu}(a_g)$	1482	293	forbidden for the point group	
$\bar{\nu}(a_g)$	1557	293	D_{2h} of the free molecule, but al-	
$\bar{\nu}(b_{3g})$	1576	293	lowed in the crystal. Due to the	
$\bar{\nu}(b_{3g})$	1634	293	space group symmetry $C_{2h}^5 -$	
$\bar{\nu}(b_{3g})$			$P2_1/a$, factor group (Davidov)	
$\bar{\nu}(b_{3g})$			splitting occurs through the in-	
			teraction of the two symmetry-	
			related molecules in the unit cell.	

(continued)

Physical property	Numerical value	Experimental conditions	Experimental method, remarks	Ref.
vibron wavenumbers (continued)				
$\bar{\nu}(a_g)$	3066	293	The asterisk denotes out-of-	
$\bar{\nu}(a_g)$	3088	293	plane vibrations.	
$\bar{\nu}(a_g)$	3108	293	For the sources of these data the	
$\bar{\nu}(b_{3u}^*)$	129/111	4.2	original reference [84K3] may be	
$\bar{\nu}(a_u^*)$	172	4.2	consulted.	
$\bar{\nu}(b_{1u})$	235	293		
$\bar{\nu}(b_{3u}^*)$	475	293		
$\bar{\nu}(a_u^*)$				
$\bar{\nu}(b_{3u}^*)$	603	293		
$\bar{\nu}(b_{2u})$	600	293		
$\bar{\nu}(b_{1u})$	650	293		
$\bar{\nu}(b_{3u}^*)$	727	293		
$\bar{\nu}(a_u^*)$				
$\bar{\nu}(b_{2u})$	808	293		
$\bar{\nu}(a_u^*)$	870	293		
$\bar{\nu}(b_{3u}^*)$	883	293		
$\bar{\nu}(b_{1u})$	903	293		
$\bar{\nu}(b_{3u}^*)$	954	293		
$\bar{\nu}(a_u^*)$				
$\bar{\nu}(b_{2u})$	998	293		
$\bar{\nu}(b_{2u})$	1068	293		
$\bar{\nu}(b_{1u})$	1145	293		
$\bar{\nu}(b_{2u})$	1163	293		
$\bar{\nu}(b_{1u})$	1270	293		
$\bar{\nu}(b_{1u})$	1314	293		
$\bar{\nu}(b_{2u})$	1346	293		
$\bar{\nu}(b_{2u})$	1398	293		
$\bar{\nu}(b_{1u})$	1447	293		
$\bar{\nu}(b_{2u})$	1462	293		
$\bar{\nu}(b_{2u})$	1533	293		
$\bar{\nu}(b_{1u})$	1616	293		
$\bar{\nu}(b_{1u})$	3024	293		
$\bar{\nu}(b_{2u})$	3050	293		
$\bar{\nu}(b_{1u})$	3050	293		
$\bar{\nu}(b_{2u})$	3093	293		
$\bar{\nu}(b_{1u})$	3108	293		

phonons: [68S3, 69B1, 69B2, 69D3, 73D, 74G1, 75N2, 78V, 79Z2, 80B3, 81D3, 82C2, 82D1]; see Fig. 5; for a comparison $C_{14}H_{10} - C_{14}D_{10}$, see [74G1]

parameters of radical ion states (see also comparison of literature results [75K1, 80S4, 67W1], also cf. [74K1, 74B4]):

I_g^a	7.40(5) eV		photoelectron spectroscopy; see also Figs. 6, 128 and 160	cf. 80S4
I_c, I_c^{th}	5.75(10) eV		photoelectron yield, cf. Fig. 7 and 143; photoelectron spectroscopy, see Fig. 8	80A3
	5.67 eV		photoelectron spectroscopy, single crystal	83S12
	5.80(34) eV		photoinjection from different metal electrodes, single crystal	67W1

(continued)

Physical property	Numerical value	Experimental conditions	Experimental method, remarks	Ref.
parameters of radical ion states (continued)				
A_g	0.55 (15) eV		literature data are reviewed in [74B4]	74B4
A_c	1.8 (2) eV		estimated value, taking into account I_c, E_g, A_g and the hole polarization energy $P^+ = I_g - I_c$, and the uncertainty of these values	74B4
$\bar{v}_{A^*+}$ (in 10^3 cm^{-1})	10.7 12.15 13.7 15.0 15.45	solution in sec-butyl-chloride at 77K	optical absorption after γ-irradiation; read from Fig. 11 of [73S2]; cf. [74K4]	73S2
$\bar{v}_{A^*-}$ (in 10^3 cm^{-1})	10.8 12.1 13.0 13.8 14.45 15.1 15.6 16.8 18.4	solution in MTHF at 77K	optical absorption after γ-irradiation; read from Fig. 1 of [73S2]; cf. [65B3] For the vibronic frequencies of the anthracene radical anion states, see [73A, 75K3]; cf. [80S4], table 2.9.	73S2
E_g (in eV)	4.0	$T = 427$ K	onset of intrinsic photoconduction, see Fig. 9	66C1
	4.1 (1)	300 K	photoinjection from metal electrodes	73B, cf. 68V
	4.00 (2)	RT	autoionization	73B
	4.08 (4)	RT	two photon excitation	74B3
	≈ 4.1	RT	two photon excitation	74K2
	4.06	RT	thermally activated dissociation of charge transfer states, see Fig. 10	80K3
	3.88 (5)	RT	photogeneration	82S6

See also [80S4], chapter 2.10.

polarization energies:

P^+	1.65 (15) eV		defined as $I_g - I_c$, cf. Fig. 161 ($=P_p$) and sections 12.3.1, 12.3.4	
P^-	1.25 (35) eV		defined as $A_c - A_g$ (values taken from the corresponding table; cf. [74B4]); see Fig. 161 ($=P_n$)	

position of valence and conduction band relative to vacuum level:

It is generally accepted that the highest valence band and the lowest conduction band are narrow (bandtwidth $\ll E_g$) and that these bands are located at the energies I_c and A_c below vacuum level, respectively, cf. Fig. 161.

Physical property	Numerical value	Experimental conditions	Experimental method, remarks	Ref.
charge transfer exciton state energies (in eV):				
E_{CT} $[\frac{1}{2}\frac{1}{2}0]$	3.45	$T = 77$ K	electromodulation spectroscopy; these results are not essentially altered on cooling down to 6 K	83S11
$[0\,1\,0]$	3.61			
$[1\,0\,0]$	3.76			
$[\frac{1}{2}\frac{3}{2}0]$ $[1\,1\,0]$	3.96			
$[0\,2\,0]$ $[\frac{3}{2}\frac{1}{2}0]$ $[1\,2\,0]$	4.15			
$[r \to \infty]$	4.40		(isotropic Coulombic law extrapolation for infinite separation with $\bar{\varepsilon} = 3.2$, interpreted as vertical ("optical") band gap E_g^{opt}, in contrast to an adiabatic gap of 4.1 eV; the difference being the relaxation energy of the two molecular ion states A^+ and A^- (A = anthracene))	
charge carrier mobilities (μ in cm^2/Vs) (a comprehensive list of references may be found in [82S3]):				
holes:				
		T [K]		
μ_{aa}	1.13 (2)	300	time-of-flight; the anisotropy and temperature dependence of the hole mobility are represented in Figs. 12…18; for pressure dependence in the range $0 < p < 3$ kbar, see Fig. 20	82K1; cf. 64K2
μ_{bb}	2.07 (4)			
μ_{c*c*}	0.73 (2)			
μ_{11}	1.28			
μ_{33}	0.57			
μ_{bb}, μ_{c*c*}		300…450		69F2
n_{aa}	-1.46 (8)	$110 \le T \le 300$	the anisotropy of the exponent n in $\mu \propto T^n$ is represented in Fig. 19	82K1
n_{bb}	-1.26 (7)			
n_{c*c*}	-1.43 (8)			
n_{11}	-1.57			
n_{33}	-1.41			
θ	-27 (2)°	300		
m_θ	0	$110 \le T \le 300$		
electrons:				
μ_{aa}	1.73 (3)	300	time-of-flight; the anisotropy and temperature dependence of the electron mobility are represented in Figs. 11…18; for pressure dependence in the range $0 < p < 3$ kbar, see Fig. 21	82K1; cf. 64K2
μ_{bb}	1.05 (2)			
μ_{c*c*}	0.39 (1)			
μ_{11}	1.74			
μ_{33}	0.38			
μ_{c*c*}		200…450		69F2
n_{aa}	-1.45 (8)	$150 \le T \le 300$	the anisotropy of the exponent n in $\mu \propto T^n$ is represented in Fig. 19	82K1
n_{bb}	-0.84 (5)			
n_{c*c*}	$+0.16$ (3)			
n_{11}	-1.36			
n_{33}	$+0.15$			
θ	-2 (1)°	300		
m_θ	-0.033°/K	$150 \le T \le 300$		

Physical property	Numerical value	Experimental conditions	Experimental method, remarks	Ref.
perdeuterated anthracene, $C_{14}D_{10}$ (isotope effect):				
$\mu_{c^*c^*}$	0.35 (1)	300	see also Fig. 14	82K1; cf. 70M2, 73M1, 77S2
$n_{c^*c^*}$	+0.10 (3)	$130 \leq T \leq 370$		82K1

The other electron mobility components and all hole mobility components are not significantly different from the values obtained for normal anthracene, $C_{14}H_{10}$, within the present experimental error span of approximately $\pm 3\%$ [82K1].

electric field dependences of mobilities $\mu_{c^*c^*}$:
The electron mobility is field-independent up to $1.6 \cdot 10^4$ V/cm at 100 K [79S1].
Field independence, up to $2.8 \cdot 10^4$ V/cm was reported later [83S7], in contrast to [80N1].
The hole mobility is field independent at room temperature up to $5 \cdot 10^5$ V/cm [82E1].

For dependence of mobilities on (intentional) impurities, see below under "carrier trapping levels".

charge carrier generation and recombination processes:
quantum yield for the intrinsic generation of one free electron plus one free hole by one photon:

η	$\approx 10^{-4}$	$\hbar\omega = 4.4$ eV $E = 0$, $T = 295$ K	cf. Fig. 10; for the electric field dependence, see Figs. 22a, b	80K3

quantum yield for excitonic charge carrier generation reactions ($T = 300$ K):

γ_{ss}	$6 \cdot 10^{-12}$ cm^3/s			73S3
	$0.9 \cdot 10^{-12.0 (4)}$ cm^3/s			68B
γ_{ST}	$1 \cdot 10^{-12}$ cm^3/s			68F
σ_{S_1}	$4 \cdot 10^{-20}$ cm^2	$\hbar\omega = 1.78$ eV		74B3
	$2 \cdot 10^{-19}$ cm^2	$2.48 \leq \hbar\omega \leq 2.95$ eV		73S3
σ_{T_1}	$5 \cdot 10^{-22}$ cm^2	$\hbar\omega = 2.35$ eV		68S1
	$1 \cdot 10^{-20}$ cm^2	$\hbar\omega \leq 2.63$ eV	flash lamp and long pass filter	67H2

For action spectrum for charge carrier generation at 300 K, cf. Fig. 23, curve a; at 5 K, cf. Fig. 25.
For action spectrum for detrapping of charge carriers by triplet excitons at 83 K, cf. Fig. 24.

spectral sensitization of quantum yield:

η	η increases by a factor of 50 (from $\eta = 2 \cdot 10^{-5}$ to $\eta = 10^{-3}$)	$\hbar\omega = 3.2$ eV surface sensitization by phenothiazine $T = 300$ K	time-of-flight pulses; excitation with (and without) a vapor-deposited thin sensitizing phenothiazine layer at the illuminated crystal surface, cf. Fig. 23	74S2, 74K1

dark conductivity:

σ_{dark}	$< 10^{-15}$ Ω^{-1} cm^{-1}	$T = 300$ K; thermal excitation by kT	contact-free ac measurement	71K1

unipolar electrode injection of charge carriers:

i_{inj}	$\leq 10^{-2}$ A/cm^2	$T = 77$ K	electrode: Li in ethylene-diamine; see also Fig. 26	69P

bipolar (double) injection of charge carriers: Fig. 27

bimolecular electron – hole recombination:

γ_{np}	$3 (1) \cdot 10^{-6}$ cm^3/s	$T = 300$ K		66H, 66K1, 67B2, 67S2

Physical property	Numerical value	Experimental conditions	Experimental method, remarks	Ref.
recombination electroluminescence: [62P]				
dependence of the emitted light intensity on the injected current density: Figs. 28, 29				
$\eta_{rec,\hbar\omega}$	0.2	$T=300$ K, $i=10^{-5}\cdots$ $2\cdot10^{-2}$ A/cm^2	for T-dependence, see [70W]	67M
	$0.01\cdots0.04$	$\lambda=410\cdots540$ nm $i\leq5\cdot10^2$ A/cm^2 at $2\cdot10^5$ V/cm	tunnel injection cathode	70D1
	$0.01\cdots0.08$	$T=300$ K, $i\leq0.1$ A/cm^2	field-assisted injection, $E=300$ V/30 μm	69D1

The electroluminescence spectra are similar to the optically excited fluorescence spectra.

For comparison of fluorescence and double injection electroluminescence spectra for neat and for tetracene-doped anthracene crystals, see Fig. 30.

dielectric tensor:

ε_{aa}	2.90 (4)	$f=2.5\cdot10^{-4}\cdots$	capacitor arrangement,	71K1
ε_{bb}	2.99 (6)	10^3 Hz,	ac bridge circuit	
ε_{c*c*}	3.72 (9)	$T=300$ K		
ε_{11}	2.51			
ε_{33}	4.11			
θ	29°			
ε_{aa}	2.90 (4)	$f=0.1\cdots5\cdot10^3$ Hz,	temperature and pressure	73M4
ε_{bb}	2.94 (3)	$T=295$ K	dependences are also given	
ε_{c*c*}	3.84 (10)			
ε_{11}	2.62 (3)			
ε_{33}	4.08 (8)			
θ	26 (2)°			

optical tensor:

n_α	1.550 (10)	$T=300$ K;	see Fig. 31 for orientation;	62N
n_β	1.775 (10)	$\lambda=589$ nm	see also [48O, 50E, 54W, 58W,	
n_γ	2.04 (8)		82J6]	

A weighted average of the 546 nm literature data is: $n_\alpha=1.559$ (7), $n_\beta=1.807$ (9), $n_\gamma=2.23$ (1), $2\,V=90.0$ (5)°.

carrier trapping levels (in eV):

$E_{t,p}$	0.43	$1\cdots40$ ppm tetracene-doped	time-of-flight mobilities, thermally activated, see Fig. 32	63H; cf. 65B1, 68O, 69S2, 74K1, 75P
	0.42	$\approx10^{-7}$ mol/mol tetracene-doped	time-of-flight mobilities, thermally activated, see Fig. 33b	74K1, 75P; cf. 68O
$E_{t,n}$	0.17	$\approx10^{-7}$ mol/mol tetracene-doped	time-of-flight mobilities, thermally activated, see Fig. 33b	74K1, 75P; cf. 68O
E_t	see Fig. 34	selected dopants	see caption of Fig. 34	74K1

Physical property	Numerical value	Experimental conditions	Experimental method, remarks	Ref.

12.1.2 Anthracene: PMDA, $C_{14}H_{10}:C_{10}H_2O_6$

molecular structure:

anthracene : pyromellitic dianhydride (PMDA)

crystal structure: triclinic, space group $C_i^1 - P\bar{1}$, $Z = 1$

lattice parameters (lengths in Å, angles in deg):

a	7.2812 (6)	$T = 153$ K	X-ray diffraction	78R1
b	10.7684 (7)			
c	7.1246 (7)			
α	117.516 (7)			
β	111.513 (8)			
γ	97.437 (8)			
a	7.37 (1)	295 K		78R1
b	10.78 (1)			
c	7.23 (1)			
α	117.5 (1)			
β	111.3 (1)			
γ	97.5 (1)			

molecular packing diagram: Fig. 35
cleavage planes: (010) and ($1\bar{1}0$), intersecting $\|$ [001]

density:

d	1.49 g/cm^3			71L2

melting point:

T_m	513 K			71L2

energy levels and lifetimes of excitons:

		T [K]		
E_{S_1} (CT)	2.271 eV	2	absorption spectrum, 0–0	77H2, 75H4, 74H1
τ_{S_1}	4 ns	300	fluorescence, pulse excitation	73H4
	5 ns	1.6		
E_{T_1}	1.921 eV	2	phosphorescence emission, 0–0	77H3, 73H4

ionization threshold for the formation of radical ion states:

I_c^{th}	6.00 (15) eV	$T = 300$ K	photoelectron spectroscopy	76I

charge carrier mobilities:

electrons:

		T [K]		
μ_{cc}	0.20 cm^2/Vs	370	time-of-flight; cf. Fig. 36	84M,
$\mu_{b^*b^*}$	0.016 cm^2/Vs	370		cf. 75K4
$\mu_{a'a'}$	0.0055 cm^2/Vs	370		84M
n_{cc}	-1.4	280$\cdots$455	thermally activated below 270 K with $E_A = 0.22$ eV	84M

holes were thermally activated ($E_A = 0.58$ eV) in the whole temperature range where time-of-flight transits were observable [75K4, 84M].

Physical property	Numerical value	Experimental conditions	Experimental method, remarks	Ref.

charge carrier generation and recombination processes:

action spectrum of photocurrent generation: Fig. 37

activation energy for photocurrent generation as a function of photon energy: Fig. 38

magnetic field effects on the photocurrent: an anisotropic magnetic field effect arises from the interaction of triplet excitons with trapped (doublet state) electrons (or holes); it is due to the anisotropy of the spin-spin interaction in the triplet state of non-spherosymmetric molecules; see Figs. 39, 40

12.1.3 Benzene, C_6H_6

molecular structure:

crystal structure: orthorhombic, space group D_{2h}^{15}-Pbca, $Z=4$

lattice parameters (in Å):

a	7.460	$T=270$ K	X-ray diffraction	58C
b	9.666			
c	7.034			
a	7.38	103 K	X-ray diffraction	55 K
b	9.57			
c	6.74			
a	7.292	78 K	X-ray diffraction	55K
b	9.471			
c	6.742			

molecular packing diagram: Fig. 41

elastic moduli (in 10^{10} dyn/cm^2):

c_{11}	6.14	$T=250$ K	ultrasonic pulse technique;	64H2
c_{22}	6.56	250 K	for the other coefficients	
c_{33}	5.83	250 K	and for other temperatures see [64H2]	

melting point:

T_m	278.6 K			81W3

vapor pressure:

p	1 Torr	$T=228.1$ K		81W3
	10 Torr	259.2 K		
	40 Torr	283.2 K		

energy levels and lifetimes of excitons:

E_{S_1}	4.688 eV	$T=4$ K; $E\|a$	0–0, absorption;	67S3
	4.693	$E\|b$	for absorption spectrum at 243 K, see Fig. 42	
E_{T_1}	3.677			70B1
τ_{T_1}	3.1 s	77 K		67S3

Physical property	Numerical value	Experimental conditions	Experimental method, remarks	Ref.
parameters of radical ion states (in eV):				
I_g^a	9.17		photoelectron spectroscopy, see also Fig. 128	82S2
I_c^{th}	7.58		photoelectron spectroscopy	82S2
A_g	-1.15		electron transmission spectrum	78J
E_g	≈ 7		estimated from $I_c^{th} - (A_g + P^+)$ with P^+ assumed as $P^+ \approx 1.7$ eV	
electron mobilities:				
μ_{aa}	2.0 cm^2/Vs	$T = 250$ K	time-of-flight; see Fig. 43	70H, 72H1
n_{aa}	-2	$173 \cdots 265$ K		70H
quantum yield of charge carrier generation processes:				
η	10^{-5}	$\hbar\omega = 4.7 \cdots 5.3$ eV	$T = 243$ K, $E = 50$ kV/cm;	70H
	$8 \cdot 10^{-5}$	5.7 eV	see Fig. 42	

12.1.4 Biphenyl, $C_{12}H_{10}$

molecular structure:

crystal structure: monoclinic, space group C_{2h}^5-P2$_1$/a, $Z = 2$ [71W]

lattice parameters:

a	8.12 (2) Å	RT	X-ray diffraction	71W
b	5.63 (1) Å			
c	9.51 (2) Å			
β	95.1 (3)°			

phase transition temperatures (in K):

T_{tr}	< 75		spectroscopic evidence	74F2
	40		Raman spectroscopy	77B5
	42 and 17		optical retardation	78B1; cf. 81I2
	38 and 21	$C_{12}D_{10}$	elastic neutron scattering,	79C6

molecular packing diagram: Fig. 44

elastic moduli:

c_{22}	$9 \cdot 10^9$ dyn/cm^2	RT, $C_{12}D_{10}$	Brillouin scattering. For the other elastic moduli [80E] may be consulted.	80E, cf. 71K2

sound velocities:

$v_{[010]}$	2600 m/s	$T = 1.8$ K	For the other elastic moduli [80E] may be consulted.	80E

melting point:

T_m	344 K			81W3

density:

d	1.19 g/cm^3			71L2

Physical property	Numerical value	Experimental conditions	Experimental method, remarks	Ref.
energy levels of excitons:				
E_{S_1}	4.1074 eV	$T = 4.2$ K	absorption, fluorescence two photon absorption	73H5, 77H5

vibrons: IR absorption: [75B2]; absorption in the visible and UV-region, fluorescence: [63B, 73H5, 74T1, 77H5]; Raman spectra: [73A, 74T1, 82B4]

phonons: inelastic neutron scattering: [81N]

parameters of radical ion states:				
I_g^v	8.14 eV	$T = 333$ K	photoelectron spectroscopy	78R2

For $\bar{\nu}_{B^*-}$, determined from absorption spectra in the visible range and from Raman spectra, see [63B, 74T1] and [73A, 74T1], respectively.

electron mobilities (μ in cm^2/Vs):

		T [K]		
μ_{aa}	0.42	300	time-of-flight; see Fig. 45	77B3
μ_{bb}	1.25			
$\mu_{c^*c^*}$	0.51			
n_{aa}	-1.0	$300\cdots130$		77B3
n_{bb}	-1.25	$300\cdots\,90$		
$n_{c^*c^*}$	-1.0	$250\cdots100$		

12.1.5 Dibenzothiophene, $C_{12}H_8S$

molecular structure: synonym: diphenylenesulfide

crystal structure: monoclinic, space group C_{2h}^5-P2$_1$/c, $Z = 4$ [70S3]

lattice parameters:

a	8.67 (1) Å	RT	X-ray diffraction	70S3
b	6.00 (1) Å			
c	18.70 (2) Å			
β	113.9°			

cleavage: $(10\bar{2}) = \textbf{\textit{rb}}$-face in Fig. 46 [71B2]
molecular packing diagram: Fig. 46

melting point:

T_m	$372\cdots373$ K			81W3

energy levels of excitons:

E_{S_1}	3.710 eV	$T = 1.3$ K	fluorescence/absorption	79G2, 71B3
E_{T_1}	3.012 eV	1.3 K	phosphorescence/absorption	79G2, 71B3

For photoelectron spectrum, see Fig. 132.
vibrons: polarized IR and Raman spectra: [71B2]

Physical property	Numerical value	Experimental conditions	Experimental method, remarks	Ref.
hole mobilities (μ in cm^2/Vs):				
		T [K]		
$\mu_{aa} = \mu_{11}$	1.34 (10)	300	time-of-flight	79C1
$\mu_{bb} = \mu_{22}$	1.18 (10)	300		
$\mu_{c^*c^*} = \mu_{33}$	1.57 (10)	300		
n	$-3.5 \cdots -4.5$	$300 \leq T \leq 360$		
$\mu_{c^*c^*}$	1.58	298	time-of-flight, see Fig. 47	75D
$n_{c^*c^*}$	-2	$300 \leq T \leq 360$		75D

optical tensor [71B2]:

optic axial plane: (010).
Angles between the optical axes: $2V = 90\,(5)°$.
Principal axes of the optical tensor: **b**, **r**, **s**, see Fig. 46.

12.1.6 1,4-Dibromonaphthalene, $C_{10}H_6Br_2$

molecular structure:

crystal structure: monoclinic, space group C_{2h}^5-$P2_1/a$, $Z = 8$

lattice parameters:

a	27.230 (13) Å	RT	X-ray diffraction;	78B4,
b	16.417 (7) Å		zone-refined material	cf. 61T
c	4.048 (2) Å			
β	91.95 (2)°			

determination of crystal orientation by optical conoscopy; the optical axes plane is perpendicular to (010), approximately ∥(100) [67C1, 80K2]; Laue diffraction [81K3].

molecular packing diagram: Fig. 48

melting point:

T_m	356 K			61T

density:

d_{theor}	2.037 g/cm^3			71L2

energy levels and lifetimes of excitons:

		T [K]		
E_{S_1}	3.763 eV	0–0, 4.2	optical absorption; lower energy sublattice	67C1, 74S3
E_{T_1}	2.504 eV	0–0, 4.2	optical absorption	67C1, 74S3
τ_{T_1}	0.4 ms	1.3	phosphorescence	74S3
	2 ms	300		

vibrons: [67C1]

Physical property	Numerical value	Experimental conditions	Experimental method, remarks	Ref.
charge carrier mobilities (in cm^2/Vs):				
holes:				
μ_{aa}	0.36	$T = 300$ K	time-of-flight	81K4
μ_{bb}	0.66			
$\mu_{c^*c^*}$	1.07		temperature dependence of	
μ_{ac^*}	-0.13		hole mobilities: Fig. 49	
electrons:				
μ_{aa}	$1.03 \cdot 10^{-2}$	$T = 300$ K	time-of-flight; temperature depen-	81K4
μ_{bb}	$1.23 \cdot 10^{-2}$		dence in the range 255$\cdots$303 K:	
$\mu_{c^*c^*}$	$4.7 \cdot 10^{-2}$		$\mu \propto T^n$, $-3.4 \leq n \leq -2.1$,	
μ_{ac^*}	$0.003 \cdot 10^{-2}$		depending on orientation	

12.1.7 9,10-Dichloroanthracene, $C_{14}H_8Cl_2$, α-form

molecular structure:

crystal structure (α-form, stable below ≈ 455 K): monoclinic, space group C_{2h}^5-P2$_1$/a, $Z = 4$ [59T, 66K2]

lattice parameters:

a	7.04 (2) Å	RT	X-ray diffraction	59T,
b	17.93 (4) Å			66K2
c	8.63 (2) Å			
β	102.93 (15)°			

molecular packing diagram: Fig. 50

melting point:

T_m	483.5 (3) K		zone-refined material; β-form; (a phase transformation from α to β occurs at 458 (1) K; for the crystal structure of the β phase, see [78B5, 79K3])	84K4

density:

d	1.525 g/cm^3			59T

energy levels and lifetime of excitons:

E_{S_1}	2.930 eV	$T = 5$ K	reflection spectroscopy	77S4
τ_{S_1}	9.2 ns	4.2 K	fluorescence	80M2, 81M1
E_{T_1}	1.75 eV	RT		76K4
	1.7287 eV	4.2 K	phosphorescence excitation spectroscopy, lower Davidov component	83N1

ionization threshold for the formation of radical ion states:

I_g^v	7.58 eV		photoelectron spectroscopy	74S4
I_c^{th}	5.8 eV		photoelectron spectroscopy	77H4
I_c	5.75 eV		photoemission yield, $\eta^{1/3}$ vs. $\hbar\omega$	77H4, cf. 70M1

Physical property	Numerical value	Experimental conditions	Experimental method, remarks	Ref.
energies of charge transfer exciton states:				
E_{CT}	3.34 eV		electroreflectance, lowest energy component	74A, 68T, cf. 77S4
hole mobilities (μ in cm^2/Vs):		T [K]		
μ_p	3.89	295	time-of-flight; error: $\pm 15\%$. Direction: 74° to a, 41° to b, and 59° to c. The relative error in the temperature dependence is about 5%	78C1
	5.30	256		
	6.31	240		
	7.82	218		
	9.63	198		
	12.2	183		
n	$-2.37\,(5)$		average of 5 crystals: $-2.4\,(2)$	78C1

12.1.8 1,4-Diiodobenzene, $C_6H_4I_2$

molecular structure: I—⟨benzene ring⟩—I

crystal structure: orthorhombic, space group D_{2h}^{15}-Pbca, $Z=4$ [59L, 60L2]

lattice parameters:

a	17.008 (2) Å	RT	X-ray diffraction	59L,
b	7.321 (2) Å			60L2
c	5.949 (2) Å			

phase transition temperature:

T_{tr}	324 (1) K		X-ray diffraction; optical polarization microscopy; the transition is reversible, but very slow at decreasing temperature	73S1, 79S3

molecular packing diagram: Fig. 51

melting point:

T_m	404···405 K			81W3

density:

d	2.79 g/cm^3			59L

hole mobilities (μ in cm^2/Vs):

μ_{aa}	11 (1)	$T=300$ K	time-of-flight	67S1
μ_{bb}	4.0 (5)	300 K		
μ_{cc}	1.7 (2)	300 K		

For the temperature dependences of the hole mobilities (determined by the time-of-flight method), see Fig. 52.

Physical property	Numerical value	Experimental conditions	Experimental method, remarks	Ref.

12.1.9 Durene, $C_{10}H_{14}$

molecular structure:

synonym: 1,2,4,5-tetramethylbenzene

crystal structure: monoclinic, space group C_{2h}^5-$P2_1/a$, $Z=2$

lattice parameters:

a	11.57 (5) Å	RT	X-ray diffraction	33R
b	5.77 (2) Å		neutron diffraction gave the	
c	7.03 (5) Å		same a, b, and c parameters,	
β	113.3°		(RT), but slightly different β	
			(112.93°) [73P2]	

molecular packing diagram: Figs. 53, 54

density:

d	1.03 g/cm³			33R

melting point:

T_m	352.4 K			81W3

energy levels of excitons:

E_{S_1}	4.441 eV	$T=20$ K	optical absorption and emission	67S3
E_{T_1}	3.4516 eV	4 K	phosphorescence	64S

vibrons, phonons: IR spectroscopy: [69H2], Raman spectroscopy: [72S2], inelastic neutron scattering: [67R, 73L]

charge carrier mobilities (μ in cm²/Vs):

holes:

μ	5	RT	time-of-flight, in a–b plane	77B2
n	−2.5		in a–b plane; see Fig. 55	77B2
$\mu_{c^*c^*}$	0.15	RT		77B2
$n_{c^*c^*}$	−2.8			77B2

electrons:

μ_{bb}	8	RT	time-of-flight, see Fig. 55	77B2
n_{bb}	−2.5			77B2

optical tensor:

the optic axes lie in the plane (010); $\measuredangle(n_\alpha, c) = -0.9°$; $(-)2V = 87.37°$ [54W]

n_β	1.6148	$\lambda = 589$ nm, RT		54W

12.1.10 Iodoform, CHI_3

molecular structure:

synonym: triiodomethane

crystal structure: hexagonal, space group C_6^6-$P6_3$, $Z=2$ [66W]

lattice parameters:

a	6.818 Å	RT(?)	X-ray diffraction	66W
c	7.524 Å			

molecular packing diagram: Fig. 56

melting point:

T_m	396 K			81W3

Physical property	Numerical value	Experimental conditions	Experimental method, remarks	Ref.
density:				
d	4.01 g/cm^3			71L2
hole mobilities (μ in cm^2/Vs):				
μ_{aa}	0.85 (10)	RT	time-of-flight	81G1,
μ_{cc}	0.20 (4)			82G1
n_{aa}	-1.10	$200 \leqq T \leqq 350$ K	obtained by a Hoesterey-Letson	81G1,
n_{cc}	-0.45		fit (cf. [63H])	82G1

12.1.11 9-Methylanthracene, $C_{15}H_{12}$

molecular structure:

crystal structure: monoclinic, space group C_{2h}^5-P2$_1$/c, $Z=4$ (arranged in two centrosymmetric pairs of molecules) [79C2, 71B1].

lattice parameters:

a	8.920 (3) Å	79C2
b	14.641 (4) Å	
c	8.078 (4) Å	
β	96.47 (3)°	

molecular packing diagram: Fig. 57

melting point:

T_m	354.7 K	81W3

energy levels and lifetimes of excitons:

		T [K]		
E_{S_1}	3.193 eV	110	dissolved in isopentane	82K7
τ_{S_1}	270 ns	4.2	two-photon excitation	83S3
	110 ns	300	fluorescence	
E_{T_1}	1.788 eV	110	phosphorescence; 0–0 transition, dissolved in isopentane/ cyclohexane	83S3

ionization threshold for the formation of radical ion states:

I_g^v	7.25 eV	$T = 409$ K	photoelectron spectroscopy	74H5
I_c^{th}	5.7 eV	RT	photoelectron spectroscopy	77H4
I_c	5.68 eV	RT	photoelectron yield, extrapolation $\eta^{1/3}$ vs. $\hbar\omega$, thin sublimed film	77H4

hole mobilities (μ in cm^2/Vs):

μ_{aa}	0.85	RT	time-of-flight; see Fig. 58	81D2
μ_{bb}	0.93			
$\mu_{c^*c^*}$	0.64			
n_{aa}	-1.7	$T = 170 \cdots 300$ K		81D2
n_{bb}	-2.2	$170 \cdots 300$ K		
$n_{c^*c^*}$	-1.1	$160 \cdots 300$ K		

Physical property	Numerical value	Experimental conditions	Experimental method, remarks	Ref.

12.1.12 Naphthalene, $C_{10}H_8$

molecular structure:

crystal structure: monoclinic, space group C_{2h}^5-$P2_1/a$, $Z=2$; perfect (001) cleavage

lattice parameters:

a	8.2606 (66) Å	$T=296$ K	X-ray diffraction	82B5
b	5.9872 (20) Å			
c	8.6816 (77) Å			
β	122.671 (80)°			
V	361.44 Å^3			

These data agree within their standard deviations with data measured independently at $T=293$ K by [81M2].

lattice parameters of perdeuterated naphthalene, $C_{10}D_8$:

a	8.096 (7) Å	$T=12$ K	neutron diffraction,	83N2;
b	5.941 (5) Å		$\lambda=0.126$ nm	cf. 82J3
c	8.648 (7) Å			
β	124.63 (5)°			

temperature dependence of lattice parameters: Fig. 59, see also [81M2, 82J3]
pressure dependence of lattice parameters: [82A, 82J3]
molecular packing diagram: Fig. 60
elastic constants: [63A, 69A, 71K2]

melting point:

T_m	353.45 (5) K			81M2

enthalpy of fusion:

ΔH_m	18.78 (25) kJ/mol			81M2

enthalpy of sublimation:

ΔH_s	72.4 kJ/mol	$T=287$ K		53B

density:

d	1.152 g/cm^3			71L2

energy levels and lifetimes of excitons:

E_{S_1}	3.902 eV	absorption/emission, $T=4.2$ K	lower Davidov component	64C1, 75W1
τ_{S_1}	115 (5) ns	$T=4.2$ K, fluorescence	lower Davidov component	67E
E_{T_1}	2.62880 eV	$T=1.6$, absorption	phosphorescence excitation, lower Davidov component, $E\|b$	69C2, 82D2
τ_{T_1}	≈ 500 ms	RT		82W2, cf. 80K2, 84K2, 84W1

Physical property	Numerical value	Experimental conditions	Experimental method, remarks	Ref.

wavenumbers of vibrons (in cm^{-1}):

Averaged experimental values and symmetry assignments in the orthorhombic D_{2h} symmetry point group of the free molecule. b_{1u} are vibrational modes polarized parallel to the short (M), b_{2u} parallel to the long (L) axis, and b_{3u} perpendicular to the plane of the molecule. These modes have been found from infrared spectroscopy. Even symmetry modes a_g and b_{3g} are Raman active. The corresponding components of the dipole moment P_i and of the polarizability tensor α_{ik} are indicated. – See also [57S, 59M, 60F, 62J, 67W2, 68S3, 69H3, 69H4, 72T, 77H6].

$\bar{\nu}$	1471	a_g	$\alpha_{xx}, \alpha_{yy}, \alpha_{zz}$	80S4
	1028			
	3060	b_{1u}	P_x	
	3029			
	1389			
	1267			
	1128			
	877			
	1510	b_{2u}	P_y	
	1369			
	1210			
	1011			
	618			
	3051	b_{3g}	α_{xy}	
	3006			
	1445			
	1242			
	968	b_{3u}	P_z	
	789			
	481			

phonons: see [68S3, 72O, 72P, 73D, 75N2, 78V, 79B3, 79S3, 79Z2, 80B3, 80D, 81D3, 81D4, 82B2, 82K4, 83D, 83S4]
Brillouin zone of naphthalene: Fig. 61
phonon dispersion curves: Fig. 62
phonon density of states: Fig. 63
Raman active phonons: Fig. 64
comparison of phonons in normal and perdeuterated naphthalene/mixed crystals: [72P]
comparison of low energy vibrons in normal and in perdeuterated naphthalene mixed crystals: [80A4]
sound velocities: [60T, 68D]

parameters of radical ion states (in eV):

I_g^a	8.12	photoelectron spectroscopy,	82S2
I_c^{th}	6.4	cf. Fig. 160	82S2
A_g	-0.19	electron transmission	78J
E_g	4.9	estimated from $I_c - A_g - P^+$	
	5.1	threshold of intrinsic photo-conduction	80B1

For $\bar{\nu}_{N^*-}$, see table 2.8 of [80S4]; spectra obtained in solution or glass matrix: [60H, 73S2, 74H2]. $\bar{\nu}_{N^*+}$: spectra obtained in solution or glass matrix, see [73S2, 74K3, 78S3].

polarization energies:

P^+	1.7 eV	$I_g^a - I_c^{th}$, cf. section 12.3.5

For position of valence and conduction band relative to vacuum level, see remark under 12.1.1 (Anthracene).

Physical property	Numerical value	Experimental conditions	Experimental method, remarks	Ref.
charge carrier mobilities (μ in cm^2/Vs): (a comprehensive list of references may be found in [82S3])				
holes:				
		T [K]		
μ_{aa}	0.94 (5)	293	time-of-flight	82W4, 84W1, 80W1
μ_{bb}	1.48 (7)	293		83W, 83S6
μ_{c*c*}	0.32 (2)	293		83S6, 80W1
μ_{ac*}	−0.023	293		84W3, cf. 80W1
μ_{11}	0.951	290		84W3
$\mu_{22} = \mu_{bb}$	1.50	290		84W3
μ_{33}	0.321	290		84W3
n_{aa}	−2.8 (2)	$78 \le T \le 300$	Fig. 65	82W4, 84W1
n_{bb}	−2.5 (1)	$78 \le T \le 300$	Fig. 66	83W, cf. 83S6
n_{c*c*}	−2.82 (15)	$78 \le T \le 300$	Fig. 67	83W, 84K2, cf. 83S6
θ	−2 (3)°		independent of temperature	84W3
electrons:				
μ_{aa}	0.62 (4)	293		83W, 80W1, 84W1
μ_{bb}	0.64 (4)	293		83W, 83S6
μ_{c*c*}	0.44 (2)	293		78S4, 80W1
	0.382 (17)	290		84W3
μ_{ac*}	−0.16 (3)	293		80W1, 84W3

For μ_{11}, $\mu_{22} (= \mu_{bb})$, μ_{33} in the temperature range 30$\cdots$300 K, see Figs. 66, 68. For the tensor ellipsoids of the electron mobility at 50, 120 and 290 K (cross section in the **a–c** plane), see Fig. 69.

n_{aa}	−1.4 (1)	$30 \le T \le 300$	Fig. 65	82W4, 84W1, 84W2
n_{bb}	−0.55 (10)	$120 \le T \le 300$	Fig. 66; it is possible to approximate the temperature dependence of several independent experimental results by $\mu_{bb} \propto \exp(E/kT)$ with $E = 9$ (1) meV in the temperature range $55 \le T \le 170$ K	83W, cf. 83S6
	−1.50 (10)	$25 \le T \le 80$		
n_{c*c*}	−1.6 (1)	25$\cdots$70	see Fig. 67; a transition from hopping to band type transport near 100 K has been inferred from such data for the electrons [78S4]	83W, 84K2, cf. 78K1 78S4, 82W1
	+0.04 (2)	120$\cdots$300		

(continued)

Physical property	Numerical value	Experimental conditions	Experimental method, remarks	Ref.
charge carrier mobilities (continued)				
$\theta(T)$	$-27°$	293	the electron mobility tensor	84W3
	$-13°$	200	rotates with temperature	
	$-6°$	100	about [010]	

For electric field dependences of mobilities determined by the time-of-flight method, see Figs. 65 (hole mobility μ_{aa}), 66 (hole mobility μ_{bb}) and 67 (electron and hole mobilities μ_{c*c*}). The charge carrier velocities tend to saturate at low temperatures for electric fields $E > 10^4$ V/cm, see Figs. 70 and 71.

charge carrier generation and recombination processes:

γ_{ss}	$2.3 \cdot 10^{-14}$ cm^3/s	RT	generation of pairs of free electrons and free holes by S_1–S_1 collisional annihilation	70B3
$\gamma_{Se_{tr}}$	$8(6) \cdot 10^{-10}$ cm^3/s	RT	release of trapped electrons by collisions of singlet excitons (singlet excitons were generated by two photon excitation)	80N3
γ_{np}	$\gtrsim 1.7 \cdot 10^{-9}$ cm^3/s	RT	bimolecular electron-hole recombination	70B3

unipolar electrode injection of holes: Fig. 72
recombination electroluminescence, $\lambda > 318$ nm: see [69L2]

dielectric tensor:

ε_{aa}	2.65 (6)	$T = 295$ K,	capacitance measurement	73M3
ε_{bb}	2.87 (5)	$f = 1592$ Hz		
ε_{c*c*}	3.21 (7)			
ε_{ac*}	-0.52 (5)			
ε_{11}	2.25 (7)			
ε_{22}	2.87 (5)			
ε_{33}	3.43 (4)			
θ	28.0 (25)°			

optical tensor:

n_α	1.525	$\lambda = 546$ nm, RT		54W
n_β	1.722			
n_γ	1.945			
θ	23.3°		with $\measuredangle(Z, c)$ of [54W] taken positive	
$(-)2V$	83°		angle between the optic axes; optical axes plane: (010)	

carrier trapping: An example of space charge-limited currents as a function of temperature is represented in Fig. 73.

12.1.13 Perylene, $C_{20}H_{12}$, α-form

molecular structure:

crystal structure (α-phase): monoclinic, space group C_{2h}^5-P2$_1$/a, $Z = 4$ (two sandwich pairs of molecules)

Physical property	Numerical value	Experimental conditions	Experimental method, remarks	Ref.
lattice parameters:				
a	11.28 Å		X-ray diffraction	64C2
b	10.83 Å			
c	10.26 Å			
β	100.55°			

molecular packing diagram: Figs. 74, 75
cleavage plane: (001)
A monoclinic β-phase, $Z=2$, is stable only above 413 K or metastable [63T].

density:				
d	1.322 g/cm^3			64C2
melting point:				
T_m	550$\cdots$552 K			81W3
	550.9 K		triple point	80W2
enthalpy of fusion:				
ΔH_m	31.87 kJ/mol			61I
enthalpy of sublimation:				
ΔH_s	126 (2) kJ/mol			61I

Vapor pressure as a function of temperature: see [61I]

energy levels and lifetime of excitons:

		T [K]		
E_{S_1}	2.632 eV	1.6	absorption spectroscopy, lower Davidov component $(\boldsymbol{E} \| \boldsymbol{b})$	74T2
τ_{S_1}	100 ns	4.2	excimer fluorescence	83S3
	60 ns	300		
E_{T_1}	1.5341 eV	2	direct absorption	69C1
	1.533 eV	77	direct absorption	69C1

vibrons: IR spectroscopy: [83K1, 65A1], vibrational analysis of $S_0 \rightarrow T_1$ transitions: [79M1], Raman spectroscopy: [83K1, 80H, 73A]

parameters of radical ion states (in eV):

I_g^a	6.90	$T=423$ K	photoelectron spectroscopy; see also Fig. 131	81S2
I_c^{th}	5.37	RT	photoelectron spectroscopy, Fig. 152	73H3
	5.2	RT	photoelectron spectroscopy	81S2
	5.12	RT	photoelectron spectroscopy, single crystal	83S12
I_c	5.39		photoemission quantum yield, Fig. 146	73H4
E_g	3.10	300 K	threshold energy for intrinsic production of electron-hole pairs	80S4

For $\bar{\nu}_{p*+}$ and $\bar{\nu}_{p*-}$, determined by absorption in glassy solutions, see [59A, 73S2] and [59A, 60H, 73S2, 73A], respectively.

For $\bar{\nu}_{p*-}$, determined by Raman spectroscopy, see [73A].

polarization energies:

P^+	1.6 (1) eV		$I_g^a - I_c^{th}$ cf. section 12.3.5	

Physical property	Numerical value	Experimental conditions	Experimental method, remarks	Ref.
charge carrier mobilities (μ in cm^2/Vs):				
holes:				
holes were always found to be thermally activated				
electrons:				
		T [K]		
μ_{aa}	2.37 (12)	300	time-of-flight; calculated from	83S5
	8.12	150	measurements in 13 different	
	24.8	80	directions, cf. Fig. 76;	
	41.4	60	for μ_{aa}, see Fig. 77	
μ_{bb}	5.53 (25)	300	for μ_{bb}, see Fig. 78	
	18.2	150		
	53.4	80		
	87.4	60		
$\mu_{c^*c^*}$	0.78 (4)	300	for $\mu_{c^*c^*}$, see Fig. 79	
	3.53	150		
	13.5	80		
	24.9	60		
μ_{ac^*}	0.23 (5)	300		
	1.21	150		
	4.94	80		
	9.22	60		
principal axes values:				
μ_{11}	2.40	300	calculated from measurements	83S5
	8.42	150	in 13 different directions,	
	26.7	80	cf. Fig. 76	
	45.5	60		
μ_{22}	5.53	300	the (010) cross sections of	
	18.2	150	the tensor ellipsoids at different	
	53.4	80	temperatures are represented	
	87.4	60	in Fig. 81	
μ_{33}	0.75	300		
	3.23	150		
	11.7	80		
	20.8	60		
θ	8.2°	300	the electron mobility tensor	83S5
	14°	150	rotates with temperature	
	21°	80	about [010]; cf. Fig. 81	
	24°	60		
n_{aa}	-1.78 (3)	300···60	valid for electric fields	83S5
			≤ 5 kV/cm	
n_{bb}	-1.72 (3)	300···60		
$n_{c^*c^*}$	-2.15 (3)	300···60		
n_{ac^*}	-2.30 (9)	300···60		

For an electron mobility in an oblique direction and shallow trapping effects at low temperatures, see Fig. 80.

For electric field dependences of charge carrier mobilities and velocities, see Figs. 77···79, and Fig.82.

For response time of perylene as a fast photodetector (time-of-flight transits), see Fig. 83.

Dye-sensitization of quantum yield for hole injection from the crystal surface: see [68G].

optical tensor:

The acute bisectrix lies in the angle between $+a$ and $+c^*$ (right handed axis system); optic axial plane is (010).

Physical property	Numerical value	Experimental conditions	Experimental method, remarks	Ref.

12.1.14 Phenazine, $C_{12}H_8N_2$, α-form

molecular structure:

crystal structure (α-form): monoclinic, space group C_{2h}^5-$P2_1/a$, $Z = 2$ [55H]

lattice parameters:

a	13.22 (1) Å	for unit cell parameters at 83K,	55H
b	5.061 (5) Å	see [56H, 57H]	
c	7.088 (7) Å		
β	109.22 (25)°		

molecular packing diagram: Fig. 84

distance between the molecular planes: 3.49 Å [55H]
cleavage plane: (100) [68C3]

melting point:

T_m	449···450 K		81W3

density:

d	1.34 g/cm^3		55H

energy levels of excitons (in eV):

		T [K]		
E_{S_1}	2.8369	4.2	absorption	62H
	2.8313	4.2	absorption	71M
E_{T_1}	1.9153	4.2	absorption	67C2
	1.91472 (6)	1.6	excitation spectrum, lower Davidov component	82D3

vibrons: IR and Raman spectroscopy [72D, 81M3], cf. [84K1]

parameters of radical ion states:

I_g^v	8.33 eV	$T = 349$ K	photoelectron spectroscopy;	75H1,
	8.44 eV		cf. Fig. 133	75M
A_c	2.3 (2) eV		estimated from A_c (anthracene) plus electron trap depth of phenazine in anthracene, 0.54 eV, cf. [75P]	

electron mobilities (μ in cm^2/Vs):

		T [K]		
$\mu_{a^*a^*}$	0.51	293	time-of-flight; $a^* \parallel b \times c$	74P1
μ_{bb}	1.1			
μ_{cc}	0.29			
$n_{a^*a^*}$	−0.1	250···370	see Fig. 85	74P1
n_{bb}	−0.65	250···350		
n_{cc}	0.05	200···350		

optical tensor:

n_α	1.73	wavelength not specified, probably 589 nm	see Fig. 86 for orientation	41W1
n_β	1.82			
n_γ	1.96			

Physical property	Numerical value	Experimental conditions	Experimental method, remarks	Ref.

12.1.15 Phenothiazine, $C_{12}H_9NS$

molecular structure:　　　　　　　　　　　synonym: thiodiphenylamine

crystal structure and lattice parameters:

a) orthorhombic phase, space group D_{2h}^{16}-Pnma, $Z = 4$ $(T > 250\ K)$; perfect (010) cleavage

a	7.916 (10) Å	RT	X-ray diffraction; the existence	76M2
b	20.974 (10) Å		of a genuine orthorhombic	
c	5.894 (10) Å		phase was questioned [77W2]	

b) monoclinic phase, space group C_{2h}^5-P2$_1$/c $(T < 250\ K)$

a	5.808 (1) Å	$T = 120\ K$	X-ray diffraction	77W2
b	7.783 (3) Å			
c	22.009 (3) Å			
β	69.87 (2)°			

phase transition temperature:

T_{tr}	245 (5) K		higher- order transition; temperature-dependent Guinier method	83K2
	249.5 (5) K	peak of Λ-shaped c_p anomaly	c_p measurements (confirming T_{tr} obtained by Raman [84N] and optical retardation [83K5] measurements)	84K5

molecular packing diagram: Fig. 87

melting point:

T_m	459···462 K			81W3

density:

d	1.352 g/cm^3	RT		76M2

vibrons: IR absorption, polycrystalline material in KBr press pellet [65M2]

ionization thresholds for the formation of radical ion states (in eV):

I_g^v	6.82		photoelectron spectroscopy	75H2
	6.72			77D
I_c	5.15		photoemission quantum yield, Fig. 149	80A3
I_c^{th}	5.18 (15)		photoelectron spectroscopy, Fig. 156	82K2

charge carrier mobilities:

electrons:

μ_{bb}	2.45 (10) cm^2/Vs	$T = 300\ K$	time-of-flight, see Fig. 88	83K2
n_{bb}	-2.8 (3)	280···370 K	reduced slope below the phase transition	83K2

holes: holes were found to be thermally activated except at the highest experimental temperatures:

μ_{bb}	$2 \cdot 10^{-2}$ cm^2/Vs	$T = 340\ K$	time-of-flight	83K2

optical tensor:

n_a	1.61	$\lambda = 589$ nm,		41W2
n_b	1.95	RT		
n_c	1.73			

Physical property	Numerical value	Experimental conditions	Experimental method, remarks	Ref.

12.1.16 Phthalocyanine, $C_{32}H_{18}N_8$, β-form

molecular structure:

crystal structure (β-form): monoclinic, space group C_{2h}^5-$P2_1/a$, $Z=2$

lattice parameters:

a	19.85 Å	X-ray diffraction; neutron	36R,
b	4.72 Å	diffraction yielded similar data	65A2
c	14.8 Å	[69H5] and disorder of the two	
β	122.25°	inner hydrogen atoms with	
		respect to the four inner N	
		atoms	

molecular packing diagram: Fig. 89
crystal habitus (sublimation-grown): Fig. 90

density:

d	1.44 g/cm^3		65A2

absorption spectrum of polycrystalline film: [77P3]

ionization energies for the formation of radical ion states (in eV):

I_g^a	6.2	$T=773$ K	photoelectron spectroscopy	79B5
I_g^v	6.41	773 K	photoelectron spectroscopy	79B5
I_c^{th}	≈ 5.5	threshold		64V1
	≈ 6.2	threshold		79B5
I_c^p	5.9 (2)	peak		81T1
E_g	≈ 2 eV	above 250 K	from conductivity measurements	77H1

charge carrier mobilities:
electrons:

		T [K]		
$\mu_{c^*c^*}$	1.2 cm^2/Vs	373		74C, 70U
$n_{c^*c^*}$	$-1.3 > n > -1.6$	300$\cdots$630	see Fig. 91	74C, 70U

holes:

$\mu_{c^*c^*}$	1.1 cm^2/Vs	373		74C, 70U
$n_{c^*c^*}$	$-1.1 > n > -1.6$	300$\cdots$630	see Fig. 91	74C, 70U

dark conduction: The electric resistivity decreases with pressure from $10^7 \,\Omega$ cm at 0 to 40 Ω cm at 570 kbar [75O1].

Physical property	Numerical value	Experimental conditions	Experimental method, remarks	Ref.

12.1.17 Pyrene, $C_{16}H_{10}$

molecular structure:

crystal structure: monoclinic, space group C_{2h}^5-$P2_1/a$, $Z=4$ (two sandwich pairs of molecules)

lattice parameters:

a	13.649 (10) Å	RT	X-ray diffraction	65C
b	9.256 (10) Å		temperature dependence of	
c	8.470 (10) Å		lattice parameters:	
β	100.28 (4)°		[78K3, 81I1]	

For neutron diffraction data at room temperature, see [72H2].
molecular packing diagram: Fig. 92
cleavage plane: (001)

phase transition temperature:

T_{tr}	110 K (300 K)	vacuum (4 kbar)		76Z
	113.5···129 K		hysteresis; by differential optical absorption	78M2

melting point:

T_m	429 K			81W3

density:

d	1.27 g/cm³			65C

enthalpy of sublimation:

ΔH_s	97.5 (20) kJ/mol			61I

For temperature dependence of vapor pressure, see [61I].

energy levels and lifetime of excitons:

E_{S_1}	3.30 eV	RT	absorption, emission; monomer	81M4
	3.16 eV	RT	excimer	81M4
τ_{S_1}	120···180 ns	RT	monomer and excimer similar	81M4
E_{T_1}	2.0893 (7) eV	$T=295$ K	excitation spectroscopy	76P2
	2.0924 eV	125 K	excitation spectroscopy	80M3, 76P2
	2.0800 (4) eV	4.2 K	excitation spectroscopy, low temp. phase	76P2, 79P
K	$8 \cdot 10^{-4}$ cm^{-1}	RT, $E\|a$, $\lambda=514.5$ nm	$S_0 \rightarrow T_1$ absorption coefficient	73E
τ_{T_1}	140 ms	RT	delayed fluorescence	73P3

vibrons: polarized absorption and fluorescence: [70C3]. IR and Raman spectroscopy of $C_{16}H_{10}$ and $C_{16}D_{10}$ at 300 K: [71B4]. Vibrational structure of $S_0 \rightarrow T_1$ absorption: [80M3].

phonons: Raman spectroscopy at $T=10···300$ K and $p=0···10$ kbar: [76Z]. IR and Raman investigations of $C_{16}H_{10}$ and $C_{16}D_{10}$ at 300 K: [71B4]

Physical property	Numerical value	Experimental conditions	Experimental method, remarks	Ref.
parameters of radical ion states (in eV):				
I_g^a	7.37		photoelectron spectroscopy	81S2
I_g^v	7.41		photoelectron spectroscopy, see Fig. 130	79C4
I_c^{th}	5.8	RT	photoelectron spectroscopy	81S2
	5.58	single crystal		83S12
I_c	5.9	RT	photoelectron yield, $\eta^{1/3}$ vs. $\hbar\omega$	73H3
E_g	4.1	$T = 400$ K	threshold for intrinsic charge carrier pair production; production by excitonic reactions was suppressed by perylene doping	68C1

For data on $\bar{v}_{P^*+}$, $\bar{v}_{P^*-}$, see [73K, 73S2] and [59A, 60H, 73S2], respectively.

polarization energies:				
P^+	1.6 eV		$I_g^a - I_c^{th}$	81S2
charge carrier mobilities:				
holes:				
$\mu_{c^*c^*}$	0.30 (5) cm^2/Vs	$T = 373$ K	time-of-flight	62L, 65B2, 75B1
$\mu_{c^*c^*}$	0.50 cm^2/Vs	300 K		83K4
$n_{c^*c^*}$	-1.52	115$\cdots$300 K	see Fig. 93	83K4

electrons:

Time-of-flight signals from which reported electron mobilities were evaluated were caused by holes generated at the rear electrode [83K4].

charge carrier generation and recombination processes:

generation kinetics:

τ_{rise}	$\leqq 5$ ns	$T = 300$ K, $\lambda_{exc} = 337$ nm, N_2 laser	time-of-flight, τ_{rise}: rise time of photocurrent pulse	72K2

electroluminescence: [75G3]

dielectric tensor:

ε_{aa}	2.83 (4)	$T = 298$ K,		76P1
$\varepsilon_{bb} = \varepsilon_{22}$	3.07 (3)	$f = 100$ kHz		
$\varepsilon_{c^*c^*}$	3.66 (6)	and 1 MHz		
ε_{ac^*}	-0.35 (7)			
ε_{11}	2.70 (5)			
ε_{33}	3.80 (8)			
θ	-20 (2)$°$		angle between the ε_{11} principal axis and $\boldsymbol{a}$	

optical tensor:

The optic axial plane is perpendicular to (010); the acute bisectrix makes an angle $\theta = 53°$ with $\boldsymbol{a}$ in the obtuse angle β [56B].

Physical property	Numerical value	Experimental conditions	Experimental method, remarks	Ref.

12.1.18 *trans*-Stilbene, $C_{14}H_{12}$

molecular structure:

crystal structure: monoclinic, space group C_{2h}^5-$P2_1/a$, $Z=4$ [74F1]

lattice parameters:

a	12.381 (3) Å		after [74F1], however with the	74F1
b	5.723 (1) Å		choice of the axes as used	
c	15.571 (4) Å		in [71W]	
β	114.11 (4)°		$(c_F = -a_W; a_F + c_F = c_W);$	
			$[000]_W \cong [0.5\ 0\ 0.5]_F$	

molecular packing diagram: Fig. 94

melting point:

T_m	397···399 K			81W3

elastic constants: [70K2, 71K2]

energy levels and lifetimes of excitons:

		T [K]		
E_{S_1}	3.6000 eV	20	absorption, emission	59P
τ_{S_1}	$3.9 \cdot 10^{-9}$ s	293	$\lambda_{exc.} = 254$ nm	67S3
E_{T_1}	2.14 eV	77	$S_0 \rightarrow T_1$ absorption in 3-methyl-pentane glass	80S5
	2.155 eV	10, crystal	delayed fluorescence excitation	78A
τ_{T_1}	10 ms	300, crystal	from delayed fluorescence	78A

vibrons: fluorescence excitation spectra of jet-cooled vapor: [83Z]; single crystal absorption, emission at 20 K: [59P].

hole mobilities:

μ	1.4 cm²/Vs	RT	"in *a-b* plane"; for temperature	64H1
$\mu_{c^*c^*}$	2.4 cm²/Vs	RT	dependence, see Fig. 95	

12.1.19 *p*-Terphenyl, $C_{18}H_{14}$

molecular structure: synonym: 1,4-diphenylbenzene

crystal structure and lattice parameters:

monoclinic, space group C_{2h}^5-$P2_1/a$, $Z=2$

room-temperature phase:

a	8.119 (5) Å	RT	X-ray diffraction; there are also	77B4
b	5.615 (3) Å		neutron diffraction data	
c	13.618 (8) Å		available, obtained with	
β	92.07 (3)°		perdeuterated *p*-terphenyl,	
			$C_{18}D_{14}$ at 200 K [77B4].	

(continued)

Physical property	Numerical value	Experimental conditions	Experimental method, remarks	Ref.

lattice parameters (continued)

low-temperature phase:

a	16.01 Å	$T = 113$ K	X-ray diffraction; the low-temperature structure is a superstructure of the room-temperature phase; a pseudo-monoclinic cell, centered on the face (001), $Z = 8$, with doubled a and b parameters has been chosen therefore, instead of the true triclinic cell, $Z = 4$.	76B2
b	11.09 Å			
c	13.53 Å			
β	92.0°			

phase transition temperature:

T_{tr}	193.55 K			83S8, cf. 79C3, 80C2

Molecular packing diagrams of the room-temperature and of the low-temperature phases are displayed in Fig. 96.

melting point:

T_m	487.0 K			83S8

density:

d	1.234 g/cm³			71L2

energy levels and lifetime of excitons:

		T [K]		
E_{S_1}	3.70 eV	5	fluorescence and fluorescence excitation spectroscopy; essentially temperature-independent between 5 and 300 K	82W3
E_{T_1}	2.53 eV	300	phosphorescence	80M4
τ_{S_1}	5 ns	300	fluorescence	67S3
	2 ns	300		81J1
	1.6 ns	77		
τ_{T_1}	100 ms	300	from delayed fluorescence	84K2, cf. 80K2, 83M1

vibrons, phonons: from fluorescence: [82W3]; FIR spectroscopy ($0 \cdots 250$ cm^{-1}, $T = 80 \cdots 300$ K) [77W4]; Raman spectroscopy: [78B2, 78G]

parameters of radical ion states:

I_g^a	7.9 eV		photoelectron spectroscopy	81S2
I_c^{th}	6.1 eV		photoelectron spectroscopy	81S2

For $\bar{v}_{T^*-}$ from absorption spectroscopy, see [60H, 63B, 65B3, 78S3].

polarization energies:

P^+	1.8 eV		$I_g^a - I_c^{th}$	81S2

Physical property	Numerical value	Experimental conditions	Experimental method, remarks	Ref.
charge carrier mobilities (μ in cm^2/Vs):				
electrons:				
		T [K]		
μ_{aa}	0.34	300	time-of-flight	78B3
μ_{bb}	1.2			
μ_{c*c*}	0.25			
n_{aa}	-0.5	$300\cdots200$	for temperature dependence,	78B3
	-2.5	$200\cdots100$	see Fig. 97	
n_{bb}	-0.7	$300\cdots180$		
	-2.3	$180\cdots\ 90$		
holes:				
μ_{c*c*}	0.80	300		78B3
μ_{aa}	0.6	170		
n_{aa}	2.9	$170\cdots110$	for temperature dependence, see Fig. 97	78B3
charge carrier generation and recombination processes:				
σ_{S_1}	$5\cdot10^{-19}\ cm^2$	$\lambda=590$ nm	photoionization of singlet excitons	83M1
$\gamma_{n,p}$	$1.3\cdot10^{-6}\ cm^3/s$	RT	bimolecular electron-hole recombination constant	83M1

12.1.20 Tetracene, $C_{18}H_{12}$

molecular structure: synonym: naphthacene, 2,3-benzanthracene

crystal structure: triclinic, space group C_i^1-$P\bar{1}$

lattice parameters:

a	7.90 Å	RT	X-ray diffraction (transition to a	62C,
b	6.03 Å		high pressure phase at ≈ 5 kbar	61R
c	13.53 Å		has been inferred from optical	
α	100.3°		measurements in a magnetic	
β	113.2°		field [76K5])	
γ	86.3°			

molecular packing diagram: see Fig. 98

melting point:

T_m	630 K			81W3

density:

d	1.24 g/cm^3			62C, 61R

enthalpy of sublimation:

ΔH_s	113 (3) kJ/mol			61I

For heat capacity in the temperature range $5\cdots350$ K, determined by adiabatic calorimetry, see [80W2]. Vapor pressure as a function of temperature in the range $360\cdots450$ K [61I].

Physical property	Numerical value	Experimental conditions	Experimental method, remarks	Ref.
energy levels of excitons:				
E_{S_1}	2.320 (6) eV	$T = 4.2 \cdots 300$ K	fluorescence	76M3
	2.317 eV	4.2 K	reflection spectrum	79K2
E_{T_1}	1.25 eV	300 K	excitation spectrum of delayed fluorescence	71T

$S_0 \rightarrow T_1$ and $S_0 \rightarrow S_1$ absorption spectra: see Fig. 162

vibrons: Raman spectra at 298 K: [71T]; fluorescence, vibrational progression at 20 K: [56S]

parameters of radical ion states (in eV):				
I_g^a	6.89		photoelectron spectroscopy, cf. Fig. 128	81S2
I_c^{th}	5.10		photoelectron spectroscopy, see Fig. 150	81S2
	5.28 (5)			73H2
I_c	5.35	RT	photoelectron emission yield, see Fig. 144	73H3
A_g	0.8 $\cdots$ 0.95			74B4
A_c	2.41 (15)		estimated from $I_g - I_c + A_g$	74B4
	2.2		estimated from $I_c - E_g$ with $I_c = 5.3$ eV and $E_g = 3.1$ eV	
$h\nu_{T^*+}$	1.44	in anthracene crystal at 83K	lowest peak; photocurrent excitation spectroscopy. For $\bar{\nu}_{T^*+}$ and $\bar{\nu}_{T^*-}$ in organic glasses, see Fig. 139	76K3
E_g	3.11 (3)		from autoionization analysis (Onsager model)	73B
	2.90 (5)			82S6
	3.0		extrapolated yield curve	66G
	3.40 (5)	10 $\cdots$ 300 K	electromodulation spectroscopy (electroabsorption), extrapolated from charge transfer transitions for $r \rightarrow \infty$. From this "optical gap" a relaxation energy of ≈ 0.3 eV has to be subtracted in order to be able to compare this result to the adiabatic gap energies listed above.	81S5
polarization energies:				
$P^+ = P^-$	1.6 eV	$T = 300$ K		65P
charge transfer exciton state energies (in eV):				
$E_{CT}[\frac{1}{2}\frac{1}{2}0]$	2.710			81S5
$[0\ 1\ 0]$	2.779			
$[1\ 0\ 0]$	2.895			
$\left.\begin{array}{l}[1\ 1\ 0]\\ [\frac{1}{2}\frac{3}{2}0]\end{array}\right\}$	3.063			
$[r \rightarrow \infty]$	3.40 (5)		for infinite electron-hole separation	
E_{CT}	2.9 eV	$T = 300$ K	photoemission from tiny crystals in Millikan apparatus	65P
hole mobilities:				
$\mu_{c^*c^*}$	0.85 (5) cm^2/Vs	RT	$c^* \| a \times b$	76B1
$n_{c^*c^*}$	$-2 > n > -3$	$T = 300 \cdots 350$ K	see Fig. 99	76B1

electroluminescence: [69V]

Physical property	Numerical value	Experimental conditions	Experimental method, remarks	Ref.

12.1.21 Tetracyanoethylene, TCNE, C_6N_4

molecular structure:

crystal structure: monoclinic, space group C_{2h}^5-$P2_1/n$, $Z=2$ [60B]

lattice parameters:

a	7.51 Å	RT	X-ray diffraction (there has been	60B
b	6.21 Å		described also a low temperature	
c	7.00 Å		cubic phase, space group Im3,	
β	97.17°		$a=9.736$ (5) Å, which is stable	
			below 292 K [71L1]; however,	
			the monoclinic phase can be	
			undercooled)	

molecular packing diagram: Fig. 100

melting point:

T_m	471···473 K			81W3

density:

d	1.31 g/cm^3			60B

vibrons: [64H2], [81C3]; phonons: inelastic neutron scattering: [81C3]

molecular electron affinity:

A_g	2.88 eV			75C2

hole mobilities (μ in cm^2/Vs):

μ_{11}	0.26 (1)	$T=273$ K	time-of-flight; monoclinic phase	80S2
μ_{22}	0.14 (1)			
μ_{33}	0.19 (1)			
θ	-24 (10)°			80S2

The temperature dependence can be described by $\mu \propto \exp(E/kT)$ between 250 K and 345 K with $E=80$ meV, see Fig. 101.

12.1.22 7,7,8,8-Tetracyanoquinodimethane, TCNQ, $C_{12}H_4N_4$

molecular structure:

crystal structure: monoclinic, space group C_{2h}^6-$C2/c$, $Z=4$ [65L]

lattice parameters:

a	8.906 (6) Å	RT	X-ray diffraction	65L
b	7.060 (4) Å			
c	16.395 (5) Å			
β	98.54 (4)°			

molecular packing diagram: Fig. 102

Physical property	Numerical value	Experimental conditions	Experimental method, remarks	Ref.
density:				
d	1.315 g/cm^3			65L
parameters of radical ion states:				
I_g^a	9.5 eV		photoelectron spectroscopy, see section 12.3.5.	
I_c^{th}	7.89 eV			76L, 74N
A_g	2.8 eV		see section 12.3.6	75C2
For $\bar{v}_{T^*-}$, see Fig. 142				
electron mobilities:				
$\mu_{c^*c^*}$	0.65 (10) cm^2/Vs	RT	time-of-flight; see Fig. 103	78S1
$n_{c^*c^*}$	-1	$T = 210 \cdots 350$ K		

12.2 Dark-conductive narrow band gap organic semiconductors

12.2.1 $(TMTSF)_2:PF_6$, (Tetramethyltetraselenafulvalene)$_2$:hexafluorophosphate, $(C_{10}H_{12}Se_4)_2:PF_6$

molecular structure:

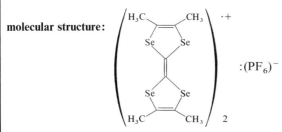

crystal structure: triclinic, space group C_i^1-P$\bar{1}$, $Z = 1$ [81J2]

lattice parameters:

a	7.297 Å	$T = 300$ K	81J2,
b	7.711 Å		81T4
c	13.522 Å		
α	83.39°		
β	86.27°		
γ	71.01°		

The flat TMTSF molecules form alternating stacks along the a axis.
molecular packing diagram: Fig.104

electrical conductivity:

σ	$10^5\ \Omega^{-1}$ cm^{-1}	$\|a$, $T = 14$ K	semiconducting or semimetal below 12 K at zero pressure, metallic above 14 K; cf. Fig. 105	80B2

Superconducting at $p = 12$ kbar, $T_c \leq 0.9$ K; first organic superconductor [80J]; cf. Figs. 106, 107.
metal/semiconductor/superconductor phase diagram: Fig. 108; cf. Fig. 107
transverse magnetoresistance as a function of magnetic field and temperature: Fig. 109

charge carrier mobility:

μ_{aa}	$10^4 \cdots 10^5$ cm^2/Vs	$T = 4.2$ K, semi-conducting state	estimated from conductivity, magneto-resistance, and Hall effect	80C1, 82C3, 82F1
	$> 10^6$ cm^2/Vs	4 K	estimate with different model assumptions	82C4

Physical property	Numerical value	Experimental conditions	Experimental method, remarks	Ref.

The electric transport data are seriously affected by minor (100 ppm) impurity concentrations [81B3, 82C3, 82F1].

TMTSF has been combined with a number of anions such as H$_2$F$_3^-$, BF$_4^-$, AsF$_6^-$, SbF$_6^-$, NbF$_6^-$, TaF$_6^-$, ClO$_4^-$, ReO$_4^-$, TcO$_4^-$, SCN$^-$, NO$_3^-$, FSO$_3^-$, F$_2$PO$_2^-$. The gross features of this "TMTSF family" are the same. This is due to the fact that the TMTSF sublattices, in which charge transport occurs, are isomorphous. But there are important differences in the details. The size, symmetry and order (or disorder) of the molecules forming the anionic sublattice influence or even suppress the metal to semiconductor, metal to superconductor, and semiconductor to superconductor transitions. The semiconducting state can arise from spin-spin interactions (spin density waves) or lattice-driven ordering. Lattice defects, nonstoichiometry, and extrinsic impurities can also strongly influence the phase transitions.

The perchlorate (ClO$_4^-$) remains highly conducting down to liquid He temperature ($\sigma > 10^5 \, \Omega^{-1} \, \mathrm{cm}^{-1}$), and superconduction is reached under ambient pressure at 1.25 K; no semiconductor phase is formed unless lattice disorder is frozen in at low temperature by quenching. The perrhenate (ReO$_4^-$), on the other hand, also metallic at room temperature, undergoes a metal insulator (semiconductor) transition already at 182 K [82J5].

A great variety of chemically modified tetraselenafulvalenes, including several of their sulfur and tellurium analogues, have been synthesized [cf. 77P2, 78M1, 79H, 80A1] and partly characterized.

The whole field is presently under very intensive study. For further information the reviews [82J1, 82J2, 82J4, 84B2, 84G2, 84J] may be consulted.

12.2.2 (Perylene)$_2$:(PF$_6$)$_{1.1}$ × 0.8 (CH$_2$Cl$_2$), C$_{40}$H$_{24}$:(PF$_6$)$_{1.1}$ × 0.8 (CH$_2$Cl$_2$)

molecular structure:

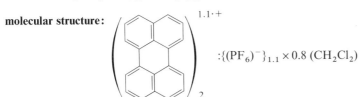

:{(PF$_6$)$^-$}$_{1.1}$ × 0.8 (CH$_2$Cl$_2$)

crystal structure: orthorhombic, space group D$_{2h}^{12}$-Pnmn, $Z = 1$ [80K4]

lattice parameters:

a	4.285 Å	RT (?)	X-ray diffraction	80K4
b	12.915 Å			
c	14.033 Å			

The perylene molecules are regularly stacked along the a axis.
molecular packing diagram: Fig. 110

electrical conductivity:

σ	900 Ω^{-1} cm^{-1}	$T = 300$ K; $E \parallel a$	four probe method; metallic above 190 K, semiconducting below 190 K; for temperature dependence, see Fig. 111	80K4

optical conductivity:

$\hbar\omega_p$	0.91 eV	RT(?)	reflectance measurements, Drude fit; cf. Fig. 112	83G1
τ	$4.4 \cdot 10^{-15}$ s			83G1
σ_{opt}	$1.9 \cdot 10^3 \, \Omega^{-1}$ cm^{-1}			83G1
m_p/m_0	0.91			83G1
$4t$	1.1 eV			83G1

Physical property	Numerical value	Experimental conditions	Experimental method, remarks	Ref.

12.2.3 $(TTT)_2:I_3$, $(Tetrathiatetracene)_2:I_3$, $(C_{18}H_8S_4)_2:I_3$

molecular structure:

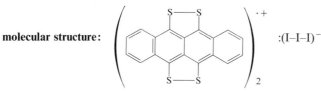

$:(I–I–I)^-$

crystal structure: orthorhombic, space group D_{2h}^{18}-Abam, $Z=2$ [77S5; 76S3, 78S5, cf. 82F2 and references given therein]

lattice parameters:

a	18.394 (12) Å	$T=298$ K	X-ray structure analysis; the	77S5
b	4.962 (5) Å		lattice parameters depend to	
c	18.319 (11) Å		some extent on the exact stoichiometry [82F2]	

The planar TTT^+ molecules are regularly stacked along b; perpendicular spacing is 3.32 Å. The linear triiodide anions lying in chains along $2b$ axes parallel to the TTT stacks form an incommensurate, disordered sublattice.
molecular packing diagram: Fig. 113
temperature dependence of crystal structure: neutron diffraction [82F2]

electrical conductivity:

σ_{bb}	$10^3\,\Omega^{-1}\,cm^{-1}$	$T=300$ K	metallic at room temperature; for temperature dependence, see Fig. 114	78I, 82S7
	$3\cdot10^3\,\Omega^{-1}\,cm^{-1}$	40 K	semiconductor below 40 K; the transition temperature depends on the exact stoichiometry and on pressure [81G2]; different samples seem to have different kind of disorder of the triiodide sublattice, cf. [82F2]; for more details the review article [82S7] may be consulted. Disorder smears out the metal to semiconductor transition [81K2]	

Comprehensive reviews on TTT radical ion salts with a number of anions and on their tetraselena and tetratelluro homologues have been given in [82S1, 82S2, 82S7].

12.2.4 $TTF:Br_{0.7}$, Tetrathialfulvalene:bromine, $C_6H_4S_4:Br_{0.7}$

molecular structure:

$:Br_{0.7}^-$

crystal structure: monoclinic, space group C_{2h}^3-C2/m, $Z=4$ [75L]

Physical property	Numerical value	Experimental conditions	Experimental method, remarks	Ref.
lattice parameters:				
a	15.617 (8) Å	RT (?)	X-ray diffraction, TTF sub-	75L
b	15.627 (8) Å		lattice; the Br sublattice is	
c	3.572 (2) Å		incommensurate;	
β	91.23 (5)°		cf. caption of Fig. 115	

The TTF molecules are arranged in two crystallographically inequivalent stacks with regular intermolecular spacing.

molecular packing diagram: Fig. 115

electrical conductivity:

σ_{cc}	$100 \cdots 500\ \Omega^{-1}\,cm^{-1}$	$T = 300$ K	four probe measurement	75W2

$\sigma_{cc}(T)$ is thermally activated below 180 K with an activation energy $E_A = 0.081$ eV (cf. Fig. 116); the high-temperature (room-temperature) phase is probably metallic; it undergoes a Peierls transition at ≈ 180 K and becomes a semiconductor below [75W2].

optical conductivity (TTF:$Br_{0.76}$):

		T [K]		
$\hbar\omega_p$	2.00 eV	295	reflection spectroscopy,	82K5
	2.05 eV	220	see Fig. 117	
	2.08 eV	120		
	2.07 eV	30		
τ	$2.9 \cdot 10^{-15}$ s	295		82K5
	$4.4 \cdot 10^{-15}$ s	220		
	$5.1 \cdot 10^{-15}$ s	120		
	$6.2 \cdot 10^{-15}$ s	30		
σ_{opt}	$2.3 \cdot 10^3\ \Omega^{-1}\,cm^{-1}$	295		82K5
	$3.8 \cdot 10^3\ \Omega^{-1}\,cm^{-1}$	220		
m_p/m_0	1.1	$30 \cdots 295$		82K5
$4t$	1.35 eV	$30 \cdots 295$		82K5

The different TTF halides have been reviewed, among other compounds, in [82M2].

12.2.5 K:TCNQ, Potassium:tetracyanoquinodimethane, K:$C_{12}H_4N_4$

molecular structure: K^+:

crystal structure and lattice parameters:

a) low-temperature modification, stable below 395 K:
monoclinic, space group C_{2h}^5-$P2_1/n$, $Z = 8$ [77K3]

a	7.0835 (7) Å	$T = 298$ K	X-ray diffraction, twinned	77K3
b	17.773 (3) Å		crystal	
c	17.859 (4) Å			
β	94.95 (1)°			

b) high-temperature modification, stable above 395 K:
monoclinic, space group $P2_1/c$, $Z = 2$ [77K3]

a	3.587 (3) Å	$T = 413$ K	X-ray diffraction	77K3
b	12.676 (5) Å			
c	12.614 (5) Å			
β	96.44 (3)°			

Physical property	Numerical value	Experimental conditions	Experimental method, remarks	Ref.

The TCNQ˙⁻ radical anions, exhibiting planar quinoid skeletons, are arranged face-to-face to form segregated stacks along the a axis. In the low-temperature phase the interplanar spacings are alternating (3.273 Å and 3.567 Å), whereas regular stacks with equal interplanar spacing of 3.497 Å are formed in the high-temperature phase. The K^+ ion displays octahedral coordination by 8 N atoms of different TCNQ⁻ ions;

molecular packing diagram: Fig. 118

A slightly different low-temperature phase has been reported independently which also undergoes a phase transition at 395 K [78R3]:

monoclinic, space group C_{2h}^5-P2$_1$/c, $Z=4$

a	3.543 (1) Å		X-ray diffraction, twinned	78R3
b	17.784(5) Å		crystal	
c	17.868 (3) Å			
β	94.96 (2)°			

The TCNQ˙⁻ radical ions form regular stacks along the a axis with interplanar distances of 3.435 Å and 3.442 Å in the two crystallographically inequivalent stacks (with long molecular axes oriented in the c and in the b direction, respectively).

The phase transition has been investigated further by measuring the temperature dependence of the intensities of superlattice and diffuse X-ray reflections [77K3, 78T2] and by Raman spectroscopy [82B3].

electrical conductivity:

σ	$3 \cdot 10^{-4} \dots$	$T = 300$ K	microwave conductivity at	82G2
	$2 \cdot 10^{-2} \, \Omega^{-1} \, cm^{-1}$		9.3 GHz; probable field	
			orientation: $\boldsymbol{E}$ parallel to the	
			stacking axis; for temperature	
			dependence, see Fig. 119	

A comprehensive review on several other alkalimetal and metal salts of TCNQ has been given in [83E]; for early work [71V] may be consulted.

12.2.6 TTF:TCNQ, Tetrathiafulvalene:tetracyanoquinodimethane, $C_6H_4S_4$:$C_{12}H_4N_4$

molecular structure:

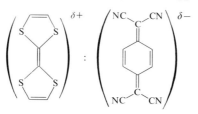

δ: degree of charge transfer; $\delta \approx 0.55$ e/molecule (at 150 K where diffuse X-ray scattering is seen) increases on cooling [79C7, 79M3]

crystal structure: monoclinic, space group C_{2h}^5-P2$_1$/c, $Z=2$ [74K5]

lattice parameters:

a	12.298 (6) Å	RT	X-ray scattering	74K5
b	3.819 (2) Å			
c	18.468 (8) Å			
β	104.46 (4)°			

The crystal structure consists of segregated regular stacks of TTF and TCNQ molecules, respectively; stacking direction is parallel $\boldsymbol{b}$. The normal of the TTF molecular plane is tilted 24.5° with respect to the b axis. The interplanar stacking distance is 3.47 Å. The TCNQ molecules are tilted opposite ($-34°$) and stacked with interplanar distance 3.17 Å.

molecular packing diagram: Fig. 120

Physical property	Numerical value	Experimental conditions	Experimental method, remarks	Ref.
electrical conductivity:				
σ_{bb}	$400\,(100)\,\Omega^{-1}\,cm^{-1}$	$T = 300$ K	four probe dc and microwave measurements	76F; cf. 74G2, 75E, 76C
	$10^4\,\Omega^{-1}\,cm^{-1}$	58 K	for $\sigma(T)$, see Fig. 121	76F, cf. 76C

The conductivity is metallic above 54 K, $\varrho = \varrho_0 + 2.4\,(5) \cdot 10^{-8}\,(T/\text{K})^2$ [75E]. At ambient pressure there are three phase transitions, at $T_H = 54$ K, $T_I = 49$ K and $T_L = 38$ K. The 54 K transition is driven by a charge density wave (CDW) Peierls instability of the TCNQ stacks; at the other two transitions ordering of the TTF stacks takes place. Between T_H and T_I there is a $2a \times 3.4b \times c$ superstructure, between T_I and T_L the a periodicity varies in a continuous way; below T_L a $4a \times 3.4b \times c$ superstructure is established. Between T_H and T_L the TTF stacks remain nearly metallic; below T_L a band gap of 0.04 eV opens. (For the derivation of these conclusions and for the experimental results on which they are based, the review article [82J1] may be consulted).

charge carrier mobility:				
$\mu_p + \mu_n$	$300 \cdots 450$ cm^2/Vs	$T = 58$ K, $E \| b$	calculated from conductivity	77C2
bandwidths:				
electron bandwidth	≈ 0.5 eV		from model calculations (TCNQ stacks)	80C3
hole bandwidth	≈ 0.5 eV		(TTF stacks)	80C3

For Hall coefficient at $50 \cdots 300$ K, see Fig. 122; for thermopower at $40 \cdots 300$ K, see Fig. 123.

optical conductivity:

optical reflectance spectrum: Fig. 124

$\hbar\omega_p$	$1.2 \cdots 1.5$ eV	$T = 4.2 \cdots 300$ K	from Drude fit; the uncertainty	82T2
τ	$(2 \cdots 6) \cdot 10^{-15}$ s		of these data is only to a minor	82T3
σ_{opt}	$800 \cdots 1700\,\Omega^{-1}\,cm^{-1}$		extent due to experimental	82T2
$4t_\|$	$0.43 \cdots 0.65$ eV		scatter of the reflectivity data;	82T2

the major part comes from interpretational problems. It has even been questioned if the optical conductivity and the dc electrical conductivity are coupled [82T2]. A more detailed analysis has been presented in [82T2, 82T3], some aspects are mentioned in the caption of Fig. 124

A great family of structurally related charge transfer complexes derives from chemically modified TTF and TCNQ molecules, see e.g. [77P2, 78M1, 79H, 80A1, 80B4, 83S9]. These complexes can be of 1:1 or other stoichiometry; the constituent molecules can pack in mixed or in segregated stacks; stack order can be regular or alternating (dimerized). Depending on geometrical aspects and molecular parameters, charge transfer can be weak, intermediate or complete, cf. [79T2]; it may vary with temperature and pressure, cf. [81T3]. Transport properties, in addition, depend on transfer integrals and scattering. The electrical properties can therefore vary over orders of magnitude, cf. [79T2]. Although progress has been made with understanding the gross features and their structural correlations, many open questions have still to be answered before substances with special desired properties can be tailor-made. It is also worth mentioning that by far more work has been invested into studying the highly conductive materials which exhibit metallic-like conductivity regimes, than into the moderately conducting semiconductor phases. However, since the electrical properties often change smoothly at metal to semiconductor transitions, knowledge of the metallic transport properties can be a useful starting point towards understanding the semiconductor transport properties.

Physical property	Numerical value	Experimental conditions	Experimental method, remarks	Ref.

12.2.7 TTF:chloranil, Tetrathiafulvalene:tetrachloro-*p*-benzoquinone, $C_6H_4S_4:C_6Cl_4O_2$

molecular structure:

crystal structure: monoclinic, space group C_{2h}^5-P2$_1$/n, $Z=2$ [79M2]

lattice parameters:

a	7.411 (1) Å	RT(?)	X-ray diffraction	79M2
b	7.621 (2) Å			
c	14.571 (3) Å			
β	99.20 (1)°			

The two component molecules form mixed stacks along the a axis;

molecular packing diagram: Fig. 125

X-ray and c_p measurements have demonstrated that the structure undergoes a phase transition at 84 K, see Fig. 126; cf. also [83A].

An abrupt phase transition at 84.4 (5) K has also been seen in temperature-dependent polarized reflection spectra [82T1].

The phase transition has been shown by infrared absorption spectroscopy [81T2], Raman spectroscopy [81T2], optical reflection spectroscopy [82T1, 83J] and thermal expansion measurements [81B4] to be from neutral to ionic. It can also be induced at room temperature by application of a pressure of 8 kbar [81T3].

charge transfer transition energy E_{CT}:

E_{CT}	0.66 eV	$T=300$ K	optical reflectance	83J
	0.55 eV	100 K		

absorption spectra in the IR and visible range: [81T2]
Raman spectra: [81T2]
optical reflexion spectra: Fig. 127; cf. also [82T1, 83J]

degree of charge transfer δ:

δ	≈ 0.25	$T=300$ K	estimated from IR and Raman	83J
	≈ 0.3	100 K	vibrational frequencies	
	≈ 0.65	below $T_{tr}=84$ K	[83G2], and from the intensities of the CT bands, and from IR vibrational spectra [83J]	

electrical conductivity:

σ	$8 \cdot 10^{-4}\,\Omega^{-1}\,cm^{-1}$	$T=300$ K	compacted powder	79T1

12.3 Comparative representation of some general properties of selected compounds

12.3.1 Molecular photoelectron spectra

A selection of gas phase vertical π-ionization potentials I_g^v (in eV) of the highest filled (1st) and next lower lying (i.e. higher ionization potential) 2nd molecular orbitals. Their difference determines the energy of the lowest electronic radical cation transition. Further data may be found under [60V, 64V1, 64V2, 66C2, 69D2, 72B1, 72C1, 72M, 74B5, 74R, 75C3, 75H1, 75H2, 75H3, 75O2, 75M, 76G1 (table 6.6), 77C4, 77D, 77S3, 78C3, 78R2, 79C4, 79C5, 81C2, and 81E2 (listing ≈ 200 compounds)].

Compound	$I_{g\ peak}^{v\ 1st}$ (in eV)	$I_{g\ peak}^{v\ 2nd}$ (in eV)	Fig. No.	Ref.
Homocycles				
Benzene	9.24	12.25	128	72C1, cf. 84G1
Naphthalene	8.15	8.88	6, 128	72C1
Anthracene	7.41	8.54	128	72C1, cf. 77S3
Tetracene	7.01	8.41	128	72C1
Pentacene	6.61	7.92	128, 129	72C1, 77S3
Hexacene	6.36	7.53		77S3
Phenanthrene	7.86	(8.15)		74B5, cf. 78R2
Chrysene	7.60	8.10		72B1
Picene	7.50	7.67		79C4
Triphenylene	7.89	8.66		74B5
Coronene	7.29	8.62		81C2
Pyrene	7.41	8.27	130	79C4, cf. 81C2
Perylene	6.97	8.54	131	77C4
Terrylene	6.42	7.79		78C3
Quaterrylene	6.11	7.17		78C3
Anthanthrene	6.92	8.08		74B5
5,6; 12,13-Dibenzopereo-pyrene	6.42	7.53		78C3
Ovalene	6.71	7.33		81C2
Biphenyl	8.38	9.14		84S3, cf. 72M 83M2
p-Terphenyl	8.04	9.03		84S3
p-Quaterphenyl	7.80	8.64		84S3
p-Quinquephenyl	7.68	8.30		84S3
p-Sexiphenyl	7.62	8.11		84S3
Fluorene	7.91	8.77		78R2, cf. 72M 83M2
Biphenylene	7.61	8.90		74B5
Fluoranthene	7.95	8.1		74B5
Azulene	7.43	8.50		74B5
1,2-Benzazulene	7.05	8.20		81C2
Benz[*f*]azulene	7.15	8.17		82S4

(continued)

Compound	$I_{g\,peak}^{v\,1st}$ (in eV)	$I_{g\,peak}^{v\,2nd}$ (in eV)	Fig. No.	Ref.
table (continued)				
Heterocycles				
Quinoline (1-Azanaphthalene)	8.62	9.07		69D2
Quinoxaline (1,4-Diazanaphthalene)	8.99	10.72		69D2
Dibenzofurane	8.09	8.34		
Carbazole	7.53			74H4
	7.60	7.99		78R2
Dibenzothiophene	7.97			66C2
	7.93	8.34	132	78R2
Acridine	7.78			60V, 64V2
	7.88	8.69		75H1
	8.13			75M
Phenazine	8.33	(9.06)	133	75H1
	8.44			75M
Phenothiazine	6.82			75H2
	6.72			77D
Phthalocyanine (metal-free)	6.41	8.75		79B5
Substituted aromatic molecules				
9-Methylanthracene	7.25	8.43		74H5
9,10-Dichloroanthracene	7.58	8.88		74S4
Hexafluorobenzene	10.12	12.77		72B2, cf. 81B6
Perfluorobiphenyl	9.80	9.96		72M
Octafluoronaphthalene	9.05 (3)	10.08 (5)		72B2
Fulvalenes				
Tetrathiafulvalene (TTF)	6.83	8.69	134	73G, 74B6
	6.70	8.47		77S6
Tetramethyltetrathia-fulvalene (TMTTF)	6.40	7.99	135	75G2
Hexamethylenetetrathia-fulvalene (HMTTF)	6.41	7.85	136	83S2
Tetraselenafulvalene (TSF)	6.90	8.32		77S6
Tetramethyltetraselena-fulvalene (TMTSF)	6.58	7.90	137	75G2
Hexamethylenetetraselena-fulvalene (HMTSF)	6.36	6.74		83S2

12.3.2 Electronic radical ion transitions

It is characteristic for the radical cation and radical anion states of aromatic molecules that they absorb at much longer wavelengths than the parent neutral molecule. The following table lists references and Fig. numbers of absorption spectra of radical ions obtained from solution or glassy matrix (if not otherwise stated).

Radical ion	Fig. No., Remarks	Ref.
Homocycles		
Benzene$^-$		73S2
Naphthalene$^-$		60H, 73S2, 74H2
Naphthalene$^+$		73S2, 74K3, 78S3
Anthracene$^-$		60H, 65B3, 73S2, 74H2, 74H3
Anthracene$^+$	138	73S2, cf. 70D3, 74K3

(continued)

Radical ion	Fig. No., Remarks	Ref.
table (continued)		
Tetracene$^-$	139	73S2, cf. 59A, 60H, 65B3, 72K1, 74H2
Tetracene$^+$	139	70D3, 72K1, 73S2, 74K4
	in an anthracene host crystal	76K3, 83K3
	140	
Pentacene$^-$	141	73S2, cf. 65B3, 72K1
Pentacene$^+$		70D3, 72K1
Biphenyl$^-$		63B, 74H2
Biphenylene$^-$		73S2
Biphenylene$^+$		73S2
p-Terphenyl$^-$		60H, 63B, 65B3, 78S3
Phenanthrene$^-$		73S2, 79K1
Phenanthrene$^+$		73S2, 74K3
Chrysene$^+$		74K4, 78K2
Pyrene$^-$		59A, 60H, 73S2
Pyrene$^+$		73K, 73S2
Perylene$^-$		59A, 60H, 73S2
Perylene$^+$		73S2
Coronene$^-$		73S2
Coronene$^+$		73S2
Heterocycles		
Quinoline$^+$ (1-Azanaphthalene$^+$)		79K1
Quinoline$^-$ (1-Azanaphthalene$^-$)		79K1
Isoquinoline$^+$ (2-Azanaphthalene$^+$)		79K1
Isoquinoline$^-$ (2-Azanaphthalene$^-$)		79K1
1,2-Diazanaphthalene$^+$		79K1
1,2-Diazanaphthalene$^-$		79K1
2,3-Diazanaphthalene$^+$		79K1
2,3-Diazanaphthalene$^-$		79K1
1,4-Diazanaphthalene$^+$		79K1
1,4-Diazanaphthalene$^-$		79K1
Carbazole$^+$		78S2
Acridine$^+$		79K1
Acridine$^-$		79K1
Phenazine$^+$		79K1
Phenazine$^-$		79K1
1-Azaphenanthrene$^+$		79K1
1-Azaphenanthrene$^-$		79K1
4-Azaphenanthrene$^+$		79K1
4-Azaphenanthrene$^-$		79K1
1,8-Diazaphenanthrene$^+$		79K1
1,8-Diazaphenanthrene$^-$		79K1
4,5-Diazaphenanthrene$^+$		79K1
4,5-Diazaphenanthrene$^-$		79K1
9,10-Diazaphenanthrene$^-$		79K1
7,7,8,8-Tetracyanoquinodimethane$^-$ (TCNQ$^-$)	142	74B6, cf. 67L
Tetrathiafulvalene$^+$ (TTF$^+$)		79B4

12.3.3 Solid state photoelectron emission yield curves

Compound	Fig. No.	I_c [eV]	Ref.
Anthracene	143	5.75	73H3, cf. 70K1, 80A3
Tetracene	144	5.35	73H3, cf. 70K1
Pentacene	145	4.98	73H3, cf. 70K1
Pyrene		5.85	75H3
Perylene	146	5.39	73H3, cf. 70K1
Quaterrylene	147	4.87	73H3
Coronene	148	5.58	73H3
Phenothiazine	149	5.15	80A3
9-Methylanthracene		5.68	77H4
9,10-Dichloroanthracene		5.75	77H4
Tetrathiatetracene (TTT)		4.42	70K1

12.3.4 Solid state photoelectron spectra

(from polycrystalline film, where not otherwise stated)

Compound	Fig. No.	I_c^{th} [eV]	Ref.
Homocycles			
Benzene	160	7.1	79G1, 84S3
p-Terphenyl		6.1	75H5
p-Sexiphenyl		5.9	84S3
Dibenzyl		7.2 (1)	80S3
Anthracene	8	5.75 (10)	82K2, cf. 73H3
Anthracene, single crystal		5.67	83S12
Anthracene:PMDA		6.00 (15)	76I
Tetracene	150	5.28 (5)	73H2
1,2-Benzanthracene		5.64	73H3
Pentacene	151	4.9 (1)	73H3, 74S2, cf. 76S2
Fluoranthene		6.14	73H3
Pyrene		5,83	73H3, 75H3
Pyrene, single crystal		5.58	83S12
Perylene	152	5.37	73H3, cf. 76S2
		5.2	74S2
Perylene, single crystal		5.12	83S12
Quaterrylene		4.83	73H3, cf. 70K1
Violanthrene A		4.95	73H3, cf. 82S5
Coronene	153	5.52	73H3, cf. 76S2
		5.3	74S2
Heterocycles			
Carbazole	154	5.87 (10)	82K2
Carbazole:PMDA	155	6.22 (10)	82K2
Acridine:PMDA		6.3 (2)	82K2
Phenothiazine	156	5.18 (15)	82K2
Phenothiazine:PMDA		5.41 (10)	82K2
Phthalocyanine (metal-free)		5.15	64V1
		6.2	79B5, cf. 73P4
Substituted aromatic molecules			
Tetraselenatetracene (TST)		4.45	76S4
Hexachlorobenzene (HCB)		7.3	81S4
Hexabromobenzene (HBB)		7.1	81S4
Hexaiodobenzene (HIB)		5.9	81S4

(continued)

Compound	Fig. No.	I_c^{th} [eV]	Ref.
table (continued)			
9-Methylanthracene		5.75	77H4
9,10-Dichloroanthracene		5.8	77H4
1,3,6,8-Tetrachloropyrene		6.3	75H3
1,3,6,8-Tetrabromopyrene		6.0	75H3
1,6-Dicyanopyrene		6.3	75H3
1,3,6,8-Tetracyanopyrene		6.2	75H3
1,3,6,8-Tetranitropyrene		6.6	75H5
Fulvalenes			
Tetrathiafulvalene (TTF)		5.0	74N
Tetraselenafulvalene (TSF)	157	4.99	83S2
Hexamethylenetetrathiafulvalene (HMTTF)	158	4.63	83S2
Hexamethylenetetraselenafulvalene (HMTSF)	159	4.75	83S2
Quinones			
Tetrachloro-*p*-benzoquinone (*p*-Chloranil)		8.1	81S4
Tetrabromo-*p*-benzoquinone (*p*-Bromanil)		7.4	81S4
Tetraiodo-*p*-benzoquinone (*p*-Iodanil)		5.6	81S4
7,7,8,8-Tetracyanoquinodimethane (TCNQ)		7.89	76L, 74N
11,11,12,12-Tetracyanonaphthoquinodi-methane (TNAP)		7.1	79I

12.3.5 Energetic comparison of gas phase and solid state photoelectron spectra; polarization energies

Adiabatic ionization potential in the gas state, I_g^a, ionization threshold of the solid, I_c^{th}, and polarization energy calculated as $P^+ = I_g^a - I_c^{th}$, for selected aromatic molecules. For the references of values labelled by a letter the original paper [81S2] should be consulted.

Compound	I_g^a [eV]	I_c^{th} [eV]	P^+ [eV]	Ref.
Homocycles				
Benzene (Fig. 160)	9.17[a]	7.58[b]	1.6	81S2
Naphthalene (Fig. 160)	8.12[a]	6.4	1.7	
Anthracene (Fig. 160)	7.36[a]	5.70[c]	1.7	
Tetracene (Fig. 160)	6.89[a]	5.10	1.8	
Pentacene	6.58[a]	4.85[c]	1.7	
Chrysene	7.51[d]	5.8	1.7	
Benz[*a*]anthracene	7.38[d]	5.64[c]	1.7	
Triphenylene	7.81[e]	6.2[f]	1.6	
Naphth[2,1-*a*]anthracene	7.2[g]	5.45[c]	1.8	
Dibenz[*a,h*]anthracene	7.35[g]	5.55[f]	1.8	
Picene	7.5[g]	5.7[f]	1.8	
Pyrene	7.37[h]	5.8[c]	1.6	
Perylene	6.90[d]	5.2[i]	1.7	
Dibenzo[*fg,op*]tetracene	7.26[e]	5.7	1.6	
Benzo[*ghi*]perylene	7.12[d]	5.4	1.7	

(continued)

Compound	I_g^a [eV]	I_c^{th} [eV]	P^+ [eV]	Ref.
table (continued)				
Coronene	7.25^e	5.52^c	1.7	81S2
Violanthrene A	6.42	4.9	1.6	
Isoviolanthrene A	6.36	4.9	1.4	
Violanthrene B	6.36	4.8	1.5	
Isoviolanthrene B	6.54	5.0	1.6	
p-Terphenyl	7.9^j	6.1^j	1.8	
Benz[a]indeno[1,2,3-cd]azulene	6.84	5.0	1.9	
Benzo[c]phenanthrene	7.6^k	6.2	1.4	
Dibenzo[g,p]chrysene	6.97^e	5.8	1.2	
Tetrabenzo[a,cd,j,lm]perylene	6.58	5.3	1.2	
Substituted homo- and heterocycles				
Toluene	8.67^l	7.3^b	1.4	
9-Methylanthracene	7.23^m	5.8^f	1.5	
9,10-Diphenylanthracene	7.05^e	5.85^n	1.2	
9,9′-Bianthryl	7.2^o	5.9^f	1.3	
Rubrene	6.41	5.3	1.1	
Hexachlorobenzene (HCB)	8.98	7.3	1.6	
Hexabromobenzene (HBB)	8.80	7.1	1.8	
Hexaiodobenzene (HIB)	7.90	5.9	2.0	
N,N-Dimethyl-N′-picrylhydrazine	6.90	5.9	1.0	
Tetrathiatetracene (TTT)	6.07	4.4	1.7	
Fulvalenes				
Tetrathiafulvalene (TTF)	6.4^p	5.0^q	1.4	
Dimethyltetrathiafulvalene (DMTTF)	6.00	5.1	0.9	
Tetramethyltetrathiafulvalene (TMTTF)	6.03	4.9	1.1	
Dibenzotetrathiafulvalene (DBTTF)	6.68	4.4	2.3	
Hexamethylenetetrathiafulvalene (HMTTF)	6.06	4.63	1.43	83S2
Bis(ethylenedithiolo)tetrathiafulvalene (BEDTTF)	6.21	4,78	1.43	
Tetrathiomethoxytetrathiafulvalene (TTMTTF)	6.29	5.00	1.29	
Tetraselenafulvalene (TSF)	6.68	4.99	1.69	
Tetramethyltetraselenafulvalene (TMTSF)	6.27	4.84	1.43	
Hexamethylenetetraselenafulvalene (HMTSF)	6.12	4.75	1.37	
Quinones				
p-Chloranil (CA)	9.74	8.1	1.6	81S2
p-Bromanil (BA)	9.59	7.4	2.2	
p-Iodanil (IA)	8.58	5.6	3.0	
7,7,8,8-Tetracyanoquinodimethane (TCNQ)	9.5^r	7.4^i	2.1	
11,11,12,12-Tetracyanonaphthoquino-dimethane (TNAP)	8.5	6.0	2.5	

12.3.6 Molecular electron affinities

Electron affinities (gas phase) of selected acceptor molecules, A, *measured* by the magnetron method (based on the thermal reaction $A^- \rightleftharpoons A + e$). These values are compared with results obtained by the alkali beam method (B), (based on the reaction $A + Cs \rightleftharpoons A^- + Cs^+$), or by photodetachment measurements (D). In the last two columns, electron affinities are derived from experimental correlation functions based on the energies of optical charge transfer transitions in complexes with selected donors (E_{CT}), and from correlations with electrochemical half-wave reduction potentials ($E_{1/2}$). For the experimental details, for information on the reliability of individual data and for the original references of several of the quoted data the cited literature should be consulted. After [75C2]. (Electron affinities in eV).

Compound	Magnetron	Other methods	E_{CT}	$E_{1/2}$
Hexacyanobutadiene	3.24 (7)		3.08 (26)	3.06 (26)
Tetracyanoethylene (TCNE)	2.88 (5)	2.03 (7) (D)	2.77 (9)	2.90 (8)
7,7,8,8-Tetracyanoquinodimethane (TCNQ)	2.83 (19)	2.8 (1) (B)	2.84 (5)	2.88 (8)
Hexacyanobenzene	2.48 (13)		2.56 (26)	
p-Bromanil (BA)		2.44 (2) (B [78C2])	2.45, cf. [78C2, table 4]	
p-Chloranil (CA)	2.40 (26)	2.76 (2) (B [78C2])	2.44 (7)	2.48 (26)
p-Fluoranil (FA)	2.27	2.92 (2) (B [78C2])		2.45 (26)
Fluorobenzoquinone	2.16		1.93 (8)	2.12 (26)
1,2,4,5-Tetracyanobenzene (TCNB)	2.15 (22)		2.00 (26)	2.05 (8)
Tetracyanopyridine	2.12 (17)			
p-Benzoquinone	1.34 (9)	1.9 (3) (B [75C1])	1.83 (12) (?)	1.98 (26) (?)
Hexafluorobenzene	1.20 (7)	$\geqq 1.8$ (3) (C)		
9,10-Anthraquinone	1.15 (10)		1.59 (26) (?)	1.55 (26) (?)
o-Dicyanobenzene	1.04 (10)			0.95 (8)
Fumaronitrile	0.75 (12)			

Negative gas phase electron affinities, obtained from electron transmission experiments [73P1, 75N1, 77M, 78J]:

Compound	A_g	Ref.
Benzene	-1.15 eV	78J
Biphenyl	-0.3 (1) eV	83M2
Naphthalene	-0.19 eV	78J, cf. 72M

Electron affinities (gas phase) *derived* from experimental correlation functions based on the energies of optical charge transfer transitions in complexes with selected donors (E_{CT}), and from correlations with electrochemical half-wave reduction potentials ($E_{1/2}$). For the experimental details, for information on the reliability of individual data and for the original references of several of the quoted data the cited literature should be consulted. After [75C2]. (Electron affinities in eV.)

Compound	$A_g (E_{CT})$	$A_g (E_{1/2})$
Tetrafluorotetracyanoquinodimethane	–	3.22
2,3-Dichloro-5,6-dicyano-p-benzoquinone (DDQ)	3.13	3.00
Hexacyanobutadiene	–	3.09

(continued)

Compound	$A_g (E_{CT})$	$A_g (E_{1/2})$
table (continued)		
Tetracyano-*p*-benzoquinone	2.87	–
7,7,8,8-Tetracyanoquinodimethane (TCNQ)	2.84	2.88
2,6-Dinitro-*p*-benzoquinone	2.82	–
2,3-Dicyano-5,6-dichloro-7-nitro-1,4-naphtoquinone	2.82	3.16
2,3-Dicyano-*p*-benzoquinone	2.78	–
Tetracyanoethylene (TCNE)	2.77	2.90
2,3-Dicyano-1,4-naphthoquinone	2.66	2.47
Hexacyanobenzene	2.56	–
Dicyanomethyleneindane-1,3-dione	2.43	2.47
Mellitic trianhydride	2.38	–
3,5-Dinitrophthalic anhydride	2.32	–
1,4,5,8-Napthalenetetracarboxylic acid dianhydride	2.28	–
Dibromopyromellitic dianhydride	2.23	2.17
Pentacyanomethylbenzene	2.19	–
2,4,7-Trinitrofluorene-9-one	2.17	2.10
Pyromellitic dianhydride (PMDA)	2.04	1.94
1,2,4,5-Tetracyanobenzene (TCNB)	2.00	2.05
p-Benzoquinone	1.83	1.98
9,10-Phenanthroquinone	1.77	1.83
Tetrachlorophthalic anhydride (TCPA)	1.72	1.63
1,4-Naphthoquinone	1.71	1.78
Tetracyanoxylene	1.65	–
Tetranitromethane	1.63	–
9,10-Anthraquinone	1.57	1.55
4-Cyanobenzonitrile (*p*-Dicyanobenzene)	1.07	1.10
Benzonitrile (Cyanobenzene)	–	0.33

Electron affinities for a number of other compounds, obtained by E_{CT} and $E_{1/2}$ correlations, may be found in [75C2].

12.3.7 Crystal electron affinity

An estimate of the crystal electron affinity (i.e. of the binding energy of an extra electron to a molecule in the crystalline state) can be obtained from the gas phase electron affinity A_g by setting $P^- = P^+$ in the relation $A_c = A_g + P^-$, and using the A_g and P^+ values of sections 12.3.6 and 12.3.5, respectively.

12.3.8 Band energy scheme of the acene class of organic photoconductors

Under the assumption that only small intermolecular interactions take place in the van der Waals solid, the (ionic) molecular states, characterized by the ionization energy I_c and the electron affinity A_c, are only moderately split into *narrow* energy bands. I_c and A_c then can be used to characterize the energy of the valence and of the conduction band, respectively. An example of the correlation of these states with the molecular size is given in Fig. 161 [74B4, 74K1].

12.3.9 Excitonic absorption spectra

Spectra for anthracene and tetracene at RT are shown in Fig. 162 as examples.

Figures for 12

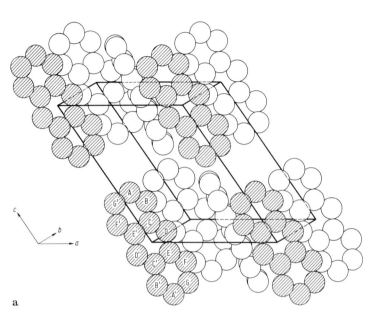

a

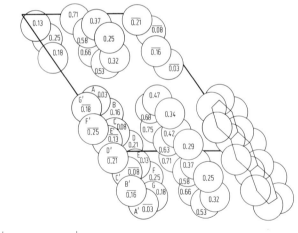

b   5 Å

Fig. 1. Anthracene. Perspective view (a) and projection of
the crystal structure along [010] (b). Fractional b coordinates of the carbon atoms are indicated; (a) after [50S],
(b) after [71W], modified.

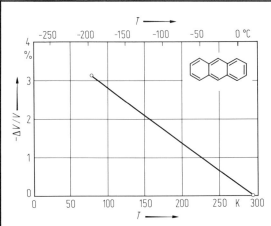

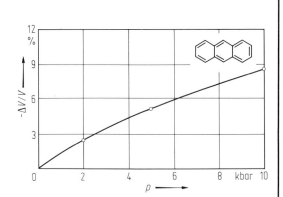

Fig. 2. Anthracene. Thermal expansion. Volume change $\Delta V/V$, vs. temperature; from [62K], after data from [53K]. Linear thermal expansion coefficients: $\alpha_{11} = 111.7 \cdot 10^{-6}\, K^{-1}$ ($15°$ to a axis); $\alpha_{22} = 13.4 \cdot 10^{-6}\, K^{-1}$ (b axis); $\alpha_{33} = 20.3 \cdot 10^{-6}\, K^{-1}$.

Fig. 3. Anthracene. Volume change vs. pressure (at room temperature), from [62K], after data from [49B].

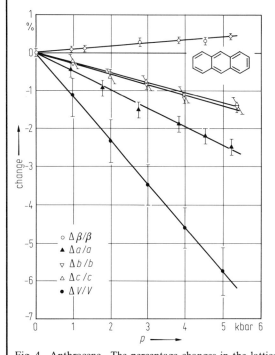

For Fig. 5, see next page.

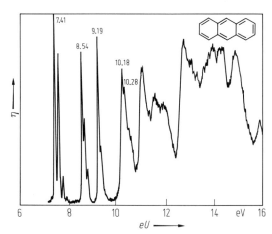

Fig. 4. Anthracene. The percentage changes in the lattice parameters a, b, c, and β vs. pressure (from neutron diffraction). $\Delta V/V$ was calculated according to $\Delta V/V = \Delta a/a + \Delta b/b + \Delta c/c + \Delta\beta \cot\beta$ [78E]. (Temperature not stated, but presumably room temperature).

Fig. 6. Anthracene. Gas phase photoelectron spectrum at $T = 370$ K, ionization quantum yield η vs. ionization potential eU [77S3].

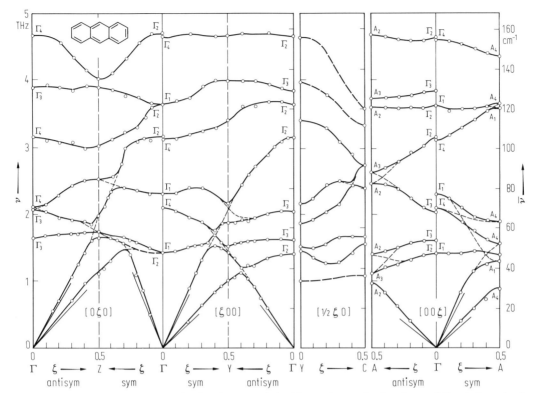

Fig. 5. Anthracene-d_{10}. The 12 intermolecular phonon and the 4 lowest intramolecular phonon (vibron) dispersion branches at $T = 12\,\text{K}$ (phonon frequency (wavenumber) vs. reduced wavevector coordinate). Strong interaction between the external and the 4 lowest internal modes is observed, cf. [82D1]. Sound velocities at 300 K are indicated by straight lines. The designation "sym" and "antisym" is with respect to the diad screw axis 2_1 and to the glide plane a of the space group symmetry $P2_1/a$. The Γ_1 and Γ_3 modes are Raman active. – From [82D1]; cf. [81D3, 82C2]; (cf. Fig. 61 for the Brillouin zone and the symmetry labels used).

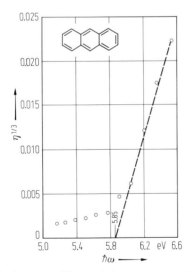

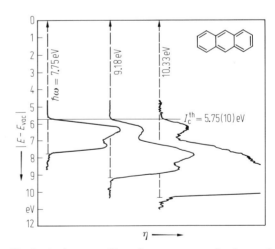

Fig.7. Anthracene. Photoemission quantum yield $\eta^{1/3}$ (from single crystal, at $T = 300\,\text{K}$), vs. photon energy [80A3].

Fig. 8. Anthracene. Photoelectron spectra of polycrystalline layers at $T = 300\,\text{K}$; the kinetic energy distributions of electrons photoemitted by light of photon energies 7.75, 9.18 and 10.33 eV are plotted, rearranged in an electron binding energy scale, relative to vacuum level [82K2].

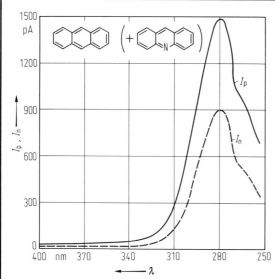

Fig. 9. Anthracene. Spectral dependence of "intrinsic" photoconduction (photocurrent vs. wavelength), due to the generation of pairs of free electrons and holes. Charge carrier generation by excitonic reactions ("extrinsic" photoconduction) is eliminated by doping with acridine, which acts as an exciton trap, but not as an electron or hole trap at the temperature of the experiment, $T = 154\,°C$. $I_{n,p}$, the photocurrent of electrons, holes respectively, has been corrected to $I_0 = 1.4 \cdot 10^{13}$ photons/s; 6.4 nm resolution [66C1].

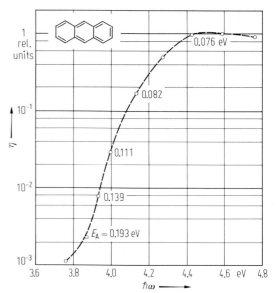

Fig. 10. Anthracene. Relative yield η of electron photocarriers at $E = 5 \cdot 10^5$ V/m and $T = 295$ K, vs. photon energy; spectral slit width 0.01 eV; (holes gave photocurrents of similar magnitude). Very low light intensity was used, generating $\approx 10^{-10}$ C/cm^2 during 1 s exposure time. The thermal activation energies indicated at the experimental points were obtained from the temperature dependence of electric field dependent yields (cf. Fig. 22), extrapolated to $E = 0$. According to Onsager's recombination model, these activation energies are indicative of the energy needed for final separation (against mutual Coulomb attraction) of the initial electron–hole pair, created at close distance. Adding the activation energy at threshold to the photon energy, a minimum total energy for free charge carrier generation of 4.06 eV is obtained. The absolute yield η at 4.4 eV is $\approx 10^{-4}$ electrons per photon (for $E = 0$) [80K3].

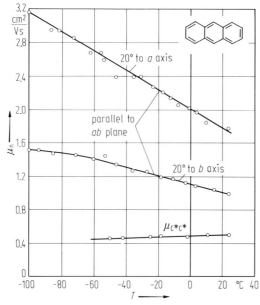

Fig. 11. Anthracene. Electron mobility vs. temperature for different orientations of the electric field with respect to the crystallographic axes [62K].

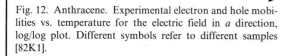

Fig. 12. Anthracene. Experimental electron and hole mobilities vs. temperature for the electric field in a direction, log/log plot. Different symbols refer to different samples [82K1].

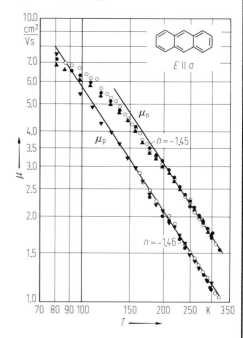

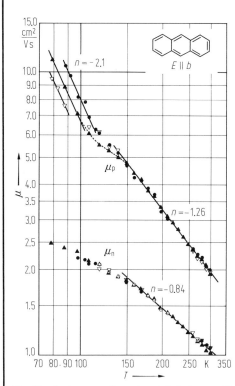

Fig. 13. Anthracene. Experimental electron and hole mobilities vs. temperature, for the electric field in b direction, log/log plot [80K1, 82K1]. One of the hole curves was presented in [78K2]. Different symbols refer to different samples.

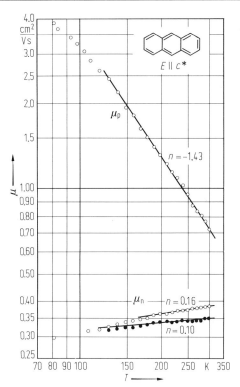

Fig. 14. Anthracene. Experimental electron and hole mobilities vs. temperature for the electric field in c^* direction, log/log plot. Open circles refer to normal anthracene [80K1], [82K1]; full circles refer to perdeuterated anthracene [82K1].

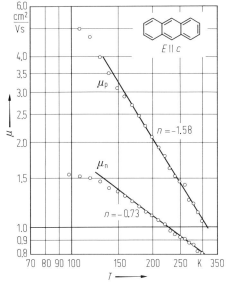

Fig. 15. Anthracene. Experimental electron and hole mobilities vs. temperature for the electric field in c direction, log/log plot [82K1].

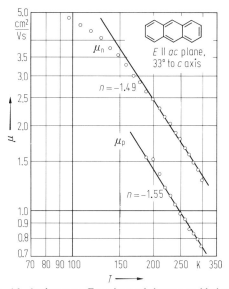

Fig. 16. Anthracene. Experimental electron and hole mobilities vs. temperature for the electric field in the $a-c$ plane, 33° from $+a$ in the direction to $+c$; log/log plot [82K1].

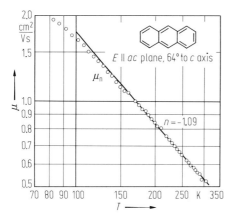

Fig. 17. Anthracene. Experimental electron mobility vs. temperature for the electric field in the $a-c$ plane, 64° from $+a$ in the direction to $+c$; log/log plot [82K1].

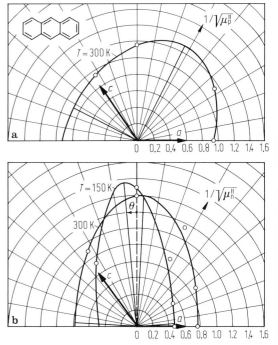

Fig. 18. Anthracene. $a-c$ cross sections through the representation quadrics (bold ellipses) of the hole (a) and the electron (b) mobility tensors (at the temperatures specified in the figures), calculated from the experimental points which are indicated. Notice that the *square root* of the *inverse* mobility parallel to the electric field direction (cf. [57N]) is plotted. After [80K1, 82K1]. μ in cm^2/Vs.

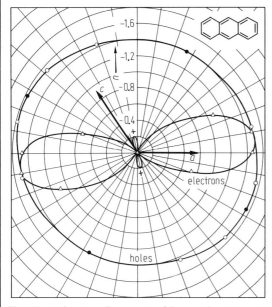

Fig. 19. Anthracene. Exponent n of the experimental $\mu \propto T^n$ temperature dependences, plotted as a function of the *internal* (true) transport direction of the charge carriers. The symbols o and $\triangle$ refer to the experimental results obtained for holes and electrons, respectively; from the μ results the exponents for the principal directions of the hole mobility tensor, marked with the symbol $\bullet$, were calculated [82K1].

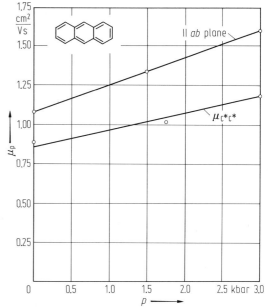

Fig. 20. Anthracene. Hole mobility (at room temperature) vs. pressure [62K].

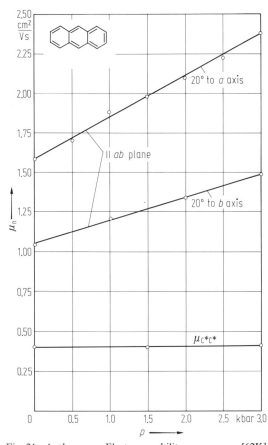

Fig. 21. Anthracene. Electron mobility vs. pressure [62K]; cf. [67K].

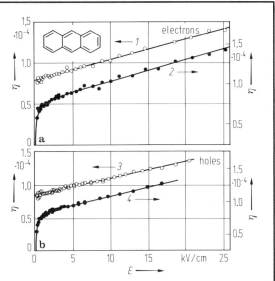

Fig. 22. Anthracene. Electron (Fig. (a)) and hole (Fig. (b)) quantum yields η at 255 nm vs. applied electric field E (at room temperature). Curves *1* and *3* for virgin crystal samples; curves *2* and *4* illustrate the reduced low-field yield which occurs when trapped holes or electrons, respectively, are present in the excited volume of the crystal [74B1]; cf. [73C1].

Fig. 23. Anthracene. Spectral sensitization of photocarrier quantum yield η vs. photon energy (at room temperature). The quantum yield under flash excitation was measured for a neat anthracene crystal under vacuum (curve *a*). The spectral region 410 to 330 nm corresponds to generation by exciton reactions; for shorter wavelengths "intrinsic" electron–hole pair generation predominates. Subsequent deposition of a thin phenothiazine layer (by evaporation) lead to the increased quantum yield in the excitonic region, whereas in the intrinsic region absorption losses in the layer reduced the yield (curve *b*) [74K1].

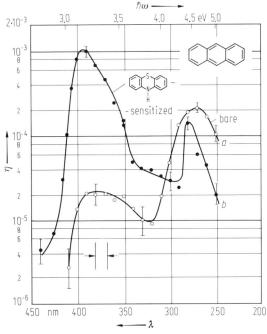

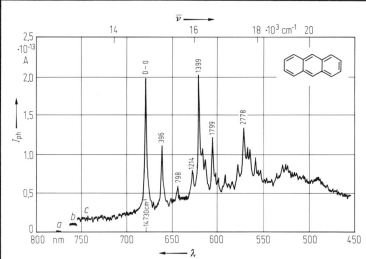

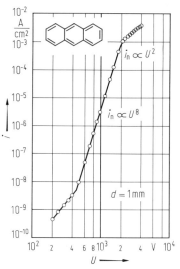

Fig. 24. Anthracene. Photocurrent vs. wavelength (wavenumber) at $T = 83$ K. The current is due to release of trapped charge carriers by mobile triplet excitons and, therefore, reflects the (weak) $S_0 \rightarrow T_1$ absorption spectrum of the anthracene crystal. The excitation spectrum was obtained after (residual) traps still present in the zone-refined crystal, were (partially) filled by capture of free charge carriers, created (via singlet excitons) by light of an energy greater than 3.15 eV ($\lambda < 393$ nm). The labels designate: a) electrical zero, b) the dark current level, and c) the photocurrent due to triplet absorption and trap release. Due to technical reasons the spectrum was taken with decreasing wavelength. The experimental set up was similar to the one used for obtaining the spectrum Fig. 140; see also [76K3]. The peaks of the vibrational progression are labelled by their energetic distance (in cm^{-1}) from the $0-0$ transition, located at 14730 cm^{-1} (1.826 eV); accuracy $\sim \pm 0.5$ nm (± 15 cm^{-1}) [75K5].

Fig. 26. Anthracene. Current-voltage characteristic of injected, space charge limited electron current. Electron injecting electrode was a solution of 1 g Li metal in 100 cm^3 ethylenediamine, saturated with anthracene [65M1]. Anthracene is otherwise a perfect insulator with a dark conductivity of less than $10^{-15} \Omega^{-1} cm^{-1}$ [71K1]. d: sample thickness.

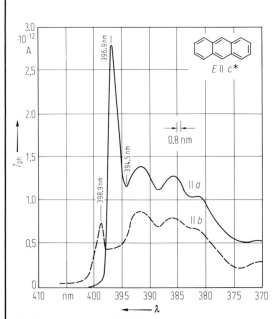

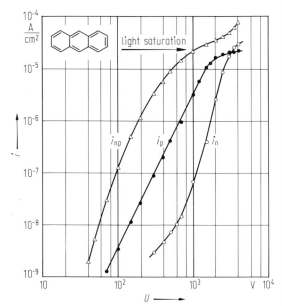

Fig. 25. Anthracene. Polariton-induced photocurrent vs. wavelength (polarized light, $T = 5$ K, crystal thickness 20 μm, applied voltage 225 V) [77C3].

Fig. 27. Anthracene. Current-voltage characteristics of injected, space-charge-limited electron and hole currents, and of space-charge-compensated double injection current (i_{np}). Electron injecting contact was Li metal dissolved in ethylenediamine, hole injecting contact was $AlCl_3$ in nitromethane [65M1].

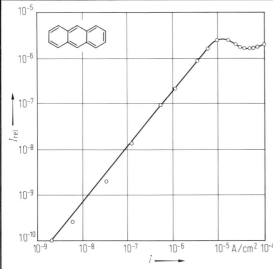

Fig. 28. Anthracene. Relative electroluminescence intensity (recombination; luminescence from excitonic levels) under double injection conditions from electrolytic contacts (cf. Fig. 27) vs. current density. The emitted light intensity is proportional to the current until saturation of the anode contact sets in [65M1]. (Temperature not stated, but presumably room temperature).

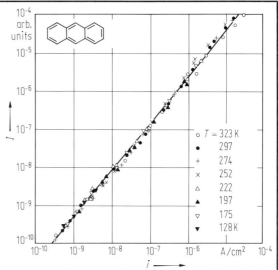

Fig. 29. Anthracene. Electroluminescence intensity (in arbitrary units) vs. current density for various temperatures between $T = 323$ K and 128 K. The figure includes results from 12 different crystals with various electrode combinations [70W].

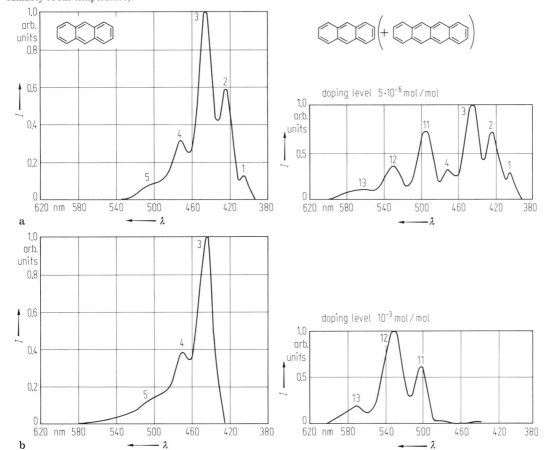

Fig. 30. Anthracene. Comparison of the surface fluorescence (a) and the bulk electroluminescence spectrum (b) of undoped (left) and of tetracene-doped (right) anthracene [67Z]. (Temperature not stated, but presumably room temperature). I: light intensity.

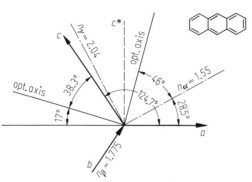

Fig. 31. Anthracene. Orientation of the optical tensor (at room temperature) [74P2], (drawn with the optical data of [62N]); $\lambda = 589$ nm, perspective representation.

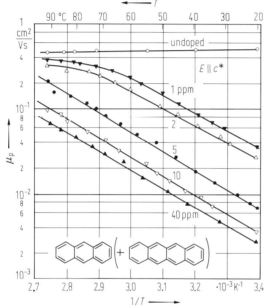

Fig. 32. Anthracene. Reduction of apparent hole mobilities $\mu_{c^*c^*}$ by (intentional) tetracene impurities at the concentrations specified in the figure, as a function of (reciprocal) temperature. From the slope of the straight lines of the doped samples an activation energy of 0.43 eV is obtained [63H].

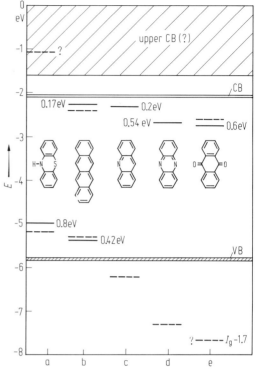

Fig. 34. Anthracene. Predicted ($---$) and measured (———) trap energies (in eV) of the molecules indicated by their symbols (from left to right: phenothiazine, tetracene, acridine, phenazine and anthraquinone). The lower levels (0.8 eV and 0.42 eV) are normally occupied states and hence act as hole traps, the upper levels are normally unoccupied states, acting as electron traps. Some of the trap energies were determined according to Fig. 33, others by thermally stimulated current spectroscopy (TSC) [74K1].

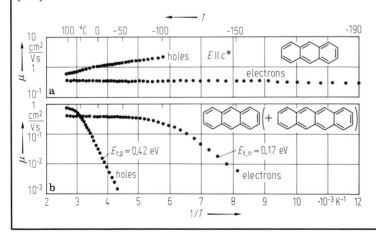

Fig. 33. Anthracene. Charge carrier mobilities vs. (inverse) temperature with the electric field in c^* direction, in ultrapure (a) and in tetracene-doped ($5 \cdot 10^{-7}$ mol/mol) (b) crystals. Tetracene forms an electron and a hole trap with depths of 0.17 and 0.42 eV, respectively [74K1, 75P].

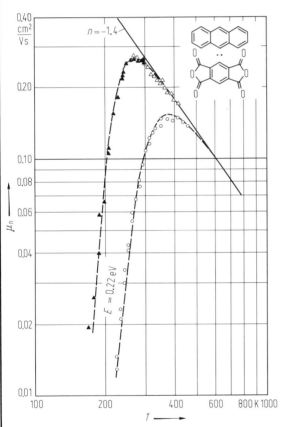

Fig. 35. Anthracene:PMDA. A stereoscopic representation of the molecular stacking (stacking direction [001]) in the donor:acceptor complex, viewed normal to the molecular planes. Drawn are the molecular skeleton (valence bonds) and the vibrational ellipsoids of the atoms [78R1].

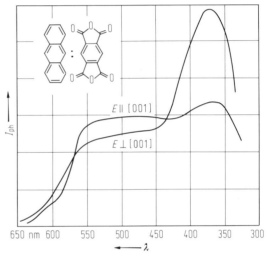

Fig. 37. Anthracene:PMDA. Photocurrent excitation spectra for the (external) electric vector parallel and perpendicular to the [001] stack direction. (Relative units, room temperature, spectra corrected for constant light intensity) [75K4].

Fig. 36. Anthracene:PMDA. Electron mobility approximately parallel to the [001] stacking direction vs. temperature, log/log plot. Different symbols refer to different crystals: triangles (full and open) refer to two parallel slices of a Bridgman crystal [84M], whereas open circles represent results obtained with less pure sublimation crystals [75K4]. The low temperature data are due to multiple shallow trapping limited transport. A Hoesterey-Letson fit (dashed curves) indicates a shallow hole trap, $E = 0.22$ eV.

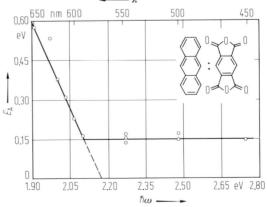

Fig. 38. Anthracene:PMDA. Activation energy of photocurrent yield vs. photon energy (wavelength). [75K4].

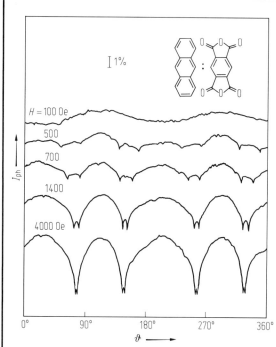

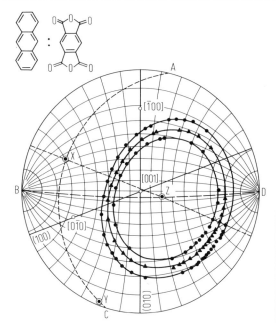

Fig. 39. Anthracene:PMDA. Magnetophotoconductivity by triplet exciton – trapped charge carrier interaction at room temperature. The modulation is due to magnetic field dependent detrapping based on triplet-doublet resonances, "TDR" (Zeeman level-(anti)-crossings in the $S=\frac{1}{2}$ charge carrier doublet and $S=1$, symmetry-split exciton triplet, spin pair state). Photocurrent vs. rotation angle in a selected (fixed) crystal plane, $\sim(1\bar{1}0)$, at various magnetic field strengths. The individual curves are shifted on the ordinate, zero point suppressed; the amplitude of a 1% modulation is indicated at the top of the figure [79Z1]. See also caption of Fig. 24.

Fig. 40. Anthracene:PMDA. Magnetophotoconductivity: angular positions of triplet-doublet resonances (dots, cf. Fig. 39), and triplet-triplet resonances (triangles) in the photocurrent at a magnetic field of 1 kG, plotted in stereographic projection in a Wulff's net. The experimental points lie on curves which for high magnetic fields merge into the tangential cone of the triplet fine structure representation quadric [57N]. At finite field the triplet-triplet resonances lie on the intersection line of the triplet representation quadric and a sphere of radius $R=g_t\mu_B H/(XYZ)^{1/2}$, where g_t is the triplet gyromagnetic ratio, μ_B the Bohr magneton, H the magnetic field, and X, Y, Z the principal values of the triplet fine structure tensor. In the example, the underlying triplet exciton is that of the unperturbed donor (anthracene) molecule with essentially no charge transfer admixture. X, Y, and Z in the figure denote the directions of the triplet fine structure tensor principal axes [79Z1]; cf. Fig. 39.

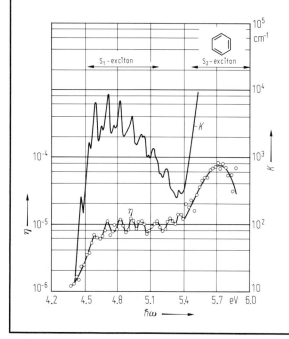

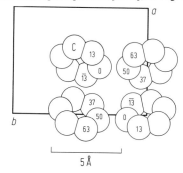

Fig. 41. Benzene. Projection of the crystal structure along [001]; the numbers are c fractional coordinates of the carbon atoms ($\times 100$) [71W].

Fig. 42. Benzene. Absorption coefficient K at $T=243$ K and photogeneration charge carrier quantum yield η (in carriers per photon) at 243 K and 50 kV/cm vs. photon energy, for the wave vector of the (unpolarized) light $k\parallel[100]$. After [70H].

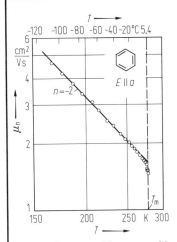

Fig. 43. Benzene. Electron mobility for the electric field $E \parallel a$ vs. temperature, log/log plot [70H].

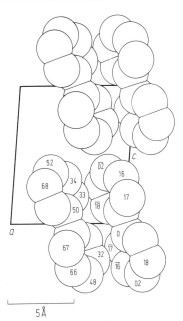

Fig. 44. Biphenyl. Projection of the crystal structure along [010]. Fractional b coordinates of the carbon atoms are indicated ($\times 100$) [71W].

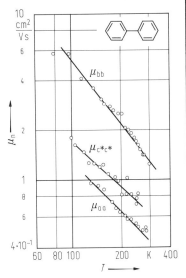

Fig. 45. Biphenyl. Electron mobility vs. temperature for three directions of the electric field, log/log plot [77B3].

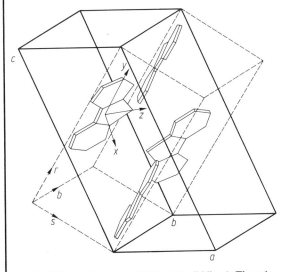

Fig. 46. Dibenzothiophene. Unit cell (solid lines). The principal optical extinction directions are indicated by the broken lines r, b, s [71B2].

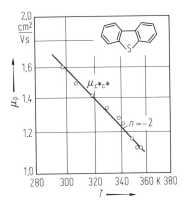

Fig. 47. Dibenzothiophene. Hole mobility vs. temperature, log/log plot, for the electric field $E \parallel c^*$ [75D].

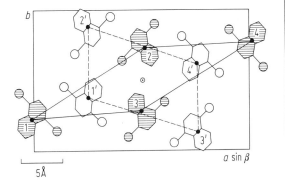

Fig. 48. 1,4-Dibromonaphthalene. Projection of the unit cell in c direction. One of the two independent molecules in the asymmetric unit is shaded [67C1].

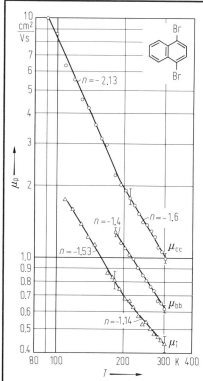

Fig. 49. 1,4-Dibromonaphthalene. Hole mobility vs. temperature, log/log plot. Hole mobilities are plotted for the electric field E approximately in c and in b direction, and for an inclined direction $\measuredangle(E, a^*) = 24°$, $\measuredangle(E,b) = 107.5°$, $\measuredangle(E,c) = 74°$ [81K4].

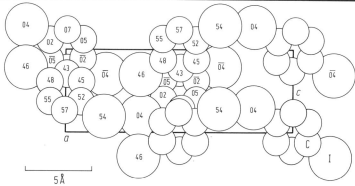

Fig. 51. 1,4-Diiodobenzene. Projection of the crystal structure along [010]. The fractional b coordinates of the carbon and iodine atoms are indicated ($\times 100$) [71W].

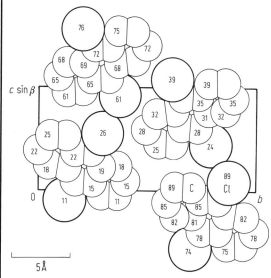

Fig. 50. 9,10-Dichloroanthracene, α-form. Projection of the crystal structure along [100]. Fractional a coordinates of the carbon and chlorine atoms are indicated ($\times 100$) [71W].

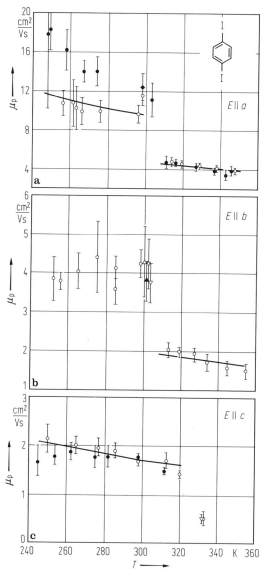

Fig. 52. 1,4-Diiodobenzene. Hole mobilities in the a, b, and c crystallographic directions of the orthorhombic crystal. Different symbols refer to different samples. A phase transition occurs at $T = 324$ K [73S1, 79S3]. After [67S1].

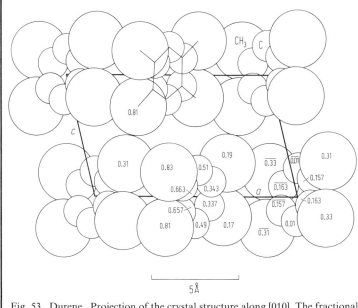

Fig. 53. Durene. Projection of the crystal structure along [010]. The fractional *b* coordinates of the carbon atoms are indicated; left hand axes [71W].

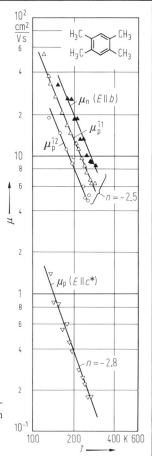

Fig. 55. Durene. Charge carrier mobilities vs. temperature for various directions, log/log plot. The intermediate direction, labelled i1, lies 45° between *a* and *b*, that labelled i2 is 35° off the *a*−*b* plane [77B2].

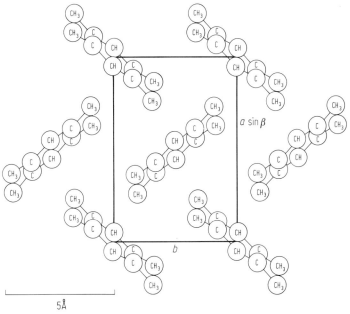

Fig. 54. Durene. Projection of the crystal structure along [001] [77B2].

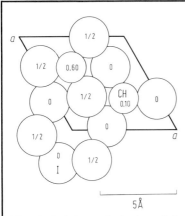

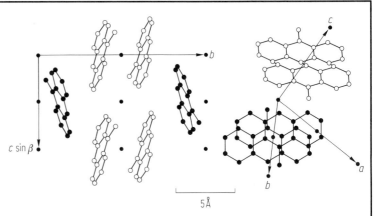

Fig. 56. Iodoform. Projection of the crystal structure along [001] with the fractional c coordinates of the carbon and iodine atoms indicated [66W].

Fig. 57. 9-Methylanthracene. Projection of the crystal structure along [100] (left), and normal to the mean plane of one of the molecular pairs (right). The two molecules of the molecular pair are related to each other by an inversion center between them [71B1].

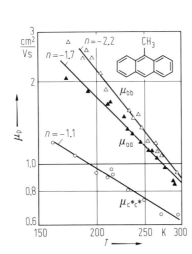

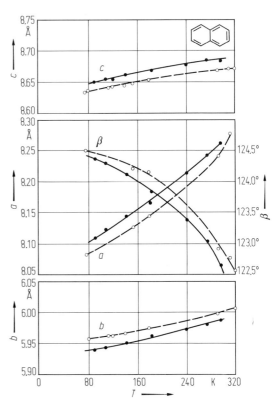

Fig. 58. 9-Methylanthracene. Hole mobilities vs. temperature, log/log plot [81D2].

Fig. 59. Naphthalene. Temperature dependence of lattice parameters. Open symbols: [68R], full symbols: [82B5].

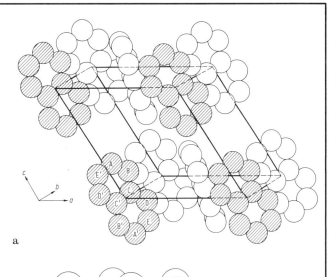

a

Fig. 60. Naphthalene. Perspective view (a), and projection of the crystal structure along [010] (b) with the fractional b coordinates of the carbon atoms indicated. The atoms labelled A,A′, etc. are related by an inversion center at [000]. (a) [49A], (b) [71W], modified.

b   5 Å

For Figs. 61, 62, see next page.

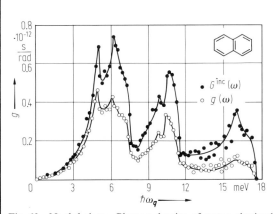

Fig. 63. Naphthalene. Phonon density of states obtained experimentally by incoherent neutron scattering (solid dots) compared with a theoretical density of states function, $g(\omega)$ (based on the experimental oscillator frequencies), and fitted with a hydrogen amplitude-weighted function $G^{\text{inc}}(\omega)$ [77W3]. See also [83D].

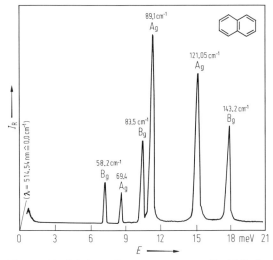

Fig. 64. Naphthalene. Raman spectrum at $T=1.3$ K. Intensity of scattered light vs. energy; arbitrary orientation and polarization [83S4]. Symmetry assignments after [83D]. Zero of the energy scale is at $\lambda = 514.54$ nm.

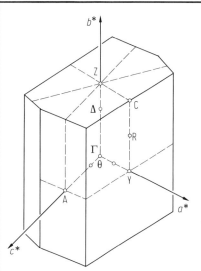

Fig. 61. Naphthalene. Brillouin zone and its symmetry points as used in Fig. 62 [80N2].

▶

Fig. 62. Naphthalene, perdeuterated. The 12 intermolecular phonon and the 4 lowest intramolecular phonon (vibron) dispersion branches at $T=6$ K, obtained by inelastic neutron scattering [80N2]; cf. [80B3, 82D1, 82K4]. Symmetry assignments as in Fig. 61.

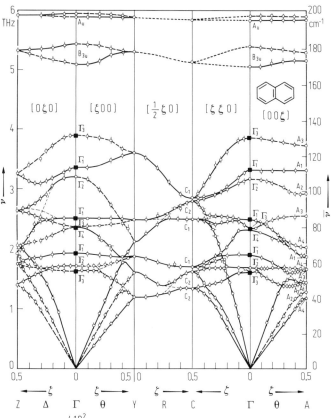

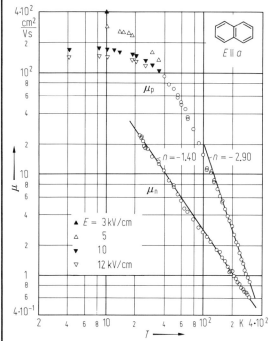

Fig. 65. Naphthalene. Electron and hole mobility vs. temperature, log/log plot, for the electric field approximately $\|a$: $\angle(E,a)=6.6°$, $\angle(E,b)=86.4°$, $\angle(E,c^*)=96.6°$. Crystal thickness was 1.01(2) mm. Mobilities $\gtrsim 100$ cm^2/Vs are electric field dependent; [82W4]; cf. [84K2, 84W1, 84W2].

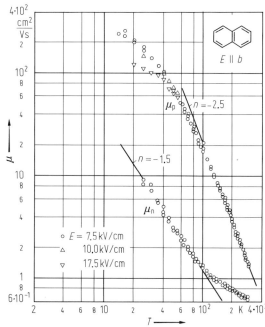

Fig. 66. Naphthalene. Electron and hole mobility vs. temperature, log/log plot, for the electric field $E\|b$ at various field strengths. These data are representative for the results obtained with several independent samples of thickness between 200 and 500 μm with the crystallographic orientations all $N\|b$ within $\pm 4°$, where N designates the normal to the individual crystal slice [83W].

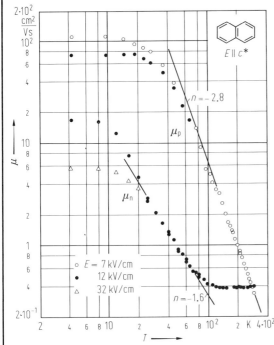

Fig. 67. Naphthalene. Electron and hole mobilities vs. temperature, log/log plot, for the electric field $E \| c^*$ (within $\pm 3°$) and various field strengths [83W]; cf. [84K2]. – For the transition of the electron mobility from nearly temperature-independent to $\mu \propto T^{-1.6}$ at $T \approx 100$ K, which has been interpreted as a hopping to band transport transition, see also [78S4, 78K1, 79S2].

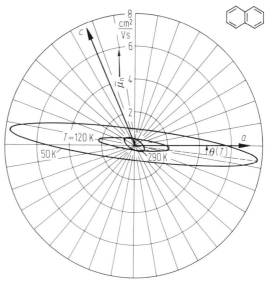

Fig. 69. Naphthalene. Tensor ellipsoids of the electron mobility at $T = 290$ K, 120 K, and 50 K, cross sections in the crystallographic $a-c$ plane (010) [84W3]. Notice that the tensor rotates with temperature about [010], cf. numerical data given in the tables. The interpretation of the figure is as follows: For any arbitrarily chosen crystal direction, the length of the radius (in this direction) from the center to the surface of the ellipsoid represents the "mobility of electrons drifting in this direction", $\tilde{\mu} = |v|/|E|$. The electric field *direction* which is necessary to cause this drift motion, however, is in general oblique to the drift direction and not obtainable from the tensor ellipsoid representation directly, except for the principal axes directions, (cf. [57N]).

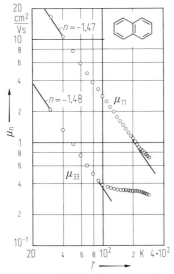

Fig. 68. Naphthalene. Principal components μ_{11} and μ_{33} of the electron mobility tensor vs. temperature, log/log plot; (for the principal component μ_{22}, which for monoclinic symmetry coincides with μ_{bb}, see Fig. 66). The points represented are best tensor values, which were calculated (for each temperature separately) from the results of 11 measurements in different crystallographic directions by a least squares matrix algorithm (cf. [57N]). After [84W3]. The tensor rotates with temperature about [010], cf. Fig. 69 and the data given in the tables.

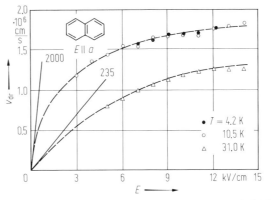

Fig. 70. Naphthalene. Hole drift velocity vs. electric field E, for $E \| a$, at three temperatures. The hole velocities tend to saturate with increasing field. The straight lines represent Ohm's law and suggest low field mobilities of 235 cm^2/Vs and 2000 cm^2/Vs, respectively; the dashed lines are a guide to the eye. These results reflect hot carrier effects [84W2]. After [82W4, 83W], cf. [84W1, 84W2].

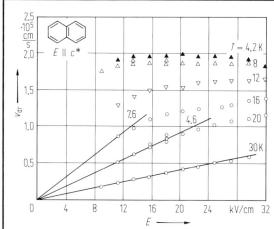

Fig. 71. Naphthalene. Electron drift velocity vs. electric field E at different temperatures, for $E\|c^*$. The low temperature electron velocities tend to saturate with increasing field. The straight lines from the origin to the 30, 20, and 16 K points represent Ohm's law with mobilities 2.1, 4.6, and 7.6 cm^2/Vs, respectively [84W3].

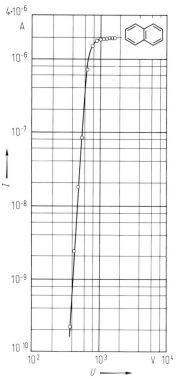

Fig. 72. Naphthalene. Hole injection current vs. voltage at room temperature, log/log plot. Hole injecting contact was formed by a $1.7 \cdot 10^{-3}$ molar solution of Ag^{2+} in 7.5 normal HNO_3. The absolute quantities refer to a crystal of thickness 48 μm and an electrode area of 0.07 cm^2 [69L2].

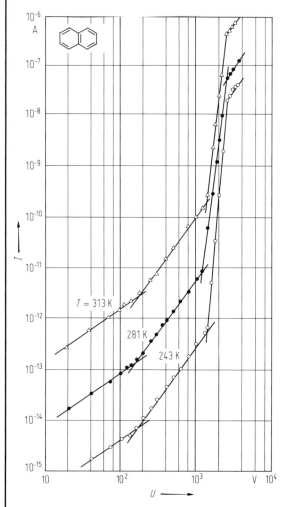

◄

Fig. 73. Naphthalene. Current-voltage characteristics at different temperatures. Electrodes were silver paint; sample dimensions typically 1 cm^2, 0.1 cm thick. The low field Ohmic current has electron and hole contributions, whereas the space charge-controlled current at the higher fields has been interpreted as being due to electron injection. Space charge limited ("SCL") current theory allows to extract from these data the concentration of free charge carriers ($2.7 \cdot 10^3$ cm^{-3}), the trap concentration ($1.6 \cdot 10^{17}$ cm^{-3}), and the trap depth (0.60 eV) [72C2].

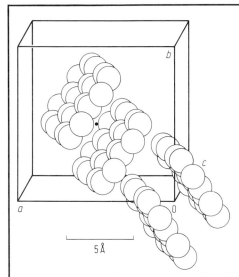

Fig. 74. Perylene, α-form. Perspective view of the crystal structure. The four almost planar perylene molecules in the unit cell are grouped in sandwich pairs about centers of symmetry at (000) and ($\frac{1}{2}\frac{1}{2}$0); the mean planes of the molecules in each pair are therefore parallel. The perpendicular distance between the planes is 3.46 Å. Plotted after the data of [64C2]. (The hydrogen atoms have been omitted for clarity; the carbon atoms have been plotted with arbitrary radius.)

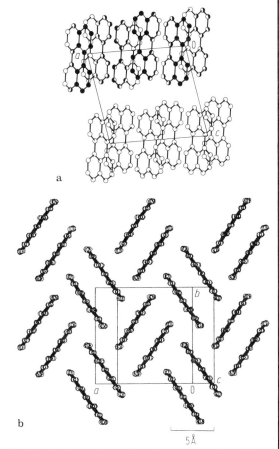

Fig. 75. Perylene, α-form. Crystal structure projected along [010], (a), and projected approximately parallel to the intersection lins of the molecular planes and perpendicular to [010], (b) (only the upper molecular layer of Fig. 75(a) is plotted). The angle between the long molecular axis of the molecules at the origin and [010] is 89.6°. In part (a) of the figure the dark atoms are closer to the observer; the hydrogen atoms have been omitted, whereas they are included in part (b). Drawn after the data of [64C2] (right hand coordinate system).

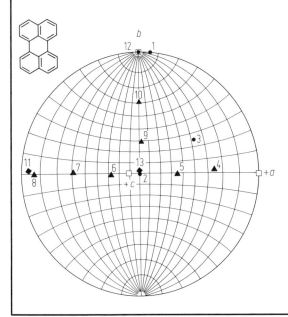

Fig. 76. Perylene, α-form. Stereographic projection of the plate normals of the 13 crystal slices used for measuring the electron mobility tensor (Figs. 77···81), full symbols; different symbols refer to samples from different crystal boules. The open squares represent the crystallographic directions a, b and c [83S5].

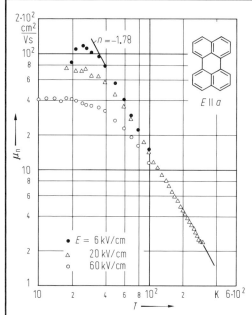

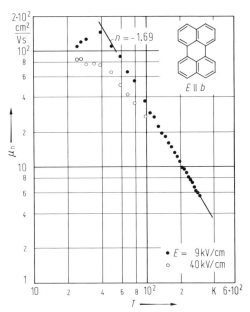

Fig. 77. Perylene, α-form. Electron mobility vs. temperature, log/log plot, for the electric field direction as indicated by point 11 in Fig. 76, ($\sim \| a$). Sample thickness was 252(3) μm [83S5, 84W1]; cf. [82K8].

Fig. 78. Perylene, α-form. Electron mobility vs. temperature, log/log plot, for the electric field direction $E \| b$ (point 12 in Fig. 76). Sample thickness was 225(5) μm [83S5]; cf. [82K8].

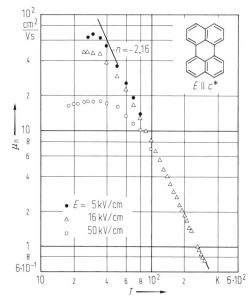

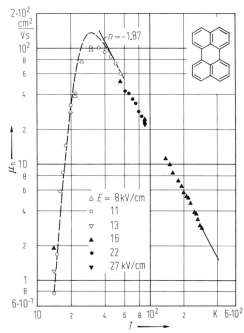

Fig. 79. Perylene, α-form. Electron mobility vs. temperature, log/log plot, for the electric field direction as indicated by point 13 in Fig. 76, ($\| c^*$) [83S5]; cf. [82K8, 84K2, 84W1]. Sample thickness was 306(3) μm.

Fig. 80. Perylene, α-form. Electron mobility vs. temperature, log/log plot, for the electric field direction as indicated by point 3 in Fig. 76; ($\angle(E, a) = 45\,(1)°$, $\angle(E, b) = 66\,(1)°$, $\angle(E, c^*) = 55\,(1)°$); sample thickness was 370(10) μm. The broken line is a fit of the Hoesterey-Letson type multiple shallow trapping model (cf. [63H]) with the parameters trap depth $= 17.5$ meV and trap concentration $= 5 \cdot 10^{-4}$ mol/mol. After [84W1]; cf. [81S3].

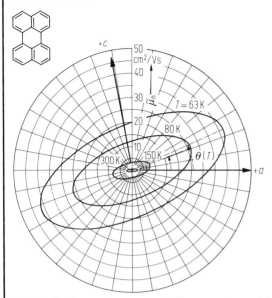

Fig. 81. Perylene, α-form. $a-c$ (010) cross sections of the electron mobility tensor ellipsoids at various temperatures, calculated from measurements in 13 different orientations, (cf. Fig. 76) by a least squares matrix formalism [57N]. Notice the tensor rotation about [010] with temperature. After [83S5]; cf. [82K8]. (For the interpretation of the tensor ellipsoids, the caption of Fig. 69 may be consulted.)

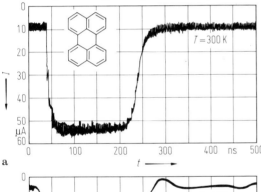

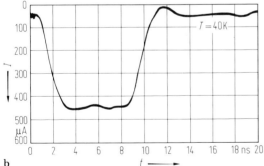

b

Fig. 83. Perylene, α-form. Electron transit current vs. time after a 0.8 ns excitation by a nitrogen laser pulse ($\lambda = 337$ nm). The time-of-flight pulse length reflects the electron transit time through the crystal slice; sample thickness was 225(5) μm, the electric field amounted to 22.2 kV/cm for the pulse at 300 K and to 31.3 kV/cm for the pulse at 40 K [83S5]. (Drawn after the original photographs of single shot oscilloscope pulses.)

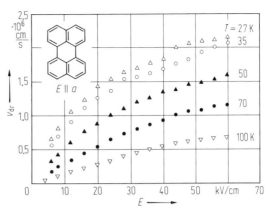

Fig. 82. Perylene, α-form. Electron drift velocity vs. electric field for $E \| a$ (orientation 11 of Fig. 76) and for various temperatures. The sublinear field dependence indicates hot electron effects [83S5, 84W1].

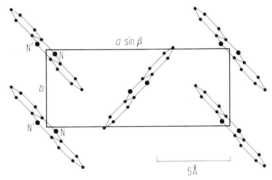

Fig. 84. Phenazine, α-form. Projection of the crystal structure along [001] [55H].

For Fig. 85, see next page.

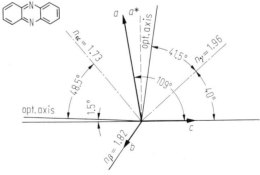

Fig. 86. Phenazine, α-form. Orientation of the optical tensor in the crystallographic axes system (at room temperature), perspective view. From [74P2], drawn after the data of [41W1].

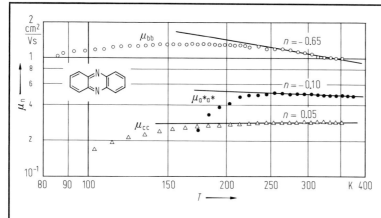

Fig. 85. Phenazine, α-form. Electron mobility vs. temperature, log/log plot, for the electric field parallel to a^*, b and c, where a^* is perpendicular to the $b-c$ cleavage plane [74P1].

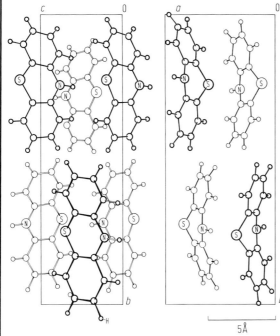

Fig. 87. Phenothiazine, room-temperature phase. Molecular packing diagram, perspective projections along [100] and along [001]. The dihedral angle between the two benzene ring moities is 158.5° [76M2].

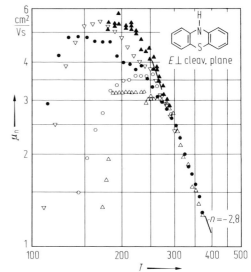

Fig. 88. Phenothiazine. Electron mobility for the electric field perpendicular to the $a-c$ cleavage plane, vs. temperature, log/log plot. Full circles: first cooling cycle; open circles: second cooling cycle after repolishing with introduction of additional defects. The other symbols refer to measurements with independent samples. Below a phase transition at $T \approx 250$ K the slopes depend on the sample history. The decreasing parts of the curves (with decreasing temperature) are due to multiple shallow trapping. After [83K2].

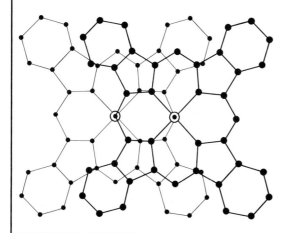

Fig. 89. Phthalocyanine, β-form. Projection of the molecular stacks on the molecular plane; stacking direction is [010]. The packing shown for metal-phthalocyanines (⊙ = metal atom) is also typical for metal-free phthalocyanine [77H1].

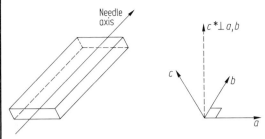

Fig. 90. Phthalocyanine, β-form. Basic shape of sublimation-grown crystals. After [74C].

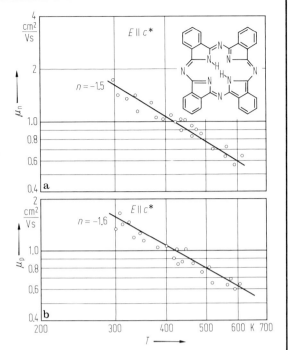

Fig. 91. Phthalocyanine, β-form. Electron and hole mobility vs. temperature, log/log plot, for the electric field normal to the $a-b$ plane, which is the pronounced face developing during sublimation growth, cf. Fig. 90 [74C].

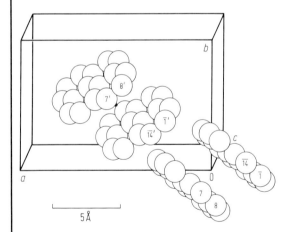

Fig. 92. Pyrene. Perspective view of the unit cell. The four nearly planar molecules, occupying general positions, are grouped in sandwich pairs about centers of symmetry at [000] and $[\frac{1}{2}\frac{1}{2}0]$ which relate atoms with the labels n and ñ, and n′ and ñ′, respectively. Glide mirror planes at $b/4$ and $3b/4$ parallel to the $a-c$ plane relate n to n′, and ñ to ñ′; twofold screw axes ∥b relate n to ñ′, and ñ to n′. The perpendicular distance between the best planes through the atoms of each pair is 3.53 Å. The labelled atoms are those closest to the observer. The carbon atoms are drawn with arbitrary radius; the same labels as in [65C] have been used. The hydrogen atoms have been omitted. – Plotted after the data of [65C].

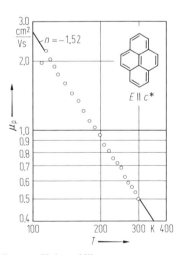

Fig. 93. Pyrene. Hole mobility $\mu_{c^*c^*}$ vs. temperature, log/log plot, for the electric field perpendicular to the $a-b$ cleavage plane [83K4]. The lowest temperature point signals the onset of the phase transition.

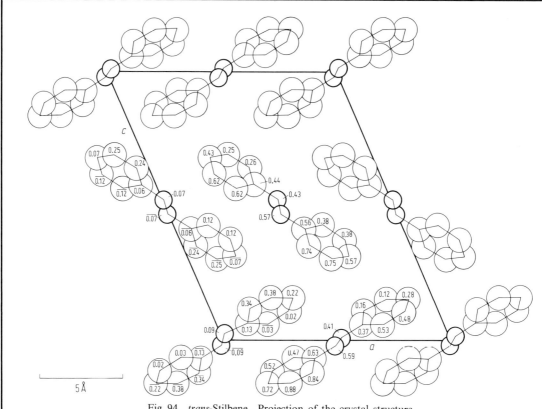

5 Å

Fig. 94. *trans*-Stilbene. Projection of the crystal structure along [010]; left hand coordinates; after [71W]. The fractional *b* coordinates are those of [74F1], where, however, a different choice of the axes has been adopted.

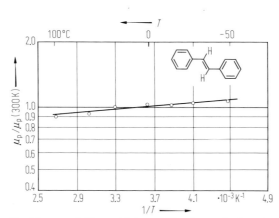

Fig. 95. *trans*-Stilbene. Hole mobility (normalized to the room temperature value), vs. (reciprocal) temperature. Sample orientation not specified [64H1].

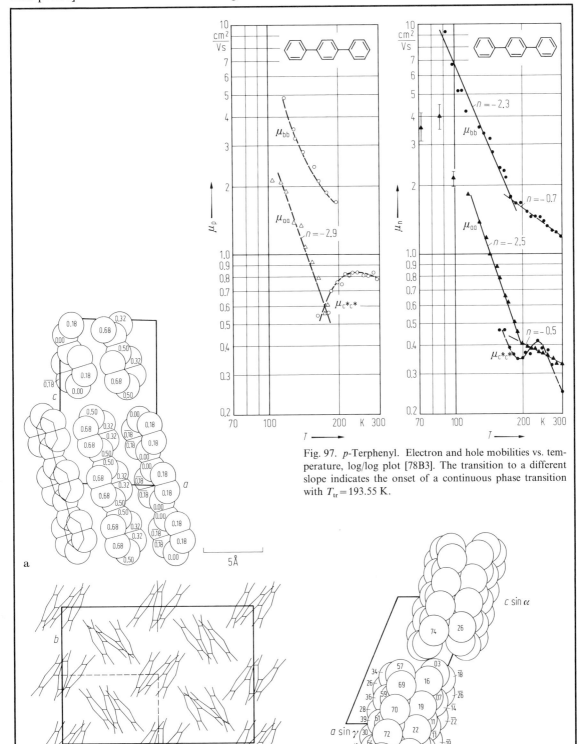

Fig. 97. p-Terphenyl. Electron and hole mobilities vs. temperature, log/log plot [78B3]. The transition to a different slope indicates the onset of a continuous phase transition with $T_{tr} = 193.55$ K.

Fig. 96. p-Terphenyl. (a) Projection of the crystal structure along [010]. Fractional b coordinates of the carbon atoms are indicated [71W], left hand axes. (b) The projection of the low-temperature phase on (001); the unit cell at room temperature is indicated by the broken lines [76B2].

Fig. 98. Tetracene. Projection of the crystal structure along [010]. Fractional b coordinates of the carbon atoms are indicated ($\times 100$) [71W].

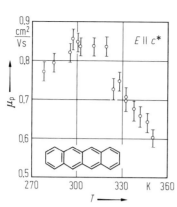

Fig. 99. Tetracene. Hole mobility vs. temperature for the electric field in c^* direction. The slope of the high temperature part of the curve is $-3 < n < -2$ in a log μ vs. log T plot [76B1].

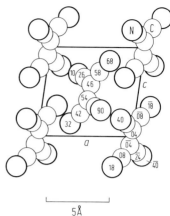

Fig. 100. Tetracyanoethylene. Projection of the crystal structure along [010] [66W]. Fractional b coordinates of the carbon and nitrogen atoms are indicated ($\times 100$).

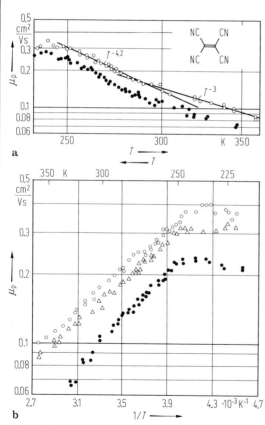

a

b

Fig. 101. Tetracyanoethylene. Hole mobility vs. temperature, (a) log/log plot, and (b) log μ vs. $1/T$ plot. Different symbols refer to different unspecified crystal orientations. In the temperature range $T > 200$ K in Fig. (b) $\mu \propto \exp(E/kT)$ fits the results with $E = 80$ meV [80S2].

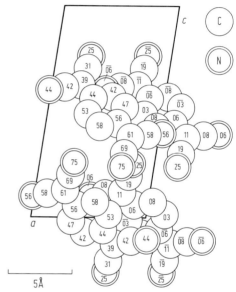

Fig. 102. 7,7,8,8-Tetracyanoquinodimethane, TCNQ. Projection of the crystal structure along [010]. The fractional b coordinates of the carbon and nitrogen atoms are indicated ($\times 100$) [71W].

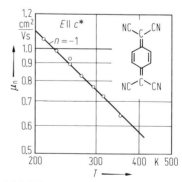

Fig. 103. 7,7,8,8-Tetracyanoquinodimethane, TCNQ. Electron mobility $\mu_{c^*c^*}$ vs. temperature, log/log plot [78S1].

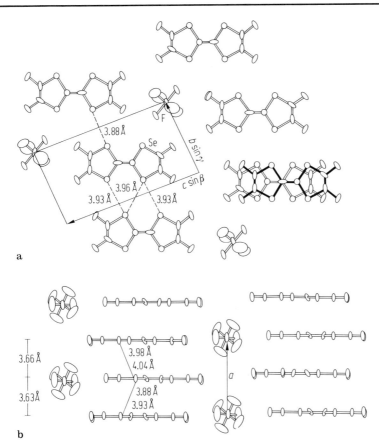

a

b

Fig. 104. $(TMTSF)_2 : PF_6$. Crystal structure viewed along a, (a), and the arrangement of the molecular stacks, (b). Se−Se distances for several intermolecular selenium contacts are indicated (in Å). The flat TMTSF molecules form alternating stacks in a direction with interplanar distances of 3.66 and 3.63 Å [81J2].

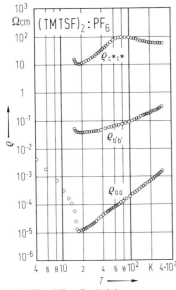

Fig. 105. $(TMTSF)_2 : PF_6$. Resistivity vs. temperature, log/log plot, for the electric field parallel to the crystallographic a, $c^* = a \times b$ and $b' = c^* \times a$ axes [81J2].

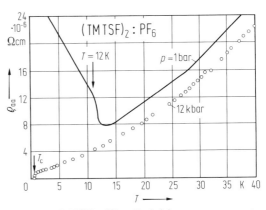

Fig. 106. $(TMTSF)_2 \cdot PF_6$. Resistivity ρ_{aa} vs. temperature at ambient pressure and under a pressure of $p = 12$ kbar. Under ambient pressure a metal to semiconductor transition occurs at $T = 12$ K, whereas at $p = 12$ kbar a superconducting state is reached at $T_c = 1$ K [82J4].

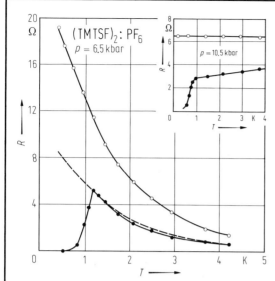

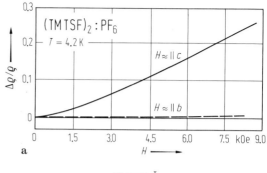

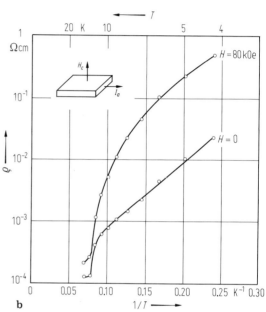

Fig. 107. $(TMTSF)_2:PF_6$. Electrical resistance for $E\|a$ vs. temperature at pressures $p=6.5$ kbar and $p=10.5$ kbar (inset). The open circles represent results obtained in a magnetic field of 41 kOe, (nearly parallel to b), the full circles were measured without a magnetic field. At 6.5 kbar coexistence of superconductivity ($T_c=1.1$ K) and a metal-semiconductor transition ($T\approx6$ K) is found [80G].

Fig. 109. $(TMTSF)_2:PF_6$. Transverse magnetoresistance $\Delta\varrho/\varrho$ at $T=4.2$ K vs. magnetic field for $H\|b$ and for $H\|c$, (a), and resistivity vs. reciprocal temperature at $H=0$ and at $H=80$ kOe, $H\|c$, (b). The electric field was always applied along the highly conducting stack (a) direction. Analysis of these results led to the conclusion that $(TMTSF)_2:PF_6$ is a two-dimensional semiconductor with high mobility at low temperature [82C3].

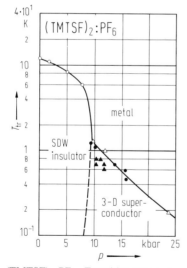

Fig. 108. $(TMTSF)_2:PF_6$. Transition temperature vs. pressure. The experimental phase transition points in the displayed phase diagram were obtained by different groups. The lines separate a spin density wave (SDW) insulator at low pressure and low temperature from a 3-dimensional superconductor at low temperature and high pressure; and a metal, existing at higher temperatures, from the insulator (more precisely: semiconductor) and from the superconductor phases [82J4].

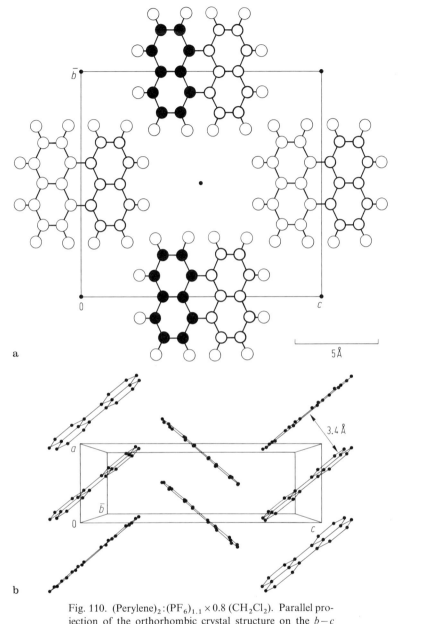

Fig. 110. (Perylene)$_2$:(PF$_6$)$_{1.1}$ × 0.8 (CH$_2$Cl$_2$). Parallel projection of the orthorhombic crystal structure on the $b-c$ plane (a), and perspective view along b (b). In Fig. a the more enhanced atoms are closer to the observer than the less enhanced. The center of the perylene molecule lies in the special position [0 $\frac{1}{2}$ 0]. The channels between the [100] oriented stacks are filled with the PF$_6^-$ ions and with the solvent molecules. These components are disordered. Fig. 110 (a) after [80K4], Fig. 110 (b) after [84S1].

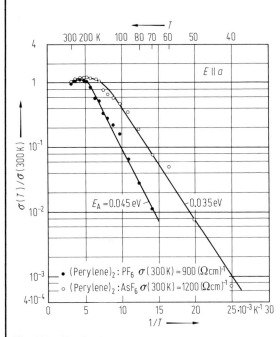

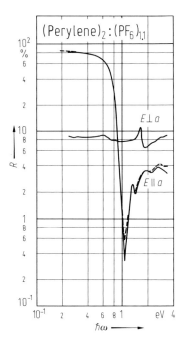

Fig. 111. $(Perylene)_2:(PF_6)_{1.1} \times 0.8(CH_2Cl_2)$. Conductivity σ_{aa} (normalized to its room temperature value) vs. (reciprocal) temperature. The figure also contains results for the isostructural AsF_6^- salt. The absolute room temperature conductivities and the thermal activation energies E_A are indicated in the figure [80K4].

Fig. 112. $(Perylene)_2:(PF_6)_{1.1} \times 0.8 (CH_2Cl_2)$. Reflectance vs. photon energy for the light E vector parallel and perpendicular to the stack axis (for an unspecified temperature, presumably 300 K). The dashed line represents a Drude fit [83G1].

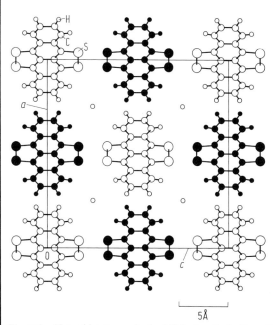

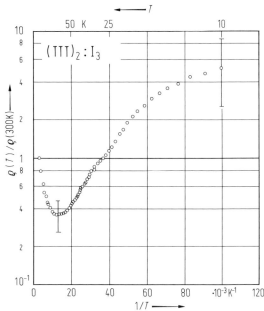

Fig. 113. $(Tetrathiatetracene)_2:I_3$, $TTT_2:I_3$. Projection of the crystal structure along [010]. The TTT interplanar spacing is 3.32 Å; the iodine sublattice, consisting of I_3^- anions, is disordered in b direction [77S5].

Fig. 114. $(Tetrathiatetracene)_2:I_3$. Resistivity vs. reciprocal temperature for the applied electric field parallel to b. Error bars indicate observed scatter; $\varrho(300\ K) = 10^{-3}\ \Omega cm$ [78I].

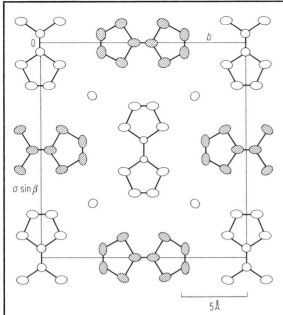

Fig. 115. TTF:$Br_{0.785}$. [001] projection of the crystal structure. Two crystallographically inequivalent TTF molecules are centered at [000] and $[0\frac{1}{2}\frac{1}{2}]$. They form separate stacks along the c axis with regular interplanar spacing of 3.572 Å. The molecular plane normals form angles of 1.4 and 1.0° with the c axis. The bromide ions also stack in columns along the c axis, forming a sublattice which is incommensurate (and dependent on the exact composition) along the c axis, but identical in its a^* and b^* reciprocal lattice vectors [75L].

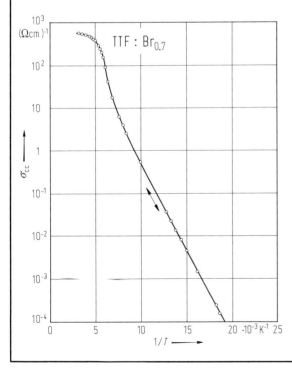

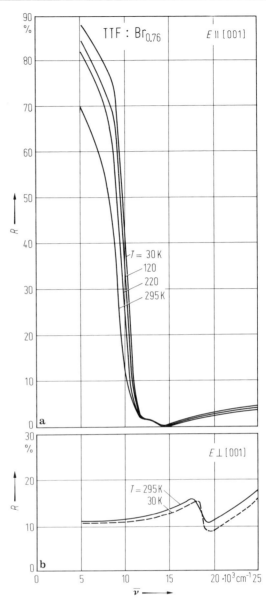

Fig. 117. TTF:$Br_{0.76}$. Reflectance vs. wavenumber for light polarization parallel (a) and perpendicular to the stack axis [001] (b) [82K5]. (Temperature not specified, but presumably room temperature.)

◄

Fig. 116. TTF:Br_n, $n \approx 0.7$. Conductivity σ_{cc} vs. reciprocal temperature; four probe measurement on needle-shaped crystals. Above $T \approx 180$ K the conductivity becomes almost temperature-independent, amounting to $100 \cdots 500$ Ω^{-1} cm^{-1} for different samples. The activation energy is 0.08 eV [75W2].

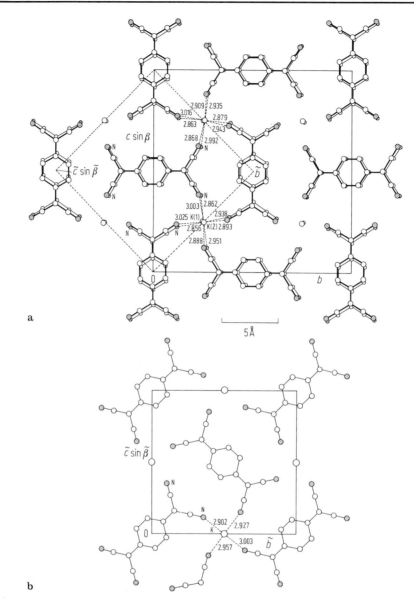

Fig. 118. K:TCNQ. Crystal structure viewed along the *a* axis, (a) low temperature modification, and (b) high temperature modification. The high temperature orientation and dimensions of the unit cell are indicated in the low temperature structure by broken lines. Both modifications are characterized by separate TCNQ and K layers (oriented parallel to the *b*−*c* plane), and by segregated TCNQ stacks in *a* direction. These stacks are dimerized (alternating) in the low temperature phase, and evenly spaced (regular) in the high temperature phase [77K3]. Atomic distances between the K ion and the adjacent N atoms are given numerically (in Å).

For Fig. 120, see next page.

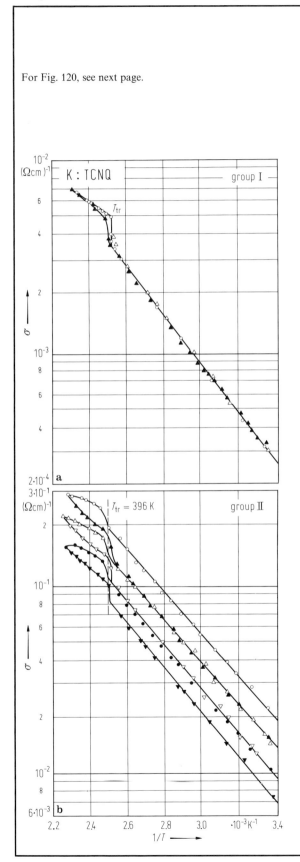

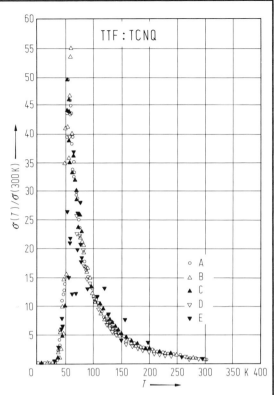

Fig. 121. TTF:TCNQ. Conductivity (normalized to its room temperature value) vs. temperature. Symbols A··· D refer to dc four probe measurements with different samples; sample D gave the microwave conductivities (9.5. GHz) which are represented by the symbol E. Current flow was in all cases parallel to the b stack (needle) axis. The absolute room temperature dc conductivities ranged about $400 \pm 100\ \Omega^{-1}\ cm^{-1}$, the enhancement at the 54 K peak (relative to the room-temperature conductivity) amounted to between 15 and 25. Microwave conductivities were lower ($170 \cdots 300\ \Omega^{-1}\ cm^{-1}$) but displayed a stronger enhancement (44 ± 6). TTF:TCNQ is metallic above 54 K and semiconductive below. After [76F]; cf. [73F].

◄

Fig. 119. K:TCNQ. Microwave conductivity at 9.3 GHz vs. reciprocal temperature. Crystals of different preparations ("group I", (a), and "group II", (b)) differ in their absolute conductivities, but both undergo a phase transition at 396 K and both exhibit similar activation energies of 0.26(1) eV and 0.13(2) eV below and above the phase transition, respectively. This behavior was taken indicative of an extrinsic semiconductor with different concentrations of donors and acceptors in the different preparations. The different symbols in the figures have the following meaning: (a) ○ first cooling cycle; ▲, second heating cycle; and △, second cooling cycle. (b) First cycle: ○, heating; ▲, cooling; second cycle: △, heating; ▽, cooling; third cycle: ●, heating; ▼, cooling [82G2]. (Probable field orientation: $E\|a$).

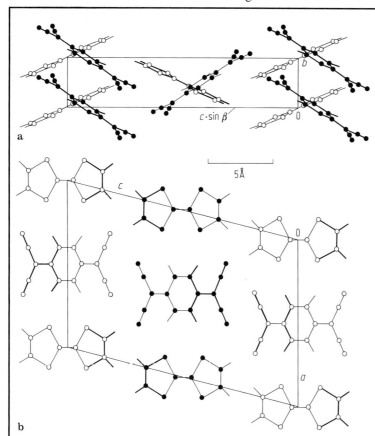

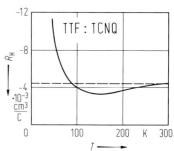

Fig. 122. TTF:TCNQ. Hall coefficient vs. temperature, full line, and expected value for metallic-like conduction, $1/nec$, calculated for a band filling of 0.58 electrons per TCNQ molecule, dashed, (the band filling factor has been derived from twice the Fermi wave vector, $2k_F$, as obtained from X-ray and neutron scattering). The measured Hall coefficient is essentially independent of temperature down to $T \approx$ 100 K, suggesting a metallic-like conduction mechanism in this temperature regime. It confirms the dominance of TCNQ carriers (electrons) in the conduction of TTF:TCNQ and the value of charge transfer of ≈ 0.6 e/molecule. The geometry of the experiment was: current along b, magnetic field along a and Hall voltage along the c^* direction [77C1]; cf. [82J1].

Fig. 120. TTF:TCNQ. Projection of the crystal structure along [100] (a), and along [010] (b). The molecules in the central vertical row in Fig. (b) have their centers at $b/2$; enhanced bonds and atoms are closer to the observer. After [74K5].

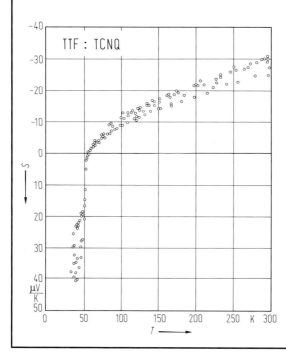

Fig. 123. TTF:TCNQ. Absolute thermopower along the b stack direction of b-elongated crystal needles vs. temperature. The large thermopower commonly observed in organic conductors (about one order of magnitude higher than in typical metals) has been ascribed to the existence of small bandwidths. A bandwidth $4t_\parallel = 0.54$ eV has been extracted from the above data. At 56 K a phase transition takes place and the metallic-like electron conduction turns over to a semiconductor-like hole conduction [73C2]; cf. [82J1].

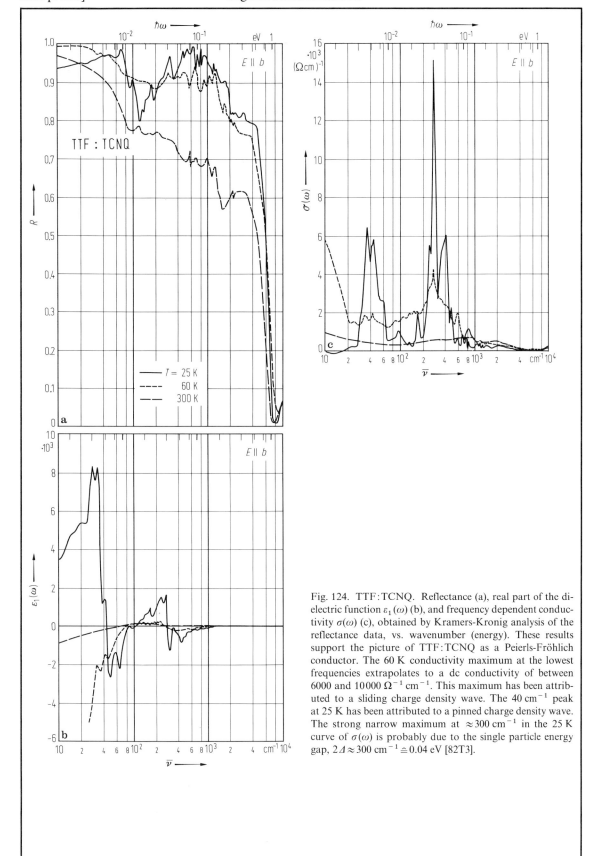

Fig. 124. TTF:TCNQ. Reflectance (a), real part of the dielectric function $\varepsilon_1(\omega)$ (b), and frequency dependent conductivity $\sigma(\omega)$ (c), obtained by Kramers-Kronig analysis of the reflectance data, vs. wavenumber (energy). These results support the picture of TTF:TCNQ as a Peierls-Fröhlich conductor. The 60 K conductivity maximum at the lowest frequencies extrapolates to a dc conductivity of between 6000 and 10000 $\Omega^{-1}\,\mathrm{cm}^{-1}$. This maximum has been attributed to a sliding charge density wave. The 40 cm^{-1} peak at 25 K has been attributed to a pinned charge density wave. The strong narrow maximum at $\approx 300\,\mathrm{cm}^{-1}$ in the 25 K curve of $\sigma(\omega)$ is probably due to the single particle energy gap, $2\varDelta \approx 300\,\mathrm{cm}^{-1} \cong 0.04$ eV [82T3].

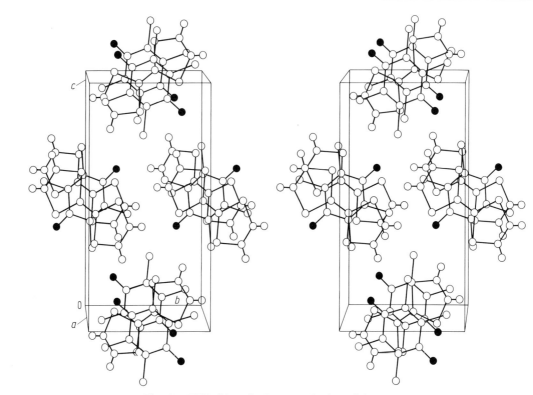

Fig. 125. TTF:chloranil. Stereoscopic view of the crystal structure in *a* direction. The constituent molecules form mixed stacks with the stack axis parallel to the crystallographic *a* direction. (Plotted with the atomic coordinates of [79M2]. The chloranil oxygen atoms have been enhanced for clarity.)

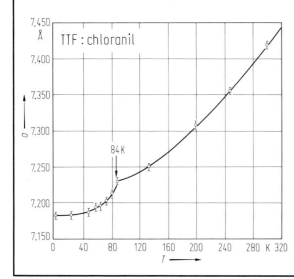

Fig. 126. TTF:chloranil. Lattice constant along stacks, *a*, vs. temperature, demonstrating the occurrence of a phase transition at 84 K [81T2].

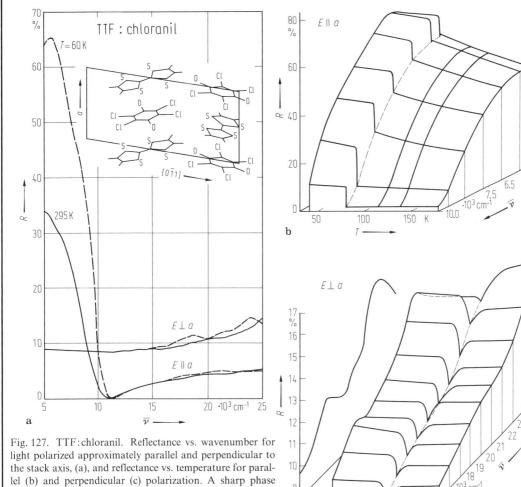

Fig. 127. TTF:chloranil. Reflectance vs. wavenumber for light polarized approximately parallel and perpendicular to the stack axis, (a), and reflectance vs. temperature for parallel (b) and perpendicular (c) polarization. A sharp phase transition is observed at 80(5) K, which has been interpreted as a neutral to ionic transition [82K6]. A phase transition temperature of 84.4(5) K has been reported by [82T1]; *several* sharp peaks were observed in [83A].

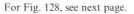

For Fig. 128, see next page.

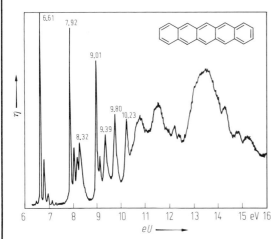

Fig. 129. Pentacene. Gas phase photoelectron spectrum at $T = 250\,°C$ [77S3].

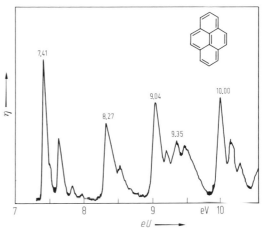

Fig. 130. Pyrene. Gas phase photoelectron spectrum at $T = 112\,°C$ [79C4].

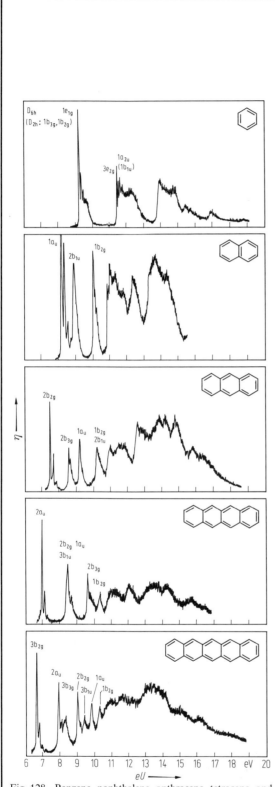

Fig. 128. Benzene, naphthalene, anthracene, tetracene, and pentacene. Photoelectron spectra (photoionization quantum yield vs. ionization potential) in the vapor phase ($T = 30 \cdots 250$ °C) [72C1].

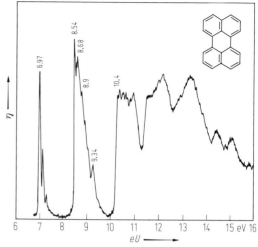

Fig. 131. Perylene. Gas phase photoelectron spectrum at $T = 150$ °C [77C4].

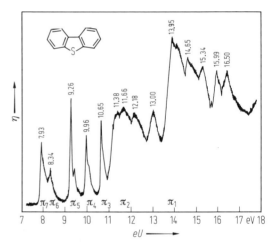

Fig. 132. Dibenzothiophene. Gas phase photoelectron spectrum at $T = 120$ °C [78R2].

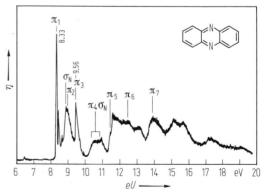

Fig. 133. Phenazine. Gas phase photoelectron spectrum at $T = 76$ °C [75H1].

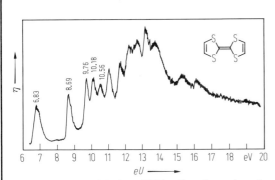

Fig. 134. Tetrathiafulvalene (TTF). Gas phase photoelectron spectrum at unspecified temperature [73G].

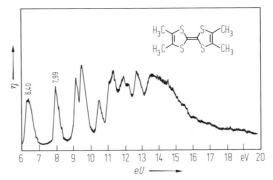

Fig. 135. Tetramethyltetrathiafulvalene (TMTTF). Gas phase photoelectron spectrum at unspecified temperature [75G2].

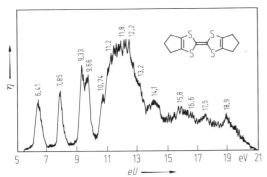

Fig. 136. Hexamethylenetetrathiafulvalene (HMTTF). Gas phase photoelectron spectrum at $T = 200$ °C [83S2].

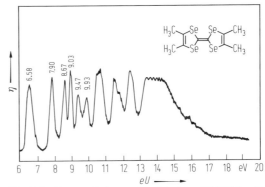

Fig. 137. Tetramethyltetraselenafulvalene (TMTSF). Gas phase photoelectron spectrum at unspecified temperature [75G2].

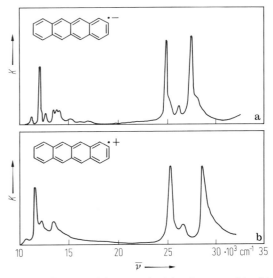

Fig. 139. Tetracene Mononegative (a) and monopositive (b) radical ion, absorption spectrum at $T = 77$ K in sec.-butyl-chloride and 2-methyltetrahydrofuran, respectively (absorption coefficient vs. photon energy) [73S2].

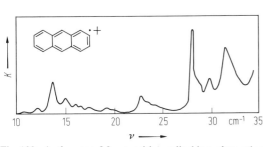

Fig. 138. Anthracene. Monopositive radical ion, absorption spectrum in sec.-butyl-chloride at 77 K (absorption coefficient vs. wavenumber) [73S2].

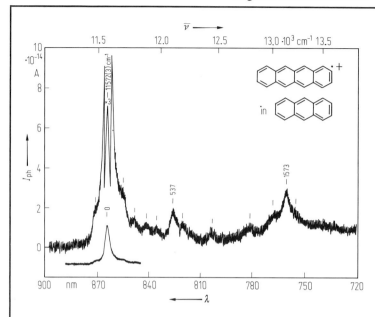

Fig. 140. Tetracene. Photocurrent excitation spectrum of the monopositive radical ion $^{1/2}D_0 \rightarrow {}^{1/2}D_1$ transition in an anthracene single crystal matrix at $T = 20$ K. The photocurrent is due to relaxation of a hole from the optically excited local radical ion state (formed in an independent preceding preparative step, by trapping of a photogenerated hole at a neutral tetracene dopant molecule) to the anthracene valence band where it contributes to conductivity [83K3]. (See also [76K3]).

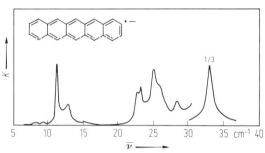

Fig. 141. Pentacene, mononegative radical ion, absorption spectrum in 2-methyl-tetrahydrofuran at $T = 77$ K (absorption coefficient vs. wavenumber) [73S2].

Fig. 142. Tetracyanoquinodimethane (TCNQ). Mononegative radical ion (alkali salt in acetonitrile solution), absorption spectrum at unspecified temperature, presumably 300 K [74B6]; cf. [67L].

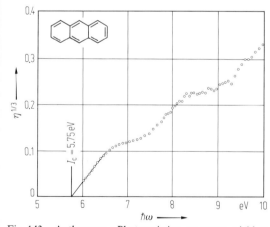

Fig. 143. Anthracene. Photoemission quantum yield η, plotted as $\eta^{1/3}$, vs. photon energy (vapor-deposited film, room temperature). The ionization threshold energy, I_c, has been extrapolated according to the relation $\eta^{1/3} \propto (\hbar\omega - I_c)$ [73H3] (modified).

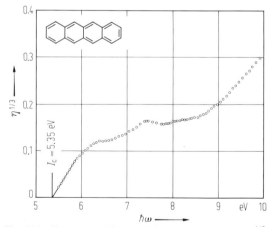

Fig. 144. Tetracene. Photoemission quantum yield, $\eta^{1/3}$, vs. photon energy (vapor-deposited film, room temperature). The ionization threshold energy I_c, has been extrapolated according to the relation $\eta^{1/3} \propto (\hbar\omega - I_c)$ [73H3] (modified).

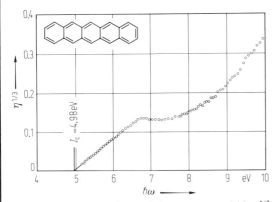

Fig. 145. Pentacene. Photoemission quantum yield, $\eta^{1/3}$, vs. photon energy (vapor-deposited film, room temperature). The ionization threshold energy, I_c, has been extrapolated according to the relation $\eta^{1/3} \propto (\hbar\omega - I_c)$ [73H3] (modified).

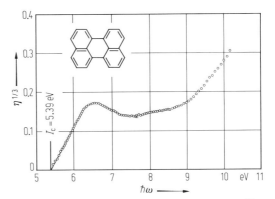

Fig. 146. Perylene, Photoemission quantum yield, $\eta^{1/3}$, vs. photon energy (vapor-deposited film, room temperature). The ionization threshold energy, I_c, has been extrapolated according to the relation $\eta^{1/3} \propto (\hbar\omega - I_c)$ [73H3] (modified).

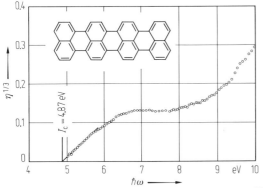

Fig. 147. Quaterrylene. Photoemission quantum yield, $\eta^{1/3}$, vs. photon energy (vapor-deposited film, room temperature). The ionization threshold energy, I_c, has been extrapolated according to the relation $\eta^{1/3} \propto (\hbar\omega - I_c)$ [73H3] (modified).

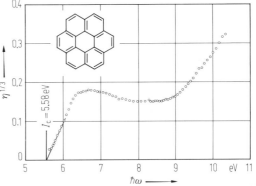

Fig. 148. Coronene. Photoemission quantum yield, $\eta^{1/3}$, vs. photon energy (vapor-deposited film, room temperature). The ionization threshold energy, I_c, has been extrapolated according to the relation $\eta^{1/3} \propto (\hbar\omega - I_c)$ [73H3] (modified).

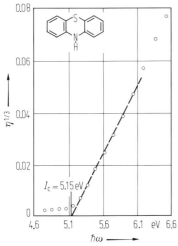

Fig. 149. Phenothiazine. Photoemission quantum yield, $\eta^{1/3}$, vs. photon energy (single crystal, room temperature). The ionization threshold energy, I_c, has been extrapolated according to the relation $\eta^{1/3} \propto (\hbar\omega - I_c)$ [80A3].

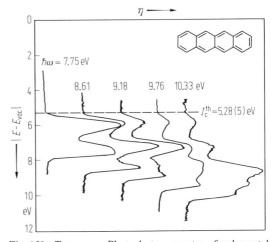

Fig. 150. Tetracene. Photoelectron spectra of polycrystalline layers, room temperature; the kinetic energy distributions of electrons photoemitted by light of the indicated photon energies are plotted rearranged in an electron binding energy scale, relative to vacuum level [73H2].

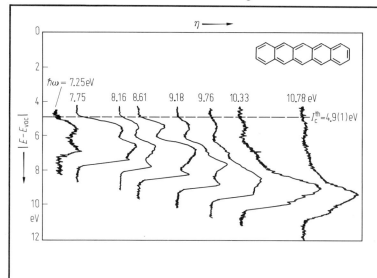

Fig. 151. Pentacene. Photoelectron spectra of polycrystalline layers, room temperature; the kinetic energy distributions of electrons photoemitted by light of the indicated photon energies are plotted rearranged in an electron binding energy scale, relative to vacuum level [73H3].

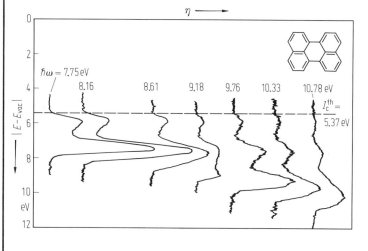

Fig. 152. Perylene. Photoelectron spectra of polycrystalline layers, room temperature; the kinetic energy distributions of electrons photoemitted by light of the indicated photon energies are plotted rearranged in an electron binding energy scale, relative to vacuum level [73H3].

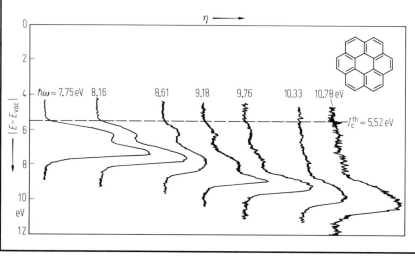

Fig. 153. Coronene. Photoelectron spectra of polycrystalline layers, room temperature; the kinetic energy distributions of electrons photoemitted by light of the indicated photon energies are plotted rearranged in an electron binding energy scale, relative to vacuum level [73H3].

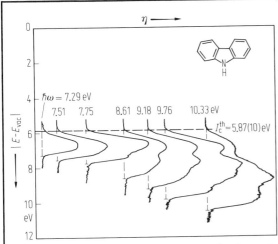

Fig. 154. Carbazole. Photoelectron spectra of polycrystalline layers, room temperature; the kinetic energy distributions of electrons photoemitted by light of the indicated photon energies are plotted rearranged in an electron binding energy scale, relative to vacuum level [82K2].

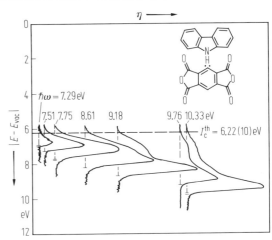

Fig. 155. Carbazole:PMDA. Photoelectron spectra of polycrystalline layers, room temperature; the kinetic energy distributions of electrons photoemitted by light of the indicated photon energies are plotted rearranged in an electron binding energy scale, relative to vacuum level [82K2].

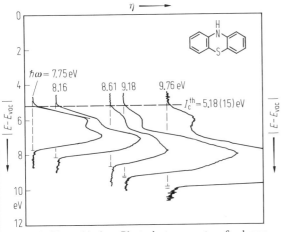

Fig. 156. Phenothiazine. Photoelectron spectra of polycrystalline layers, room temperature; the kinetic energy distributions of electrons photoemitted by light of the indicated photon energies are plotted rearranged in an electron binding energy scale, relative to vacuum level [82K2].

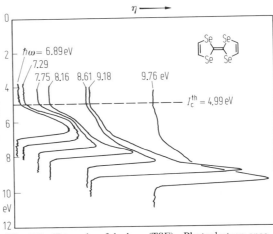

Fig. 157. Tetraselenafulvalene (TSF). Photoelectron spectra of polycrystalline layers, room temperature; the kinetic energy distributions of electrons photoemitted by light of the indicated photon energies are plotted rearranged in an electron binding energy scale, relative to vacuum level [82S2].

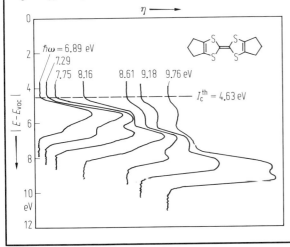

Fig. 158. Hexamethylenetetrathiafulvalene (HMTTF). Photoelectron spectra of polycrystalline layers, room temperature; the kinetic energy distributions of electrons photoemitted by light of the indicated photon energies are plotted rearranged in an electron binding energy scale, relative to vacuum level [83S2].

For Fig. 159, see next page.

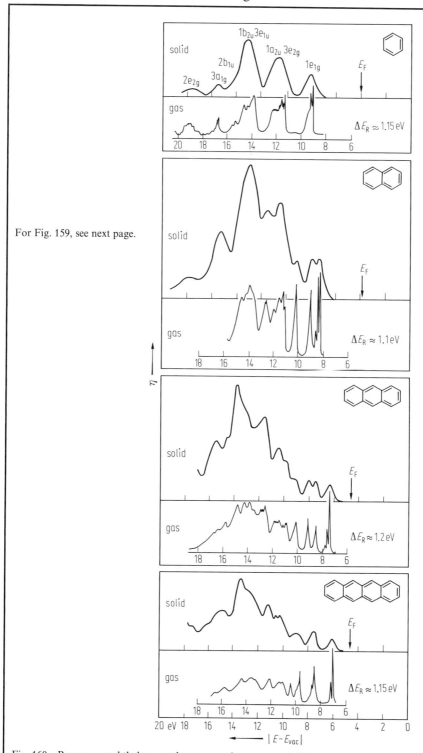

Fig. 160. Benzene, naphthalene, anthracene, and tetracene. Solid state and gas phase photoelectron spectra (quantum yield, arbitrary units, vs. energy). The original spectra have been rearranged in electron binding energy scales, and the gas phase (weak) curves have been shifted relative to the solid state spectra (bold lines), as indicated, by an increment ΔE_R (relaxation shift), in such a way that coincidence of most of the prominent peaks was obtained; [79G1].

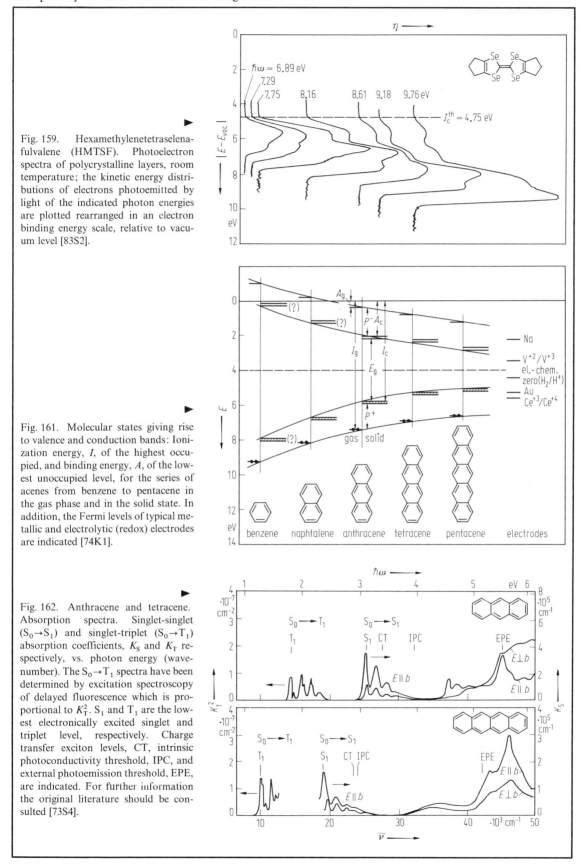

Fig. 159. Hexamethylenetetraselena-fulvalene (HMTSF). Photoelectron spectra of polycrystalline layers, room temperature; the kinetic energy distributions of electrons photoemitted by light of the indicated photon energies are plotted rearranged in an electron binding energy scale, relative to vacuum level [83S2].

Fig. 161. Molecular states giving rise to valence and conduction bands: Ionization energy, I, of the highest occupied, and binding energy, A, of the lowest unoccupied level, for the series of acenes from benzene to pentacene in the gas phase and in the solid state. In addition, the Fermi levels of typical metallic and electrolytic (redox) electrodes are indicated [74K1].

Fig. 162. Anthracene and tetracene. Absorption spectra. Singlet-singlet $(S_0 \rightarrow S_1)$ and singlet-triplet $(S_0 \rightarrow T_1)$ absorption coefficients, K_S and K_T respectively, vs. photon energy (wavenumber). The $S_0 \rightarrow T_1$ spectra have been determined by excitation spectroscopy of delayed fluorescence which is proportional to K_T^2. S_1 and T_1 are the lowest electronically excited singlet and triplet level, respectively. Charge transfer exciton levels, CT, intrinsic photoconductivity threshold, IPC, and external photoemission threshold, EPE, are indicated. For further information the original literature should be consulted [73S4].

References for 12

33R   Robertson, J.M.: Proc. Roy. Soc. (London) **141 A** (1933) 594; **142 A** (1933) 659.

36R   Robertson, J.M.: J. Chem. Soc. **1936** (1936) 1195.

41W1   Wood, R.G., Williams, G.: Philos. Mag. **31** (1941) 115.

41W2   Wood, R.G., McCale, C.H., Williams, G.: Philos. Mag. **31** (1941) 71.

48O   Obreimov, I.W., Prichotkov, A.V., Rodnikova, I.W.: Zh. Eksp. Teor. Fiz. **18** (1948) 409.

49A   Abrams, S.C., Robertson, J.M., White, J.G.: Acta Crystallogr. **2** (1949) 233.

49B   Bridgman, P.W.: Proc. Am. Acad. Arts Sci. **76** (1949) 20.

50E   Eitchis, A.Yu.: Zh. Eksp. Teor. Fiz. **20** (1950) 471.

50M   Mathieson, A.McL., Robertson, J.M., Sinclair, V.C.: Acta Crystallogr. **3** (1950) 245.

50S   Sinclair, V.C., Robertson, J.M., Mathieson, A.McL.: Acta Crystallogr. **3** (1950) 251.

53B   Bradley, R.S., Cleasby, T.G.: J. Chem. Soc. **1953** (1953) 1690.

53K   Kozhin, V.M., Kitaigorodskii, A.I.: Zh. Fiz. Khim. **27** (1953) 1676.

54W   Winchell, A.N.: "The Optical Properties of Organic Compounds"; New York: Academic Press, **1954**.

55H   Herbstein, F.H., Schmidt, G.M.J.: Acta Crystallogr. **8** (1955) 399, 406.

55K   Kozhin, V.M., Kitaigorodskii, A.I.: Zh. Fiz. Khim. **29** (1955) 2074.

56B   Porter, M.W., Spiller, R.C. (eds.): "The Barker Index of Crystals II, 3"; Cambridge: Heffer a. Sons, **1956**.

56H   Hirshfeld, F.L., Schmidt, G.M.J.: Acta Crystallogr. **9** (1956) 233.

56K   Krishna Murti, G.S.R.: Indian J. Phys. **30** (1956) 537.

56S   Sidman, J.M.: J. Chem. Phys. **25** (1956) 122.

57H   Hirshfeld, K.F.L., Schmidt, G.M.J.: J. Chem. Phys. **26** (1957) 923.

57N   Nye, J.F.: "Physical Properties of Crystals, Their Representation by Tensors and Matrices"; Oxford: Clarendon Press; **1957** and **1972**.

57S   Scully, D.B., Whiffen, D.H.: J. Mol. Spectrosc. **1** (1957) 257.

58C   Cox, E.G., Cruickshank, D.W.J., Smith, J.A.S.: Proc. Roy. Soc. (London) **A 247** (1958) 1.

58W   Wolf, H.C.: Z. Naturforsch. **13a** (1958) 414.

59A   Aalbersberg, W.Ij., Hoijtink, G.J., Mackor, E.L., Weijland, W.P.: J. Chem. Soc. (London) **1959** (1959) 3049, 3055.

59L   Liang, T.C., Struchkov, Y.I.: Izv. Akad. Nauk SSSR, Otd. Khim. Nauk **12** (1959) 2095.

59M   Mitra, M.M., Bernstein, H.J.: Can. J. Chem. **37** (1959) 553.

59P   Prikhot'ko, A.F., Fugol, I.Ya.: Opt. Spectrosc. (USSR) **7** (1959) 35; Engl. transl. **7** (1959) 19.

59T   Trotter, J.: Acta Crystallogr. **12** (1959) 54.

60B   Bekoe, D.A., Trueblood, K.N.: Z. Kristallogr. **113** (1960) 1.

60F   Freeman, D.E., Ross, I.G.: Spectrochim. Acta **16** (1960) 1393.

60H   Hoijtink, K.J., Zandstra, P.J.: Mol. Phys. **3** (1960) 371.

60K   Kepler, R.G.: Phys. Rev. **119** (1960) 1226.

60L1   LeBlanc, O.H., Jr.: J. Chem. Phys. **33** (1960) 626.

60L2   Liang, T.C., Struchkov, Y.T.: Izv. Akad. Nauk SSSR, Otd. Khim. Nauk **1960** (1960) 1010.

60T   Teslenko, V.F.: Sov. Phys. Crystallogr. **12** (1960) 946.

60V   Vilesov, F.I.: Dokl. Akad. Nauk SSSR **132** (1960) 632.

61I   Inokuchi, H., Akamatu, H.: Solid State Phys. **12** (1961) 93.

61R   Robertson, J.M., Sinclair, V.C., Trotter, J.: Acta Crystallogr. **14** (1961) 697.

61T   Trotter, J.: Can. J. Chem. **39** (1961) 1574.

62B   Brothy, J.J., Buttrey, J.W. (eds.): "Organic Semiconductors", New York: The Macmillan Comp. **1962**.

62C   Campbell, R.B., Robertson, J.M., Trotter, J.: Acta Crystallogr. **15** (1962) 289.

62H   Hochstrasser, R.M.: J. Chem. Phys. **36** (1962) 1808.

62J   Jeljashevich, M.J.: "Atomic and Molecular Spectroscopy", (in Russian); Izd. Fiz. Mat. Lit., Moscow **1962**.

62K   Kepler, R.G.: in Brothy, J.J., Buttrey, J.W. (eds.): "Organic Semiconductors", New York: The Macmillan Comp., **1962**, p. 1.

62L   LeBlanc, O.H.: J. Chem. Phys. **37** (1962) 916.

62N   Nakada, I.: J. Phys. Soc. Jpn. **17** (1962) 113.

62P   Pope, M., Kallmann, H.P., Magnante, P.: J. Chem. Phys. **38** (1962) 2042.

63A	Aleksandrov, K.S., Belikova, G.S., Ryznenkov, A.P., Teslenko, V.R., Kitaigorodskii, A.I.: Sov. Phys. Crystallogr. **8** (1963) 164.
63B	Buschow, K.H.J., Dieleman, J., Hoijtink, G.J.: Mol. Phys. **7** (1963) 1.
63H	Hoesterey, D.C., Letson, G.M.: J. Phys. Chem. Solids **24** (1963) 1609.
63L	Lyons, L.E. in Fox, D., Labes, M.M., Weissberger, A. (eds.): "Physics and Chemistry of the Organic Solid State" Vol. I; New York: Interscience, **1963**, p. 745.
63T	Tanaka, J.: Bull. Chem. Soc. Jpn. **36** (1963) 1237.
64C1	Craig, P., Wolf, H.C.: J. Chem. Phys. **40** (1964) 2057.
64C2	Camerman, A., Trotter, J.: Proc. Roy. Soc. (London) **279 A** (1964) 129.
64H1	Hoesterey, D.C., Letson, G.M.: J. Chem. Phys. **41** (1964) 675.
64H2	Heseltine, J.C.W., Elliott, W., Wilson, O.B. Jr.: J. Chem. Phys. **40** (1964) 2584.
64K1	Kearns, D.R.: Adv. Chem. Phys. **8** (1964) 282.
64K2	Kepler, R.G. in Bak, T.A. (ed.): "Phonons and Phonon Interactions"; New York: Benjamin Inc., **1964**, p. 578.
64K3	Krainov, E.P.: Opt. Spectrosc. **16** (1964) 532.
64M	Mason, R.: Acta. Crystallogr. **17** (1964) 547.
64O	Okamoto, Y., Brenner, W.: "Organic Semiconductors", New York: Reinhold Publ. Comp. **1964**.
64S	Sponer, H., Kanda, Y.: J. Chem. Phys. **40** (1964) 778.
64V1	Vilesov, F.I., Zagrubskii, A.A., Garbuzov, D.Z.: Sov. Phys. Solid State **5** (1964) 1460.
64V2	Vilesov, F.I.: Sov. Phys. Usp. **6** (1964) 888.
65A1	Ambrosino, F., Califano, S.: Spectrochim. Acta **21** (1965) 1401.
65A2	Assour, J.M., Khan, W.K.: J. Chem. Soc. **87** (1965) 207.
65B1	Bogus, C.: Z. Physik **184** (1965) 219.
65B2	Bepler, W.: Z. Physik **185** (1965) 507.
65B3	Buschow, K.H.J., Dieleman, J., Hoijtink, G.J.: J. Chem. Phys. **42** (1965) 1993.
65C	Camerman, A., Trotter, J.: Acta Crystallogr. **18** (1965) 636.
65F	Feil, D., Linck, M.H., McDowell, J.J.H.: Nature **207** (1965) 285.
65K	Kommandeur, J. in Fox, D., Labes, M.M., Weissberger, A. (eds.): "Physics and Chemistry of the Organic Solid State" Vol. 2; New York: Interscience **1965**, p. 1.
65L	Long, R.E., Sparks, R.A., Trueblood, K.N.: Acta Crystallogr. **18** (1965) 932.
65M1	Mehl, W., Büchner, W.: Z. Phys. Chem. **47** (1965) 76.
65M2	Mecke, R., Langenbucher, F.: "Infrared Spectra of Selected Chemical Compounds"; London: Heyden, **1965**.
65P	Pope, M., Burgos, J., Giachino, J.: J. Chem. Phys. **43** (1965) 3367.
66C1	Chaiken, R.F., Kearns, D.R.: J. Chem. Phys. **45** (1966) 3966.
66C2	Cooper, A.R., Crowne, G.W.P., Farell, P.G.: Trans. Faraday Soc. **62** (1966) 18.
66G	Geacintov, N., Pope, M., Kallmann, H.: J. Chem. Phys. **45** (1966) 2639.
66H	Helfrich, W., Schneider, W.G.: J. Chem. Phys. **44** (1966) 2902.
66K1	Kepler, R.G., Coppage, F.N.: Phys. Rev. **151** (1966) 610.
66K2	Kuwano, H., Kondo, M.: Bull. Chem. Soc. Jpn. **39** (1966) 2779.
66T	Toptschijew, A.W. (Rexer, E., German (ed.)): "Organische Halbleiter"; Berlin: Akademieverlag, **1966**.
66W	Wyckoff, R.W.G.: "Crystal Structures" Vol. 5; New York: Interscience Publ. 2nd ed., **1966**.
67A	Afanaseva, K.S., Aleksandrov, K.S., Kitaigorodskii, A.I.: Phys. Status Solidi **24** (1967) K61.
67B1	Bourdon, J., Schnuriger, B., in Fox, D., Labes, M.M., Weissberger, A. (eds.): "Physics and Chemistry of the Organic Solid State" Vol 3; New York: Interscience Publ., **1967**, p. 60.
67B2	Bogus, C.: Z. Phys. **207** (1967) 281.
67C1	Castro, G., Hochstrasser, R.M.: J. Chem. Phys. **47** (1967) 2241.
67C2	Clarke, R.H., Hochstrasser, R.M.: J. Chem. Phys. **47** (1967) 1915.
67E	El-Kareh, T.B., Wolf, H.C.: Z. Naturforsch. **22a** (1967) 1242.
67G	Gutman F., Lyons, L.E.: "Organic Semiconductors"; New York: J. Wiley, **1967**.
67H1	Helfrich, W., in Fox, D., Labes, M.M., Weissberger, A. (eds.): "Physics and Chemistry of the Organic Solid State" Vol. 3; New York: Interscience Publ. **1967**, p. 2.
67H2	Holzman, P., Morris, R., Jarnagin, R.C., Silver, M.: Phys. Rev. Lett. **19** (1967) 506.
67K	Kajiwara, T., Inokuchi, H., Minomura, S.: Bull. Chem. Soc. Jpn. **40** (1967) 1055.
67L	Lowitz, D.A.: J. Chem. Phys. **46** (1967) 4698.
67M	Mehl, W., Funk, B.: Phys. Lett. **25 A** (1967) 364.
67P	Pope, M.: Sci. Am., Jan. **1967**, 86.

67R	Rush, J.J.: J. Chem. Phys. **47** (1967) 3936.
67S1	Schwartz, L.M., Ingersoll, H.G. Jr., Hornig, J.F.: Mol. Cryst. **2** (1967) 379.
67S2	Silver, M., Sharma, R.: J. Chem. Phys. **46** (1967) 692.
67S3	Schmillen, A., Legler, R.: "Luminescence of Organic Substances"; Landolt-Börnstein, New Series, Vol. II/3, Berlin, Heidelberg, New York: Springer, **1967**.
67W1	Williams, R., Dresner, J.: J. Chem. Phys. **46** (1967) 2133.
67W2	Weulersse, P.: C. R. Acad. Sci. (Paris) **B264** (1967) 327.
67Z	Zschokke-Gränacher, I., Schwob, H.P., Baldinger, E.: Solid State Commun. **5** (1967) 825.
68A	Avakian, P., Merrifield, R.E.: Mol. Cryst. **5** (1968) 37.
68B	Braun, C.L.: Phys. Rev. Lett. **21** (1968) 215.
68C1	Chaiken, R.F., Kearns, D.R.: J. Chem. Phys. **49** (1968) 2846.
68C2	Chafik, A., Mecke, R.: Z. Naturforsch. **23a** (1968) 716.
68C3	Clarke, R.H., Hochstrasser, R.M.: J. Chem. Phys. **48** (1968) 1745.
68D	Danno, T., Inokuchi, H.: Bull. Chem. Soc. Jpn. **41** (1968) 1783.
68F	Fourny, J., Delacote, G., Schott, M.: Phys. Rev. Lett. **21** (1968) 1085.
68G	Gerischer, H., Michel-Beyerle, M.E., Rebentrost, F., Tributsch, H.: Electrochim. Acta **13** (1968) 1509.
68K	Katon, J.E. (ed.): "Organic Semiconducting Polymers"; New York: Marcel Dekker, **1968**.
68O	Oyama, K., Nakada, I.: J. Phys. Soc. Jpn. **24** (1968) 792.
68R	Ryzhenkov, A.P., Kozhin, V.M.: Sov. Phys. Crystallogr. **12** (1968) 943.
68S1	Strome, G.: Phys. Rev. Lett **20** (1968) 3.
68S2	Schechtman, B.H.: Ph.D. Thesis, Stanford Univ. (1968).
68S3	Suzuki, M., Yokoyama, T., Ito, M.: Spectrochim. Acta **24a** (1968) 1091.
68T	Tanaka, J., Shibata, M.: Bull. Chem. Soc. Jpn. **41** (1968) 34.
68V	Vaubel, G., Baessler, H.: Phys. Lett. **27A** (1968) 328.
69A	Afanaseva, G.K.: Sov. Phys. Crystallogr. **13** (1969) 892.
69B1	Bree, A., Kydd, R.A.: Chem. Phys. Lett. **3** (1969) 357.
69B2	Bree, A., Kydd, R.A.: J. Chem. Phys. **51** (1969) 989.
69C1	Clarke, R.H., Hochstrasser, R.M.: J. Mol. Spectrosc. **32** (1969) 309.
69C2	Castro, G., Robinson, G.W.: J. Chem. Phys. **50** (1969) 1159.
69D1	Dresner, J.: RCA Rev. **30** (1969) 322.
69D2	Dewar, M.J.S., Worley, S.D.: J. Chem. Phys. **51** (1969) 263.
69D3	Durocher, G., Williams, D.F.: J. Chem. Phys. **51** (1969) 5405.
69F1	Foster, R.: "Organic Charge–Transfer-Complexes"; Academic Press, **1969**.
69F2	Fourny, J., Delacôte, G.: J. Chem. Phys. **50** (1969) 1028.
69G	Geacintov, N., Pope, M.: J. Chem. Phys. **50** (1969) 814.
69H1	Huntington, H.B., Gangoli, S.G., Mills, J.L.: J. Chem. Phys. **50** (1969) 3844.
69H2	Hadni, A., Wyncke, B., Morlot, G., Gerbaux, X.: J. Chem. Phys. **51** (1969) 3514.
69H3	Hanson, D., Gee, A.R.: J. Chem. Phys. **51** (1969) 5052.
69H4	Hanson, D.: J. Chem. Phys. **51** (1969) 5063.
69H5	Hoskins, B.F., Mason, S.A., White, J.C.B.: Chem. Commun. **1969** (1969) 554.
69K	Korn, A.I., Arndt, R.A., Damask, A.C.: Phys. Rev. **186** (1969) 938.
69L1	Lipsett, F.R.: "Energy Transfer in Polyacene Solid Solutions VIII, a Bibliography for 1968", Mol. Cryst. Liq. Cryst. **6** (1969) 175; see also references for 1956···1967, given in the introduction.
69L2	Lohmann, F., Mehl, W.: J. Chem. Phys. **50** (1969) 500.
69P	Pott, G.T., Williams, D.F.: J. Chem. Phys. **51** (1969) 1901.
69S1	Smith, G.C.: Phys. Rev. **185** (1969) 1133.
69S2	Schmillen, A., Falter, W.W.: Z. Phys. **218** (1969) 401.
69V	Vaubel, G.: Phys. Status Solidi **35** (1969) K67.
70B1	Birks, J.B.: "Photophysics of Aromatic Molecules"; London: Wiley-Interscience, **1970**.
70B2	Boguslavskii, L.I., Vannikov, A.V.: "Organic Semiconductors and Biopolymers"; New York: Plenum Press, **1970**.
70B3	Braun, C.L., Dobbs, G.M.: J. Chem. Phys. **53** (1970) 2718.
70C1	Caywood, J.M.: Mol. Cryst. Liq. Cryst. **12** (1970) 1.
70C2	Clark, L.B., Philpott, M.R.: J. Chem. Phys. **53** (1970) 3790.
70C3	Chandhuri, M.K., Gauguly, S.C.: J. Phys. **C3** (1970) 1791.
70D1	Dresner, J., Goodman, A.M.: Proc. IEEE **1970** (1970) 1868.

70D2	Dresner, J.: J. Chem. Phys. **52** (1970) 6343.
70D3	Distler, D., Hohlneicher, G.: Ber. Bunsenges. Phys. Chem. **74** (1970) 960.
70H	Hirth, H.: Dissertation, Univ. Karlsruhe, **1970**.
70K1	Kochi, M., Harada, Y., Hirooka, T., Inokuchi, H.: Bull. Chem. Soc. Jpn. **43** (1970) 2690.
70K2	Krupnyi, A.I., Aleksandrov, K.S., Belikova, G.S.: Kristallografiya **15** (1970) 589; Sov. Phys. Crystallogr. (English. Transl.) **15** (1970) 507.
70M1	Marchetti, A.P., Kearns, D.R.: Mol. Cryst. Liq. Cryst. **6** (1970) 299.
70M2	Munn, R.W., Nicholson, J.R., Siebrand, W., Williams, D.F.: J. Chem. Phys. **52** (1970) 6442.
70S1	Sharp, J.H., Smith, M.: "Organic Semiconductors" in Eyring, H., Henderson, D., Jost, W. (eds.): "Physical Chemistry"; Academic Press **1970**, p. 435.
70S2	Schadt, M., Williams, D.F.: Phys. Status Solidi **39** (1970) 223.
70S3	Schaffrin, R.M., Trotter, J.: J. Chem. Soc. A **1970** (1970) 1561.
70U	Usov, N.N., Benderskii, V.A.: Phys. Status Solidi **37** (1970) 535.
70W	Williams, D.F., Schadt, M.: J. Chem. Phys. **53** (1970) 3480.
71B1	Bart, J.C.J., Schmidt, G.M.J.: Isr. J. Chem. **9** (1971) 429.
71B2	Bree, A., Zwarich, R.: Spectrochim. Acta **27 A** (1971) 599.
71B3	Bree, A., Zwarich, R.: Spectrochim. Acta **27 A** (1971) 621.
71B4	Bree, A., Kydd, R.A., Misra, T.N., Vilkos, V.V.B.: Spectrochim. Acta **27 A** (1971) 2315.
71C	Castro, G.: IBM J. Res. Dev. **15** (1971) 27.
71K1	Karl, N., Rohrbacher, H., Siebert, D.: Phys. Status Solidi **a 4** (1971) 105.
71K2	Krupnyi, A.I., Al'chinkov, V.V., Aleksandrov, K.S.: Kristallografiya **16** (1971) 801; Sov. Phys. Crystallogr. (English Transl.) **16** (1972) 692.
71L1	Little, R.G., Pautler, D., Coppens, P.: Acta Crystallogr. **B27** (1971) 1493.
71L2	Landolt-Börnstein, New Series, Vol. III/5; Berlin, Heidelberg, New York: Springer, **1971**.
71M	Mikami, N.: J. Mol. Spectrosc. **37** (1971) 147.
71P1	Pope, M., Kallmann, H.: in "Electrical Conduction in Organic Solids", Discuss. Faraday Soc. **51** (1971) 7.
71P2	Pethig, R., Morgan, K.: Phys. Status Solidi **b 43** (1971) K119.
71S	Spielberg, D.H., Korn, A.I., Damask, A.C.: Phys. Rev. **B3** (1971) 2012.
71T	Tomkiewicz, Y., Groff, R.P., Avakian, P.: J. Chem. Phys. **54** (1971) 4504.
71V	Vegter, J.G., Kuindersma, P.I., Kommandeur, J.: in "Conduction in Low Mobility Materials", Klein, N., Tanner, D.S., Pollak, M., (eds.), London: Taylor & Francis, **1971**.
71W	Wyckoff, R.W.G.: "Crystal Structures", 2nd ed., Vol. 6, Part 2, New York: J. Wiley, **1971**.
72B1	Boschi, R., Murell, J.N., Schmidt, W.: Faraday Discuss. Chem. Soc **54** (1972) 116.
72B2	Brundle, C.R., Robin, M.B., Kuebler, N.A.: J. Am. Chem. Soc. **94** (1972) 1466.
72C1	Clark, P.A., Brogli, F., Heilbronner, E.: Helv. Chim. Acta **55** (1972) 1415.
72C2	Campos, M.: Mol. Cryst. Liq. Cryst. **18** (1972) 105.
72D	Durnick, T.J., Wait, S.C., Jr.: Mol. Spectrosc. **42** (1972) 211.
72H1	Hirth, H.: Dissertation, Univ. Karlsruhe, **1970**; see also: Hirth, H., Stöckmann, F.: Phys. Status Solidi **b 51** (1972) 691.
72H2	Hazell, A.C., Larsen, F.K., Lehmann, M.S.: Acta Crystallogr. **B28** (1972) 2977.
72K1	Katchkurova, I.Ya.: Dokl. Akad. Nauk SSSR **206** (1972) 568.
72K2	Karl, N.: Deutsche Phys. Gesellschaft, Spring Meeting Freudenstadt, **1972**, HL 16.
72L	Lehmann, M.S., Pawley, G.S.: Acta Chem. Scand. **26** (1972) 1996.
72M	Maier, J.P., Turner, D.W.: Faraday Discuss. Chem. Soc. **54** (1972) 149.
70O	Ostertag, R.: Dissertation, Univ. Stuttgart, **1972**.
72P	Prasad, P.N., Kopelman, R.: J. Chem. Phys. **57** (1972) 863.
72S1	Shchegolev, I.F.: Phys. Status Solidi **a 12** (1972) 4.
72S2	Sanquer, M., Meinnel, J.: C.R. Acad. Sci. (Paris) **274** (1972) 1241.
72T	Tsuji, K., Yamada, H.: J. Phys. Chem. **76** (1972) 260.
73A	Alexandrov, I.V., Bobovich, Ya.S., Maslov, V.G., Sidorov, A.N.: Opt. Spectrosc. **35** (1973) 264.
73B	Baessler, H., Killesreiter, H.: Mol. Cryst. Liq. Cryst. **24** (1973) 21.
73C1	Chance, R.R., Braun, C.L.: J. Chem. Phys. **59** (1973) 2269.
73C2	Chaikin, P.M., Kwak, J.F., Jones, T.E., Garito, A.F., Heeger, A.J.: Phys. Rev. Lett. **31** (1973) 601.
73D	Dows, D.A., Hsu, L., Mitra, S.S., Brafman, O., Hayek, M., Daniels, W.B., Crawford, R.K.: Chem. Phys. Lett. **22** (1973) 595.
73F	Ferraris, J., Cowan, D.O., Walatka, V., Jr., Perlstein, J.H.: J. Am. Chem. Soc. **91** (1973) 948.

73E   Ern, V., Bouchriha, H., Bisceglia, M., Arnold, S., Schott, M.: Phys. Rev. **B8** (1973) 6038.

73G   Gleiter, R., Schmidt, E., Cowan, D.O., Ferraris, J.P.: J. Electron Spectrosc. Relat. Phenom. **2** (1973) 207.

73H1   Hanson, D.M.: CRC Crit. Rev. Solid State Sci. **3** (1973) 243.

73H2   Hirooka, T., Tanaka, K., Kuchitsu, K., Fujihira, M., Inokuchi, H., Harada, Y.: Chem. Phys. Lett. **18** (1973) 390.

73H3   Hirooka, T.: Thesis, Univ. Tokyo, **1973**.

73H4   Haarer, D., Karl, N.: Chem. Phys. Lett. **21** (1973) 49.

73H5   Hochstrasser, R., McAlpine, R.D., Whiteman, J.D.: J. Chem. Phys. **58** (1973) 5078.

73K   Khan, Z.H., Khanna, B.N.: J. Chem. Phys. **59** (1973) 3015.

73L   Livingstone, R.C., Grant, D.M., Pugmire, R.J., Strong, K.A., Brugger, R.M.: J. Chem. Phys. **58** (1973) 1438.

73M1   Mey, W., Sonnostine, T.J., Morel, D.L., Hermann, A.M.: J. Chem. Phys. **58** (1973) 2542.

73M2   Malaspina, M., Gigli, R., Bardi, G.: J. Chem. Phys. **59** (1973) 387.

73M3   Munn, R.W., Williams, D.F.: J. Chem. Phys. **59** (1973) 1742.

73M4   Munn, R.W., Nicholson, J.R., Schwob, H.P., Williams, D.F.: J. Chem. Phys. **58** (1973) 3828.

73P1   Pisanias, M.N., Christophorou, L.G., Carter, J.G., McCorkle, D.L.: J. Chem. Phys. **58** (1973) 2110.

73P2   Prince, E., Schroeder, L.W., Rush, J.J.: Acta Crystallogr. **B29** (1972) 184.

73P3   Peter, L., Vaubel, G.: Chem. Phys. Lett. **18** (1973) 531; **21** (1973) 158.

73P4   Pong, W., Smith, J.A.: J. Appl. Phys. **44** (1973) 174.

73R   Räsänen, J., Stenman, F., Penttinen, E.: Spectrochim. Acta **29a** (1973) 395.

73S1   Soltzberg, L.J., Ash, B.M., McKay, P.C.: Mol. Cryst. Liq. Cryst. **21** (1973) 283.

73S2   Shida, T., Iwata, S.: J. Am. Chem. Soc. **95** (1973) 3473.

73S3   Schott, M., Berrehar, J.: Phys. Status Solidi **b59** (1973) 175.

73S4   Swenberg, E., Geacintov, N.E.: in Birks, J.B. (ed.): "Organic Molecular Photophysics", Vol. I, London: J. Wiley, **1973**, p. 489.

74A   Abbi, S.C., Hanson, D.M.: J. Chem. Phys. **60** (1974) 319.

74B1   Braun, C.L., Chance, R.R.: in Masuda, K., Silver, M. (eds.): "Energy and Charge Transfer"; New York: Plenum Press, **1974**. p. 17.

74B2   Burland, D.M.: Phys. Rev. Lett. **33** (1974) 833.

74B3   Bergman, A., Jortner, J.: Phys. Rev. **B9** (1974) 4560.

74B4   Belkind, I.A., Grechov, V.V.: Phys. Status Solidi **a26** (1974) 377.

74B5   Boschi, R., Clar, E., Schmidt, W.: J. Chem. Phys. **60** (1974) 4406.

74B6   Biber, A., Andre, J.J.: Chem. Phys. **5** (1974) 166.

74C   Cox, G.A., Knight, P.C.: J. Phys. **C7** (1974) 146.

74F1   Finder, C.J., Newton, M.G., Allinger, N.L.: Acta Crystallogr. **B30** (1974) 411.

74F2   Friedman, P.S., Kopelman, R., Prasad, P.N.: Chem. Phys. Lett **24** (1974) 15.

74G1   Glockner, E.: Dissertation, Univ. Stuttgart **1974**.

74G2   Groff, R.P., Suna, A., Merrifield, R.E.: Phys. Rev. Lett. **33** (1974) 418.

74H1   Haarer, D.: Chem. Phys. Lett. **27** (1974) 91.

74H2   Hoytink, G.J.: Chem. Phys. Lett. **26** (1974) 318.

74H3   Heinze, J., Serafimov, O., Zimmermann, H.W.: Ber. Bunsenges. Phys. Chem. **78** (1974) 652.

74H4   Haink, H.J., Adams, J.E., Huber, J.R.: Ber. Bunsenges. Phys. Chem. **78** (1974) 436.

74H5   Hino, S., Inokuchi, H.: Chem. Lett. **1974** (1974) 363.

74K1   Karl, N.: "Organic Semiconductors", Festkörperprobleme/Adv. Solid State Phys. **14** (1974) 261.

74K2   Kepler, R.G.: Phys. Rev. **B9** (1974) 4468.

74K3   Khan, Z.H., Khanna, B.N., Zaidi, Z.H.: Indian J. Pure Appl. Phys. **12** (1974) 66.

74K4   Khan, Z.H., Khanna, B.N.: Can. J. Chem. **52** (1974) 827.

74K5   Kistenmacher, T.J., Philips, T.E., Cowan, D.O.: Acta Crystallogr. **B30** (1974) 763.

74M1   Meier, H.: "Organic Semiconductors, Dark and Photoconductivity of Organic Solids"; Weinheim: Verlag Chemie, **1974**.

74M2   Masuda, K., Silver, M.: "Energy and Charge Transfer in Organic Semiconductors"; New York: Plenum Press, **1974**.

74N   Nielsen, P., Epstein, A.J., Sandman, D.S.: Solid State Commun. **15** (1974) 53.

74P1   Probst, K.-H., Karl, N.: unpubl.; Probst, K.-H.: Dissertation, Univ. Stuttgart, **1974**; cf. [74K1] (where, however, the labelling of the phenazine *a* and *c* axes has been interchanged for better comparison with anthracene).

74P2	Probst, K.-H.: Dissertation, Univ. Stuttgart, **1974**.
74R	Robinson, J.W. (ed.): "Handbook for Spectroscopy"; CRC Press **1974**.
74S1	Soos, Z.G.: "Theory of π-Molecular Charge-Transfer Crystals", Annual Rev. Phys. Chem. **25** (1974) 121.
74S2	Seki, K.: Thesis, Univ. Tokyo, **1974**.
74S3	Schmidberger, R.: Dissertation, Univ. Stuttgart, **1974**.
74S4	Streets, D.G., Williams, T.A.: J. Electron Spectrosc. **3** (1974) 71.
74T1	Takahashi, Ch., Maeda, S.: Chem. Phys. Lett. **24** (1974) 584.
74T2	Tanaka, J., Kishi, T., Tanaka, M.: Bull. Chem. Soc. Jpn. **47** (1974) 2376.
75B1	Berrehar, J., Karl, N., Schott, M.: unpublished, **1975**.
75B2	Bree, A., Edelson, M., Kydd, R.A.: Spectrochim. Acta **A31** (1975) 1569.
75C1	Cooper, C.D., Naff, W.T., Compton, R.N.: J. Chem. Phys. **63** (1975) 2752.
75C2	Chen, E.C.M., Wentworth, W.E.: J. Chem. Phys. **63** (1975) 3183.
75C3	Clar, E., Schmidt, W.: Tetrahedron **31** (1975) 2263.
75D	DiMarco, P., Giro, G.: Mol. Cryst. Liq. Cryst. **29** (1975) 179.
75E	Etemad, S., Penny, T., Engler, E.M., Scott, B.A., Seiden, P.E.: Phys. Rev. Lett. **34** (1975) 741.
75G1	Goodings, E.P.: Endeavour **34** (1975) 123.
75G2	Gleiter, R., Kobayashi, M., Spanget-Larsen, J., Ferraris, J.P., Bloch, A.N., Bechgaard, K., Cowan, D.O.: Ber. Bunsenges. Phys. Chem. **79** (1975) 1219.
75G3	Gonzalez-Basurto, J., Burshtein, Z.: Mol. Cryst. Liq. Cryst. **31** (1975) 211.
75H1	Hush, N.S., Cheung, A.S., Hilton, P.R.: J. Electron Spectrosc. Relat. Phenom. **7** (1975) 385.
75H2	Haink, H.J., Huber, J.R.: Chem. Ber. **108** (1975) 1118.
75H3	Hino, S., Hirooka, T., Inokuchi, H.: Bull. Chem. Soc. Jpn. **48** (1975) 1133.
75H4	Haarer, D.: Chem. Phys. Lett. **31** (1975) 192.
75H5	Hino, S., Seki, K., Inokuchi, H.: Chem. Phys. Lett. **36** (1975) 335.
75K1	Karl, N.: "Halbleitereigenschaften des organischen Molekülkristalls Anthracen", Habilitationsschrift, Univ. Stuttgart, **1975**.
75K2	Keller, H.H. (ed.): "Low-Dimensional Cooperative Phenomena"; New York: Plenum Press, **1975**.
75K3	Kachkurova, I.Ya.: Zh. Prikl. Spectrosk. **22** (1975) 689.
75K4	Karl, N., Ziegler, J.: Chem. Phys. Lett. **32** (1975) 438.
75K5	Karl, N.: unpublished; presented in: Solid Organic State Letters **10** (1976).
75L	LaPlaca, S.J., Corfield, P.W.R., Thomas, R., Scott, B.A.: Solid State Commun. **17** (1975) 635.
75M	Maier, J.P., Muller, J.-F., Kubota, T., Yamakawa, M.: Helv. Chim. Acta **58** (1975) 1641.
75N1	Nenner, I., Schulz, G.J.: J. Chem. Phys. **62** (1975) 1747.
75N2	Nicol, M., Vernon, M., Woo, J.T.: J. Chem. Phys. **63** (1975) 1992.
75O1	Onodera, A., Kawai, N., Kobayashi, T.: Solid State Commun. **17** (1975) 775.
75O2	Oberland, S., Schmidt, W.: J. Am. Chem. Soc. **97** (1975) 6633.
75P	Probst, K.-H., Karl, N.: Phys. Status Solidi **a27** (1975) 499.
75S1	Schuster, H.G. (ed.): "One-Dimensional Conductors"; Berlin, Heidelberg, New York: Springer **1975**.
75S2	Soos, Z.G., Klein, D.J.: "Charge Transfer in Solid State Complexes" in Foster, R. (ed.): "Molecular Association", Vol. 1; Academic Press, **1975**, p. 1.
75W1	Wolf, H.C., Port, H.: in "Molecular Spectroscopy of Dense Phases, Proceedings 12th Europ. Congr. Mol. Spectrosc. Strasbourg 1975; Amsterdam: Elsevier Publ. Comp., **1976**, p. 31.
75W2	Warmack, R.J., Callcott, T.A., Watson, C.R.: Phys. Rev. **B12** (1975) 3336.
76B1	Berrehar, J., Delannoy, P., Schott, M.: Phys. Status Solidi **b77** (1976) K119.
76B2	Baudur, J.L., Delugeard, Y., Cailleau, H.: Acta Crystallogr. **B32** (1976) 150.
76C	Cohen, M.J., Coleman, L.B., Garito, A.F., Heeger, A.J.: Phys. Rev. **B13** (1976) 5111.
76F	Ferraris, J.P., Finnegan, T.F.: Solid State Commun. **18** (1976) 1169.
76G1	Gill, W.D.: "Polymeric Photoconductors", in Mort, J., Pai, D.M., (eds.): "Photoconductivity and Related Phenomena"; Amsterdam: Elsevier Publ. Comp., **1976**, p. 303.
76G2	Goodings, E.P.: Chem. Soc. Rev. **5** (1976) 95.
76I	Ishii, K., Sakamoto, K., Seki, K., Sato, N., Inokuchi, H.: Chem. Phys. Lett. **41** (1976) 154.
76K1	Kepler, R.G.: "Organic Molecular Crystals: Anthracene", in Hannay, N.B. (ed.): "Treatise on Solid State Chemistry", Vol. 3; New York: Plenum Press, **1976**, p. 615.
76K2	Kurik, M.B. (ed.): "Organic Semiconductors"; Akad. Nauk USSR, Inst. Fiziki, Kijev 1976 (in Russian).
76K3	Karl, N., Feederle, H.: Phys. Status Solidi **a34** (1976) 497.

76K4	Kotani, M.: Chem. Phys. Lett. **43** (1976) 603.
76K5	Kalinowski, J., Godlewski, J., Jankowiak, R.: Chem. Phys. Lett. **43** (1976) 127.
76L	Lipari, N.O., Nielsen, P., Ritsko, J.J., Epstein, A.J., Sandman, D.J.: Phys. Rev. **B14** (1976) 2229.
76M1	Mort, J., Pai, D.M. (eds.): "Photoconductivity and Related Phenomena", especially chapter 5, Inokuchi, H., Maruyama, Y.: "Molecular Crystals"; New York: Elsevier Publ. Comp., **1976**.
76M2	McDowell, J.J.H.: Acta Crystallogr. **B32** (1976) 5.
76M3	Müller, H., Bässler, H.: J. Lumin. **12/13** (1976) 259.
76P1	Price, A.H., Williams, J.O., Munn, R.W.: Chem. Phys. **14** (1976) 413.
76P2	Port, H., Mistelberger, K.: J. Lumin. **12/13** (1976) 351.
76S1	Soos, Z.G., Klein, D.J.: in N.B. Hannay (ed.): "Treatise on Solid State Chemistry", Vol. 3, New York: Plenum Press, **1976**, p. 679.
76S2	Seki, K., Hirooka, T., Kamura, Y., Inokuchi, H.: Bull. Chem. Soc. Jpn. **49** (1976) 904.
76S3	Shibaeva, R.P., Kaminskii, V.F.: Kristallografiya **23** (1976) 499.
76S4	Shirotani, I., Kamura, Y., Inokuchi, H.: Chem. Phys. Lett. **40** (1976) 257.
76T	Tanner, D.B., Jacobsen, C.S., Garito, A.F., Heeger, A.J.: Phys. Rev. **B13** (1976) 3381.
76Z	Zallen, R., Griffiths, C.H., Slade, M.L., Hayek, M., Brafman, O.: Chem. Phys. Lett. **39** (1976) 85.
77B1	Burland, D.M., Konzelmann, U.: J. Chem. Phys. **67** (1977) 319.
77B2	Burshtein, Z., Williams, D.F.: Phys. Rev. **B15** (1977) 5769.
77B3	Burshtein, Z., Williams, D.F.: J. Chem. Phys. **67** (1977) 3592.
77B4	Baudour, J.L., Cailleau, H., Yelon, W.B.: Acta Crystallogr. **B33** (1977) 1773.
77B5	Bree, A., Edelson, M.: Chem. Phys. Lett. **46** (1977) 500.
77C1	Cooper, J.R., Miljak, M., Deplanque, G., Jérome, D., Weger, M., Fabre, J.M., Giral, L.: J. Phys. (Paris) **38** (1977) 1097.
77C2	Conwell, E.M.: Phys. Rev. Lett. **39** (1977) 777.
77C3	Coret, A., Fort, A.: Il Nuovo Cimento **39B** (1977) 544.
77C4	Clar, E., Schmidt, W.: Tetrahedron **33** (1977) 2093.
77D	Domelsmith, L.N., Munchausen, L.L., Houk, K.N.: J. Am. Chem. Soc. **99** (1977) 6506.
77F	Fort, A., Coret, A.: Phys. Status Solidi **b83** (1977) 599.
77H1	Hamann, C.: Krist. Tech. **12** (1977) 651.
77H2	Haarer, D.: J. Chem. Phys. **67** (1977) 4076.
77H3	Haarer, D., Keijzers, C.P., Silbey, R.: J. Chem. Phys. **66** (1977) 563.
77H4	Hino, S., Hirooka, T., Inokuchi, H.: Bull. Chem. Soc. Jpn. **50** (1977) 620.
77H5	Hochstrasser, R.M., Sung, H.N.: J. Chem. Phys. **66** (1977) 3265.
77H6	Hochstrasser, R.M., Sung, H.N.: J. Chem. Phys. **66** (1977) 3276.
77K1	Keller, H.J. (ed.): "Chemistry and Physics of One-Dimensional Metals"; New York: Plenum Press, **1977**.
77K2	Kurik, M.B. (ed.): "Photonics of Organic Semiconductors"; Akad. Nauk USSR, Inst. Fiziki, Kijev, **1977** (in Russian).
77K3	Konno, M., Ishii, T., Saito, Y.: Acta Crystallogr. **B33** (1977) 763.
77L	LeBlanc, O.H., Jr.: "Conductivity", in Fox, D., Labes, M.M., Weissberger, A. (eds.) "Physics and Chemistry of the Organic Solid State", Vol. III, New York: Interscience Publ., **1977**, p. 133.
77M	Mathur, D., Hasted, J.B.: Chem. Phys. Lett. **48** (1977) 50.
77O	Ong, N.P., Portis, A.M.: Phys. Rev. **B15** (1977) 1782.
77P1	Pál, L., Grüner, G., Jánossy, A., Sólyom, J. (eds.): "Organic Conductors and Semiconductors", Siofok Conference, Berlin, Heidelberg, New York: Springer, **1977**.
77P2	Perlstein, J.H.: Angew. Chem. **89** (1977) 534.
77P3	Popovic, Z.D., Sharp, J.H.: J. Chem. Phys. **66** (1977) 5076.
77S1	Silinsh, E.: "Electronic States of Organic Molecular Crystals"; Monograph. Zinatne, Riga 1977 (in Russian); a revised and complemented English version has appeared later; Berlin, Heidelberg, New York: Springer, **1980**; see also [80S4].
77S2	Schein, L.B.: Chem. Phys. Lett. **48** (1977) 571.
77S3	Schmidt, W.: J. Chem. Phys. **66** (1977) 828.
77S4	Syassen, K., Philpott, M.R.: Chem. Phys. Lett. **50** (1977) 14.
77S5	Smith, D.L., Luss, H.R.: Acta Crystallogr. **B33** (1977) 1744.
77S6	Schweig, A., Thon, N., Engler, E.M.: J. Electron Spectrosc. **12** (1977) 335.
77W1	Walker, E.I.P., Marchetti, A.P., Young, R.H.: J. Chem. Phys. **68** (1977) 4134.
77W2	van de Waal, B.W., Feil, D.: Acta Crystallogr. **B33** (1977) 314.

77W3	White, J.W.: in Lovesey, S.W., Springer, T. (eds.): "Dynamics of Solids and Liquids by Neutron Scattering"; Berlin, Heidelberg, New York: Springer, **1977**.
77W4	Wyncke, B., Brehat, F., Hadni, A.: J. Phys. (Paris) **38** (1977) 1171.
78A	Aimé, J.P., Ern, V., Fave, J.L., Schott, M.: Mol. Cryst. Liq. Cryst. **46** (1978) 169.
78B1	Bree, A., Edelson, M.: Chem. Phys. Lett. **55** (1978) 319.
78B2	Bolton, B.A., Prasad, P.N.: Chem. Phys. **35** (1978) 331.
78B3	Burshtein, Z., Williams, D.F.: J. Chem. Phys. **68** (1978) 983.
78B4	Bellows, J.C., Stevens, E.D., Prasad, P.N.: Acta Crystallogr. **B34** (1978) 3256.
78B5	Burshtein, Z., Hanson, A.W., Ingold, C.F., Williams, D.F.: J. Phys. Chem. Solids **39** (1978) 1125.
78C1	Cipollini, N., Braun, C.L., Chance, R.R.: Chem. Phys. Lett. **53** (1978) 404.
78C2	Cooper, C.D., Frey, W.F., Compton, R.N.: J. Chem. Phys. **69** (1978) 2367.
78C3	Clar, E., Schmidt, W.: Tetrahedron **34** (1978) 3219.
78E	Elnahwy, S., ElHamamsy, Y., Damask, A.C., Cox, D.E., Daniels, W.B.: J. Chem. Phys. **68** (1978) 1161.
78G	Girard, A., Cailleau, H., Marqueton, Y., Ecolivet, C.: Chem. Phys. Lett. **54** (1978) 479.
78H	Haman, C. (ed.): "Organische Festkörper und Organische dünne Schichten"; Leipzig: Akad. Verlagsges. Geest u. Portig, **1978**.
78I	Isett, L.C., Perez-Albuerne, E.A.: in Miller, J.S., Epstein, A.J. (eds.): "Synthesis and Properties of Low-Dimensional Materials"; New York: Academy of Science, **1978**, p. 395.
78J	Jordan, K.D., Burrow, P.D.: Acc. Chem. Res. **11** (1978) 341.
78K1	Karl, N., Warta, W.: in "Electrical and Related Properties of Organic Solids", Karpacz Conference; Wroclaw Techn. Univ. **1978**, p. 45.
78K2	Khan, Z.H.: Can. J. Spectrosc. **23** (1978) 8.
78K3	Kai, Y., Hama, F., Yasuoka, N., Kasai, N.: Acta Crystallogr. **B34** (1978) 1263.
78M1	Miller, J.S., Epstein, A.J. (eds.): "Synthesis and Properties of Low-Dimensional Materials", Ann. New York Acad, Sci. **313**; New York: Academy of Science **1978**.
78M2	Matsui, A., Tomioka, K., Tomotika, T.: Solid State Commun. **25** (1978) 237.
78R1	Robertson, B.E., Stezowski, J.J.: Acta Crystallogr. **B34** (1978) 3005.
78R2	Ruščić, B., Kovač, B., Klasink, L., Güsten, H.: Z. Naturforsch. **A33** (1978) 1006.
78R3	Richard, P., Zanghi, J.-C., Guédon, J.-F.: Acta Crystallogr. **B34** (1978) 788.
78S1	Schein, L.B., Nigrey, P.J.: Phys. Rev. **B18** (1978) 2929.
78S2	Shida, T., Nosaka, Y., Kato, T.: J. Phys. Chem. **82** (1978) 695.
78S3	Shida, T.: J. Phys. Chem. **82** (1978) 991.
78S4	Schein, L.B., Duke, C.B., McGhie, A.R.: Phys. Rev. Lett. **40** (1978) 197.
78S5	Shibaeva, R.P., Kaminskii, V.F.: Sov. Phys. Crystallogr. **23** (1978) 277.
78S6	Syassen, K., Philpott, M.R.: J. Chem. Phys. **68** (1978) 4870.
78T1	Toombs, G.A.: "Quasi One Dimensional Conductors", Phys. Rep. **C40** (1978) 181.
78T2	Terauchi, H.: Phys. Rev. **B17** (1978) 2446.
78V	Vovelle, F., Chedin, M.-P., Dumas, G.G.: Mol. Cryst. Liq. Cryst. **48** (1978) 261.
79B1	Barišić, S., Bjeliš, A., Cooper, J.R., Leontić, B. (eds.): "Quasi One Dimensional Conductors I, II", Dubrovnik Conference; Berlin, Heidelberg, New York: Springer, **1979**.
79B2	Blythe, A.R.: "Electrical Properties of Polymers"; Cambridge: Cambridge Univ. Press, **1979**.
79B3	Bellows, J.C., Prasad, P.N.: J. Chem. Phys. **70** (1979) 1864.
79B4	Bozio, R., Zanon, I., Girlando, A., Pecile, C.: J. Chem. Phys. **71** (1979) 2282.
79B5	Berkowitz, J.: J. Chem. Phys. **70** (1979) 2819.
79C1	Cehak, A.: in "Electrical and Related Properties of Organic Solids", Supplement, Karpacz Conference; Wroclaw Techn. Univ. **1979**, and private commun.
79C2	Cox, P.J., Sim, G.A.: Acta Crystallogr. **B35** (1979) 404.
79C3	Cailleau, H., Dworkin, A.: Mol. Cryst. Liq. Cryst. **50** (1979) 217.
79C4	Clar, E., Schmidt, W.: Tetrahedron **35** (1979) 1027.
79C5	Clar, E., Schmidt, W.: Tetrahedron **35** (1979) 2673.
79C6	Cailleau, H., Moussa, F., Mons, J.: Solid State Commun. **31** (1979) 521.
79C7	Comès, R., Shirane, G.: in [79D], p. 17.
79D	Devreese, J.T., Evrard, R.P., van Doren, V.E. (eds.): "Highly Conducting One-Dimensional Solids", New York: Plenum Press, **1979**.
79G1	Grobman, W.D., Koch, E.E.: "Photoemission from Organic Molecular Crystals", Topics Appl. Phys. **27** (1979) 261; Berlin, Heidelberg, New York: Springer.
79G2	Goldacker, W., Schweitzer, D., Zimmermann, H.: Chem. Phys. **36** (1979) 15.

79H	Hatfield, W.E. (ed.): "Molecular Metals", New York: Plenum Press, **1979**.
79I	Ikemoto, I., Sato, Y., Sugano, T., Kosugi, N., Koruda, H., Ishii, K., Sato, N., Seki, K., Inokuchi, H., Takahashi, T., Harada, Y.: Chem. Phys. Lett. **61** (1979) 50.
79J	Jakubowski, B.: Krist. Tech. **14** (1979) 991.
79K1	Kato, T., Shida, T.: J. Am. Chem. Soc. **101** (1979) 6869.
79K2	Kolendritskii, D.D., Kurik, M.V., Piryatinski, Yu.P.: Phys. Status Solidi **b91** (1979) 741.
79K3	Krauss, R., Schulz, H., Nesper, R., Thiemann, K.H.: Acta Crystallogr. **B35** (1979) 1419.
79M1	Mistelberger, K.: Dissertation, Univ. Stuttgart **1979**.
79M2	Mayerle, J.J., Torrance, J.B., Crowley, J.I.: Acta Crystallogr. **B35** (1979) 2988.
79M3	Megtert, S., Pouget, J.P., Comès, R.: in [79H] p. 87.
79P	Port, H., Mistelberger, K., Rund, D.: Mol. Cryst. Liq. Cryst. **50** (1979) 11.
79S1	Schein, L.B., McGhie, A.R.: Chem. Phys. Lett. **62** (1979) 356.
79S2	Schein, L.B., McGhie, A.R.: Phys. Rev. **B20** (1979) 1631.
79S3	Soltzberg, L.J., Armstrong, E.C., Kelley, P.A.: Mol. Cryst. Liq. Cryst. **50** (1979) 179.
79T1	Torrance, J.B., Mayerle, J.J., Lee, V.Y., Bechgaard, K.: J. Am. Chem. Soc. **101** (1979) 4747.
79T2	Torrance, J.B.: Acc. Chem. Res. **12** (1979) 80.
79Z1	Ziegler, J., Karl, N.: Chem. Phys. **40** (1979) 207.
79Z2	Zallen, R., Conwell, E.M.: Solid State Commun. **31** (1979) 557.
80A1	Alcácer, L. (ed.): "The Physics and Chemistry of Low Dimensional Solids"; Dodrecht: Reidel Publ. Comp., **1980**.
80A2	Andres, K., Wudl, F., McWhan, D.B., Thomas, G.A., Nalewajek, D., Stevens, A.L.: Phys. Rev. Lett. **45** (1980) 1449.
80A3	Anthonj, R., Karl, N., Robertson, B.E., Stezowski, J.J.: J. Chem. Phys. **72** (1980) 1244.
80A4	Ahlgren, D.C., Kopelman, R.: Chem. Phys. **48** (1980) 47.
80B1	Braun, C.L.: "Organic Semiconductors", in Keller, S.P. (ed.) "Handbook of Semiconductors", Vol. 3; Amsterdam: North Holland, **1980**.
80B2	Bechgaard, K., Jakobsen, C.S., Mortensen, K., Pedersen, H.J., Thorup, N.: Solid State Commun. **33** (1980) 1119.
80B3	Broude, V.L., Sheka, E.F.,: Mol. Cryst. Liq. Cryst. **57** (1980) 145.
80B4	Bechgaard, K., Andersen, J.R., in: Alcácer, L. (ed.), "The Physics and Chemistry of Low Dimensional Solids", Dodrecht: Reidel Publ. Comp., **1980**.
80C1	Chaikin, P.M., Grüner, G., Engler, E.M., Greene, R.L.: Phys. Rev. Lett. **45** (1980) 1874.
80C2	Chen, M.-C., Cullick, A.S., Gerkin, R.E., Lundstedt, A.P.: Chem. Phys. **46** (1980) 423.
80C3	Conwell, E.: Phys. Rev. **B22** (1980) 1761.
80D	Dumas, G.G., Vovelle, F., Chédin, M.-P.: C.R. Acad. Sci. (Paris) **B290** (1980) 373.
80E	Ecolivet, C., Sanquer, M.: J. Chem. Phys. **72** (1980) 4145.
80G	Greene, R.L., Engler, E.M.: Phys. Rev. Lett. **45** (1980) 1587.
80H	Hochstrasser, R.M., Nyi, C.: J. Chem. Phys. **72** (1980) 2591.
80J	Jérome, D., Mazaud, A., Ribault, M., Bechgaard, K.: J. Phys. (Paris) Lett. **41** (1980) L95.
80K1	Karl, N.: 9th Molecular Crystal Symposium, Mittelberg, **1980**, Conference Proceedings, p. 149.
80K2	Karl, N.: "High Purity Organic Molecular Crystals" in H.C. Freyhardt (ed.), "Crystals", Vol. 4, Berlin Heidelberg, New York: Springer, **1980**, p. 1.
80K3	Kato, K., Braun, C.L.: J. Chem. Phys. **72** (1980) 172.
80K4	Keller, H.J., Nöthe, D., Pritzkow, H., Wehe, D., Werner, M., Koch, P., Schweitzer, D.: Mol. Cryst. Liq. Cryst. **62** (1980) 181.
80M1	Mort, J.: Adv. Phys. **29** (1980) 367.
80M2	Mayer, U.: Dissertation, Univ. Stuttgart, **1980**.
80M3	Mistelberger, K., Port, H.: Mol. Cryst. Liq. Cryst. **57** (1980) 203.
80M4	Morikawa, E., Kotani, M.: Z. Naturforsch. **35a** (1980) 823.
80N1	Nakano, S., Maruyama, Y.: Solid State Commun. **35** (1980) 671.
80N2	Natkaniec, I., Bokhenkov, E.L., Dorner, B., Kalus, J., Mackenzie, G.A., Pawley, G.S., Schmelzer, U., Sheka, E.F.: J. Phys. **C13** (1980) 4265.
80N3	Nakagawa, K., Kotani, M., Tanaka, H.: Phys. Status Solidi **b102** (1980) 403.
80R1	Ribault, M., Benedek, G., Jérome, D., Bechgaard, K.: J. Phys. (Paris) Lett. **41** (1980) L397.
80R2	Ribault, M., Pouget, J.-P., Jérome, D., Bechgaard, K.: J. Phys. (Paris) Lett. **41** (1980) L607.
80R3	Roberts, G.G., Aspley, N., Munn, R.W.: "Temperature-Dependent Electronic Conduction in Semiconductors", Phys. Rep. **60** (1980) 59.
80S1	Schein, L.B., Warta, W., McGhie, A.R., Karl, N.: Chem. Phys. Lett. **75** (1980) 267.

80S2	Samoć, M., Zboiński, Z., Mierzejewski, A.: Chem. Phys. **48** (1980) 209.
80S3	Seki, K.: unpublished, **1980**.
80S4	Silinsh, E.A.: "Organic Molecular Crystals, their Electronic States"; Berlin, Heidelberg, New York: Springer, **1980**.
80S5	Saltiel, J., Khalil, G.-E., Schanze, K.: Chem. Phys. Lett. **70** (1980) 233.
80S6	Saito, G., Inokuchi, H.: "Organic Conductors in Electroorganic Chemistry", Kagaku-Zokan **86** (1980) 45 (in Japanese).
80W1	Walker, E.I.P., Marchetti, A.P., Young, R.H.: J. Chem. Phys. **72** (1980) 3426.
80W2	Wong, W.-K., Westrum, E.F., Jr.: Mol. Cryst. Liq Cryst. **61** (1980) 207.
81B1	Bernasconi, J., Schneider, T. (eds.): "Physics in One Dimension"; Berlin, Heidelberg, New York: Springer, **1981**.
81B2	Bechgaard, K., Carneiro, K., Olsen, M., Rasmussen, F.B., Jakobsen, C.S.: Phys. Rev. Lett. **46** (1981) 852.
81B3	Bouffard, S., Zuppiroli, L.: Mater. Sci. **7** (1981) 101; Techn. Univ. Wroclaw, Poland.
81B4	Batail, P., LaPlaca, S.J., Mayerle, J.J., Torrance, J.B.: J. Am. Chem. Soc. **103** (1981) 951.
81B5	Bässler, H.: Phys. Status Solidi **(b)107** (1981) 9.
81B6	Bieri, G., Åsbrink, L., v. Niessen, W.: J. Electron. Spectrosc. **23** (1981) 281.
81C1	The Internat. Conf. Low Dim. Synth. Metals, Helsingør 1980: Chem. Scripta **17** (1981).
81C2	Clar, E., Robertson, J.M., Schlögl, R., Schmidt, W.: J. Am. Chem. Soc. **103** (1981) 1320.
81C3	Chaplot, S.L., Mierzejewski, A., LeFebvre, J., Pawley, G.S., Luty, T.: J. Phys. (Paris) **42**, Suppl. **C6** (1981) 584.
81D1	Dehlhaes, P., Amiell, J., Manceau, J.P., Keryer, G., Flandrois, S., Fabre, J.M., Giral, L.: C.R. Acad. Sci. (Paris) Ser. II **293** (1981) 347.
81D2	DiMarco, P.G., Giro, G.: Mol. Cryst. Liq. Cryst. **69** (1981) 193.
81D3	Dorner, B., Bokhenkov, E.L., Sheka, E.F., Chaplot, S.L., Pawley, G.S., Kalus, J., Schmelzer, U., Natkaniec, I.: J. Phys. (Paris) Colloque C6, supplement to Vol. **42** (12), **1981**, p. C6–602.
81D4	Duppen, K., Hesp, B.M.M., Wiersma, D.A.: Chem. Phys. Lett. **79** (1981) 399.
81E1	Epstein, A.J., Conwell, E.M. (eds.): Boulder Conf., Mol. Cryst. Liq. Cryst. **77** (1981) and **79, 81, 83, 85, 86** (1982).
81E2	Eilefeld, P., Schmidt, W.: J. Electron Spectrosc. Relat. Phenom. **24** (1981) 101.
81G1	Giermańska, J., Samoć, A., Sworakowski, J., Zboiński, Z.: Mater. Sci. **7** (1981) 153, Tech. Univ. Wroclaw, Poland.
81G2	Gorelov, B.M., Laukhin, V.N., Schegolev, I.F.: Chem. Scr. **17** (1981) 23.
81H	Hamann, C., Heim, J., Burghardt, H.: "Organische Leiter, Halbleiter und Photoleiter"; Braunschweig: Vieweg, **1981**.
81I1	Ivanov, Yu.P., Antipin, M.Yu., Pertsin, A.J., Struchkov, Yu.T.: Mol. Cryst. Liq. Cryst. **71** (1981) 181.
81I2	Ishibashi, Y.: J. Phys. Soc. Jpn. **50** (1981) 1255.
81J1	Jones, A.C., Styrcz, K.J., Elliot, D.A., Williams, J.O.: Phys. Lett. **80** (1981) 413.
81J2	Jacobsen, C.S., Mortensen, K., Thoroup, N., Tanner, D.B., Weger, M., Bechgaard, K.: Chem. Scri. **17** (1981) 103.
81K1	Kao, K.C., Hwang, W.: "Electrical Transport in Solids – with particular reference to organic semiconductors"; Oxford: Pergamon Press, **1981** (2600 references with titles).
81K2	Khanna, S.K., Fuller, W.W., Gruner, G., Chaikin, P.M.: Phys. Rev. **B24** (1981) 2958.
81K3	Karl, N., Port, H., Schrof, W.: Mol. Cryst. Liq. Cryst. **78** (1981) 55.
81K4	Karl, N., Jahnel, J.: 3rd Conference on Electrical and Related Porperties of Organic Solids, (Wdzydze, 1981); cf. Mater. Sci. (Techn. Univ. Wroclaw, Polen) **7** (1981) 193.
81M1	Mayer, U., Auweter, H., Braun, A., Wolf, H.C., Schmid, D.: Chem. Phys. **59** (1981) 449.
81M2	Meresse, A.: Thèse d'Etat; Univ. Bordeaux I, **1981**.
81M3	Mitchell, M.B., Smith, G.R., Guillory, W.A.: J. Chem. Phys. **75** (1981) 44.
81M4	Matsui, A., Iemura, M., Nishimura, H.: J. Lumin. **24/25** (1981) 445.
81N	Natkaniec, I., Bielushkin, A.V., Wasiutynski, T.: Phys. Status Solidi **b105** (1981) 413.
81P1	Pigoń, K., Chojnacki, H.: "Electrical Conductivity of Solid Molecular Complexes", in Ratajczak, H., Orville-Thomas, W.J. (eds.), "Molecular Interactions, Vol. 2, New York: John Wiley, **1981**, p. 451.
81P2	Port, H., Rund, D., Wolf, H.C.: Chem. Phys. **60** (1981) 81.
81S1	Sato, N., Inokuchi, H., Shirotani, I.: Chem. Phys. **60** (1981) 327.
81S2	Sato, N., Seki, K., Inokuchi, H.: J. Chem. Soc., Faraday Trans. II, **77** (1981) 1621.

81S3	Stehle, R., Karl, N.: Presented as a post deadline poster at the 3rd Conference on Electrical and Related Properties of Organic Solids, Wdzydze, Poland, **1981**.

81S3   Stehle, R., Karl, N.: Presented as a post deadline poster at the 3rd Conference on Electrical and Related Properties of Organic Solids, Wdzydze, Poland, **1981**.

81S4   Sato, N., Seki, K., Inokuchi, H.: J. Chem. Soc., Faraday Trans. II, **77** (1981) 47.

81S5   Sebastian, L., Weiser, G., Baessler, H.: Chem. Phys. **61** (1981) 125.

81T1   Tegeler, E., Iwan, M., Koch, E.-E.: J. Electron Spectrosc. Relat. Phenom. **22** (1981) 297.

81T2   Torrance, J.B., Girlando, A., Mayerle, J.J., Crowley, J.I., Lee, V.Y., Batail, P., LaPlaca, S.J.: Phys. Rev. Lett. **47** (1981) 1747.

81T3   Torrance, J.B., Vazquez, J.E., Mayerle, J.J., Lee, V.Y.: Phys. Rev. Lett. **46** (1981) 253.

81T4   Thorup, N., Rindorf, G., Soling, M., Bechgaard, K.: Acta Crystallogr. **B37** (1981) 1236.

81W1   Willig, F.: "Electrochemistry at the Organic Molecular Crystal/Electrolyte Interface", in Gerischer, H., Tobias, C. (eds.): "Adv. Electrochemistry a. Electrochemical Engineering" Vol. **12**; New York: John Wiley, **1981**, p. 1.

81W2   Warta, W., Karl, N.: unpublished, **1981**.

81W3   Weast, R.C., Astle, M.J. (eds.): "CRC Handbook of Chemistry and Physics" 62nd ed.; Boca Raton: CRC Press, **1981**.

82A   Alt, H., Kalus, J.: Acta Crystallogr. **B38** (1982) 2595.

82B1   Braun, A., Mayer, U., Auweter, H., Wolf, H.C., Schmid, D.: Z. Naturforsch. **37a** (1982) 1013.

82B2   Backer, M., Häfner, W., Kiefer, W.: J. Raman Spectrosc. **13** (1982) 247.

82B3   Bandrauk, A.D., Truong, K.D., Jandl, S.: Mol. Cryst. Liq. Cryst. **85** (1982) 297.

82B4   Bree, A., Zwarich, R., Taliani, C.: Chem. Phys. **70** (1982) 257.

82B5   Brock, P.C., Dunitz, J.D.: Acta Crystallogr. **B38** (1982) 2218.

82C1   Chaplot, S.L., Lehner, N., Pawley, G.S.: Acta Crystallogr. **B38** (1982) 483.

82C2   Chaplot, S.L., Pawley, G.S., Dorner, B., Jindal, V.K., Kalus, J., Natkaniec, I.: Phys. Status Solidi **b110** (1982) 445.

82C3   Chaikin, P.M., Mu-Yong Choi, Haen, P., Engler, E.M., Greene, R.L.: Mol. Cryst. Liq. Cryst. **79** (1982) 95.

82C4   Conwell, E.M., Banik, N.C.: Mol. Cryst. Liq. Cryst. **79** (1982) 95.

82D1   Dorner, B.: "Coherent Inelastic Neutron Scattering in Lattice Dynamics", Berlin, Heidelberg, New York: Springer, **1982**, p. 44.

82D2   Doberer, U., Port, H., Benk, H.: Chem. Phys. Lett. **85** (1982) 253.

82D3   Doberer, U.: Dissertation, Univ. Stuttgart, **1982**.

82E1   Eichhorn, M., Willig, F., Charlé, K.P., Bitterling, K.: Proc. 7th Int. Conf. Conduct. a Breakdown in Dielectr. Liquids, Berlin 1981 (IEEE, New York 1981), p. 6; J. Chem. Phys. **76** (1982) 4648.

82E2   Etemad, S., Heeger, A.J., MacDiarmid, A.G.: Ann. Rev. Phys. Chem. **33** (1982) 443.

82F1   Forro, L.: Mol. Cryst. Liq. Cryst. **79** (1982) 315.

82F2   Filhol, A., Gaultier, J., Hauw, C., Hilti, B., Mayer, C.W.: Acta Crystallogr. **B38** (1982) 2577.

82G1   Giermańska, J., Samoć, A., Samoć M., Sworakowski, J., Zboiński, Z.: Chem Phys. Lett. **87** (1982) 71.

82G2   Goudard, J.-L., Lakhani, A.A., Hota, N.K.: Solid State Commun. **41** (1982) 423.

82H   Haddon, R.C., Kaplan, M.L., Wudl, F.: in Kirk-Othmer, Encyclop. Chem. Technol. (3rd ed.), New York: J. Wiley, **20** (1982) 674.

82J1   Jérome, D., Schulz, H.J.: "Organic Conductors and Superconductors", Adv. Phys. **31** (1982) 299.

82J2   Jérome, D.: "Organic Superconductors: A Survey of Low Dimensional Phenomena", Mol. Cryst. Liq. Cryst. **79** (1982) 155.

82J3   Jordan, J.F.J., Axmann, A., Egger, H., Kalus, J.: Phys. Status Solidi **a71** (1982) 457.

82J4   Jérome, D.: Physica **109/110B** (1982) 1447.

82J5   Jacobsen, C.S., Pedersen, H.J., Mortensen, K., Torrance, J.B., Bechgaard, K.: J. Phys. **C15** (1982) 2651.

82J6   Julian, M.E., Bloss, F.D.: Acta Crystallogr. **A38** (1982) 167.

82K1   Karl, N., Warta, W.: to be published; presented in part at the 9th Molecular Crystal Symposium, Mittelberg 1980; cf. Conference Proceedings, p. 149; Warta, W.: Diplomarbeit, Univ. Stuttgart, **1978**.

82K2   Karl, N., Sato, N., Seki, K., Inokuchi, H.: J. Chem. Phys. **77** (1982) 4870.

82K3   Kagoshima, S., Sambongi, T., Nagasawa, H.: "One Dimensional Conductors"; Tokyo: Syokabo, **1982** (in Japanese).

82K4   Kalus, J., Dorner, B., Jindal, V.K., Karl, N., Natkaniec, I., Pawley, G.S., Press, W., Sheka, E.F.: J. Phys. **C15** (1982) 6533.

82K5	Kuroda, H., Yakushi, K., Cao, Y.: Mol. Cryst. Liq. Cryst. **85** (1982) 325.
82K6	Kikuchi, K., Yakushi, K., Kuroda, H.: Solid State Commun. **44** (1982) 151.
82K7	Karbach, H.J., Nickel, B.: unpublished results; private communication, **1982**.
82K8	Karl, N., Stehle, R.: partial results were presented at the 10th Molecular Crystal Symposium (St. Jovite, Canada **1982**) cf. Conference Proceedings, p. 148.
82M1	Miller, J.S.: "Extended Linear Chain Compounds", Vol. I, II; New York: Plenum Press, **1982**, Vol. III (1983).
82M2	Marks, T.J., Kalina, D.W.: in Miller, J.S. (ed.) "Extended Linear Chain Compounds", Vol. I; New York: Plenum Press, **1982**, p. 197.
82M3	Mort, J., Pfister, G. (eds.): Electronic Properties of Polymers; New York: J. Wiley, **1982**.
82P1	Pope, M., Swenberg, Ch.E.: "Electronic Processes in Organic Crystals"; Oxford: Clarendon Press, **1982**.
82S1	Sandman, D.J., Stärk, J.C., Hamill, G.P., Burke, W.A., Foxman, B.M.: Mol. Cryst. Liq. Cryst. **86** (1982) 79.
82S2	Shibaeva, R.P.: in Miller, J.S. (ed.) "Extended Linear Chain Compounds", Vol. II; New York: Plenum Press, **1982**, p. 435.
82S3	Schein, L.B., Brown, D.W.: Mol. Cryst. Liq. Cryst. **87** (1982) 1.
82S4	Schmidt, W.: private communication, **1982**.
82S5	Sato, N., Seki, K., Inokuchi, H., Harada, Y., Takahashi, T.: Solid State Commun. **41** (1982) 759.
82S6	Silinsh, E.A., Kolesnikov, V.A., Muzikante, I.J., Balode, D.R.: Phys. Status Solidi **b**113 (1982) 379.
82S7	Shchegolev, I.F., Yagubskii, E.B.: in Miller, J.S. (ed.) "Extended Linear Chain Compounds", Vol. II; New York: Plenum Press **1982**, p. 385.
82T1	Tokura, Y., Koda, T., Mitani, T., Saito, G.: Solid State Commun. **43** (1982) 757.
82T2	Tanner, D.B.: in Miller, J.S. (ed.) "Extended Linear Chain Compounds", Vol. II; New York: Plenum Press, **1982**, p. 205.
82T3	Tanner, D.B., Jacobsen, C.S.: Mol. Cryst. Liq. Cryst. **85** (1982) 137.
82W1	Warta, W., Karl, N.: unpublished, **1982**.
82W2	Wakayama, N.: Mol. Cryst. Liq. Cryst. **89** (1982) 1.
82W3	Wakayama, N.I.: J. Lumin. **27** (1982) 299.
82W4	Warta, W., Karl, N.: presented at the 10th Molecular Crystal Symposium (St. Jovite, Canada **1982**) as a post deadline poster.
83A	Ayache, C., Torrance, J.B.: Solid State Commun. **47** (1983) 789.
83D	DellaValle, R.G., Fracassi, P.E., Righini, R., Califano, S.: Chem. Phys. **74** (1983) 179.
83E	Enders, H.: in Miller, J.S. (ed.): "Extended Linear Chain Compounds", Vol. III; New York: Plenum Press, **1983**, p. 263.
83G1	Geserich, H.P., Wilckens, R., Ruppel, W., Enkelmann, V., Wegner, G., Wieners, G., Schweitzer, D., Keller, H.J.: Mol. Cryst. Liq. Cryst. **93** (1983) 385.
83G2	Girlando, A., Marzola, F., Pecile, C., Torrance, J.B.: J. Chem. Phys. **79** (1983) 1075.
83G3	Gutman, F., Keyzer, H., Lyons, L.E., Somoano, R.B.: "Organic Semiconductors, Part B", Malabar, Krieger Publ. **1983**.
83J	Jacobsen, C.S., Torrance, J.B.: J. Chem. Phys. **78** (1983) 112.
83K1	Kosic, Th.J., Schosser, C.L., Dlott, D.D.: Chem. Phys. Lett. **96** (1983) 57.
83K2	Karl, N., Schmid, E., Stehle, R., Kotani, M., Isono, Y., Tanaka, H.: to be published in Chem. Phys.
83K3	Karl, N.: unpublished, **1983**.
83K4	Karl, N., Stehle, R.: to be published.
83K5	Kotani, M., Isono, Y.: private communication, **1983**.
83M1	Morikawa, E., Isono, Y., Kotani, M.: J. Chem. Phys. **78** (1983) 2691.
83M2	Modelli, A., Distefano, G., Jones, D.: Chem. Phys. **82** (1983) 489.
83N1	Nissler, H., Port, H.: to be published.
83N2	Natkaniec, I., Belushkin, V.A., Dyck, W., Fuess, H., Zeyen, C.M.E.: Z. Kristallogr. **163** (1983) 285.
83P1	Ponomarev, V.I., Shilov, G.V.: Sov. Phys. Crystallogr. **28** (1983) 397.
83P2	Parkin, S.S.P., Engler, E.M., Schumaker, R.R., Lagier, R., Lee, V.Y., Scott, J.C., Greene, R.L.: Phys. Rev. Lett. **50** (1983) 270.
83S1	Sheka, E.F.: "Spectroscopy of Molecular Crystals, a Bibliography for 1980", Mol. Cryst. Liq. Cryst. **91** (1983) 197, and references given in the introduction for the years 1969⋯1979.

83S2	Sato, N., Saito, G., Inokuchi, H.: Chem. Phys. **76** (1983) 79.
83S3	Schrof, W.: private communication, **1983**.
83S4	Schweitzer, D.: unpublished, **1983**.
83S5	Stehle, R., Karl, N.: to be published.
83S6	Schein, L.B., Warta, W., McGhie, A.R., Karl, N.: Chem. Phys. Lett. **100** (1983) 34.
83S7	Schein, L.B., Narang, R.S., Anderson, R.W., Meyer, K.E., McGhie, A.R.: Chem. Phys. Lett. **100** (1983) 37.
83S8	Shu-Sing Chang: J. Chem. Phys. **79** (1983) 6229.
83S9	Saito, G., Yamaji, K.: "TTF-TCNQ and Related Compounds", in "Chemistry of Low-Dimensional Conductive Materials"; Kagabu Sosetu **42** (1983), p. 59 (in Japanese).
83S10	Saito, G., Inokuchi, H.: "Fundamental Research of Organic Metals and their Applications", in "Function and Materials", CMC **3** (1983) 38 (in Japanese).
83S11	Sebastian, L., Weiser, G., Peter, G., Bässler, H.: Chem. Phys. **75** (1983) 103.
83S12	Schmid, B.M., Sato, N., Inokuchi, H.: Chem. Lett. Jpn. **1983** (1983) 1897.
83T	Turlet, J.M., Kottis, Ph., Philpott, M.R.: "Polariton and Surface Exciton State Effects in the Photodynamics of Organic Molecular Crystals", in Prigogine, I., Rice, S.A. (eds.), Adv. Chem. Phys. **54**; New York: J. Wiley, **1983**, p. 303.
83W	Warta, W., Karl, N.: unpublished, **1983**.
83Y	Yagubskii, E.B., Shchegolev, J.F., Laukhin, V.N., Kononovich, P.A., Karatsovnik, M.V., Zvarykina, A.V., Buravov, L.I.: Pis'ma Zh. Eksp. Teor. Fiz. **39** (1984) 12; JETP Lett. (engl. Transl.): **39** (1984) 12.
83Z	Zwier, T.S., Carrasquillo, E., Levy, D.H.: J. Chem. Phys. **78** (1983) 5493.
84B1	Bässler, H.: Philos. Mag. **B50** (1984) 347.
84B2	Buzdin, A.I., Bulaevskii, L.N.: Usp. Fiz. Nauk **144** (1984) 415.
84G1	Grubb, S.G., Whetten, R.L., Albrecht, A.C., Grant, E.R.: Chem. Phys. Lett. **108** (1984) 420.
84G2	Greene, R.L., Chaikin, P.M.: Physica B+C **126** (1984) 431.
84H	Hodina, A.J.: Synth. Met. **10** (1984) B55; B99; B155.
84J	Jérome, D.,: in Acrivos, J.V., Mott, N.F., Yoffe, A.D. (eds.): "Physics and Chemistry of Electrons and Ions in Condensed Matter"; Dodrecht: Reidel **1984**, p. 595; also: idem: in Reineker, P., Haken, H., Wolf, H.C. (eds.): "Organic Molecular Aggregates"; Berlin, Heidelberg, New York, Tokyo: Springer **1983**, p. 252.
84K1	Kessler, R.J., Fischer, M.R., Tripathi, G.N.R.: Chem. Phys. Lett. **112** (1984) 575.
84K2	Karl, N.: J. Mater. Sci. **10** (1984) 365; Techn. Univ. Wroclaw, Poland.
84K3	Krivenko, T.A., Dementjev, V.A., Bokhenkov, E.L., Kolesnikov, A.I., Sheka, E.F.: Mol. Cryst. Liq. Cryst. **104** (1984) 207.
84K4	Karl, N., Rommel, H., Warth, M.: unpublished (measured by DSC at different heating rates S and extrapolated to S → 0).
84K5	Karl, N., Rommel, H.; unpublished (measured by DSC, temperature steps method).
84M	Massa, D., Karl, N.: unpublished, **1984**.
84N	Nakayama, H., Ishii, K., Chijiwa, E., Wada, M., Sawada, A.: Solid State Commun., to be published.
84S1	figure provided by Schweitzer, D., **1984**.
84S2	Sheka, E.F., Makarova, V.S., Krivenko, T.A.: Spectroscopy of Molecular Crystals; a Bibliography for 1981: in Mol. Cryst. Liq. Cryst. **104** (1984) 1; for 1982: in Mol. Cryst. Liq. Cryst. **114** (1984) 305.
84S3	Seki, K., Karlsson, U.O., Engelhardt, R., Koch, E.-E., Schmidt, W.: Chem. Phys. **91** (1984) 459.
84W1	Warta, W., Stehle, R., Karl, N.: Appl. Phys. **A36** (1984) to be published.
84W2	Warta, W., Karl, N.: Phys. Rev. **B** (submitted 1984).
84W3	Warta, W., Karl, N.: to be published.

B. Special topics

13 Space charge layers at surfaces and interfaces

13.0 List of symbols

Symbol	Property	Unit
$\boldsymbol{B}$	magnetic induction ($|\boldsymbol{B}|=B$)	T
C	capacitance	F
d_i	thickness of insulator	nm
D_{it}	density of surface states	$cm^{-2}eV^{-1}$
D	implantation dose	cm^{-2}
e	electron charge	As
E_V	energy of top of the valence band	eV
E_C	energy of bottom of the conduction band	eV
E_F	Fermi energy	eV
E_i	intrinsic energy level	eV
E_i	eigenvalues of energy from Schrödinger's equation	eV
$\boldsymbol{E}$	electric field ($|\boldsymbol{E}|=E$)	$V\,cm^{-1}$
E_z	electric field perpendicular to the surface	$V\,cm^{-1}$
f	frequency	Hz
f	factor, fraction	
g^*	electron g-factor	
g_s	spin degeneracy factor	
g_v	valley degeneracy factor	
G	conductance	Ω^{-1}
h	Planck's constant	J s
$\boldsymbol{H}$	magnetic field strength ($|\boldsymbol{H}|=H$)	$A\,m^{-1}$
I	electric current	A
I_D	electric current along the surface channel	A
k_B	Boltzmann's constant	$J\,K^{-1}$
$\boldsymbol{k}$	wave vector ($|\boldsymbol{k}|=k$), k_x and k_y are parallel to the surface	cm^{-1}
L	geometric channel length	µm
m_0	free electron mass	kg
m^*	effective electron mass	kg
n	electron concentration	cm^{-3}
n_i	intrinsic charge carrier concentration	cm^{-3}
n_s	carrier concentration of free electrons in space charge layers	cm^{-2}
n_{tot}	total induced carrier concentration in space charge layers	cm^{-2}
N	bulk doping concentration	cm^{-3}
N_a, N_d	acceptor and donor doping concentrations, respectively	cm^{-3}
N_{depl}	depletion layer concentration	cm^{-2}
N_{it}	surface state density	cm^{-2}
N_{ox}	oxide charge density	cm^{-2}
p	hole concentration	cm^{-3}
p_s	hole concentration in space charge layers	cm^{-2}
Q	electric charge	C
s	spin quantum number	
T	temperature	K
V_G	gate voltage	V
V_T	threshold voltage	V
V_D	drain voltage, voltage along the surface channel	V
w	geometric channel width	µm
w_D	depletion layer width	µm
x, y	coordinate directions parallel to the surface	
X	uniaxial stress (compression or tension)	$N\,mm^{-2}$
z	coordinate direction perpendicular to the surface	

Symbol	Property	Unit
ε_i	insulator permittivity	F m^{-1}
ε_0	free space permittivity	F m^{-1}
ε_s	semiconductor permittivity	F m^{-1}
ε_r	relative permittivity	
μ_B	Bohr magneton	J T^{-1}
μ_B	bulk mobility	cm^2 V^{-1}s^{-1}
μ_c	conductivity mobility	cm^2 V^{-1}s^{-1}
μ_{FE}	field effect mobility	cm^2 V^{-1}s^{-1}
μ_H	Hall mobility	cm^2 V^{-1}s^{-1}
μ_0	permeability of vacuum	H m^{-1}
ϱ	resistivity	Ω cm
$\varrho(z)$	charge density perpendicular to the surface	As cm^{-1}
σ	conductivity	Ω^{-1}cm^{-1}
τ	relaxation time	s
π_{ij}	Piezoresistance coefficient	mm^2 N^{-1}
φ_i, ξ_i	wave functions	
ω_c	cyclotron frequency	Hz
ψ_B	bulk potential	V
ψ_c	contact potential	V
ψ_{MS}	metal-semiconductor contact potential	V
ψ_s	surface potential	V
Φ	work function	eV
$\Delta\Phi_{MS}$	work function difference between metal and semiconductor	eV

13.1 General remarks

Semiconductor surfaces (interface to the vacuum) and other interfaces occur in the metal-semiconductor interface (Schottky barrier) or in the metal-insulator-semiconductor (MIS) structure. In all cases, the electronic band structure and the basic semiconductor properties are modified by space charges induced by the electric field present in the interface and by the structural change. Structural changes also occur in semiconductor-semiconductor interfaces which will be discussed elsewhere.

The difference in the electrostatic potentials between the surface and the underlying bulk defines the potential barrier, ψ_c, (contact potential), which is the most fundamental parameter controlling the electronic processes at the surface. Fig. 1, p. 228, shows the electronic energy band diagrams for an ideal contact between a metal and an n-type semiconductor, and a metal-insulator-n-type semiconductor structure, respectively. When the materials are not in contact, the vacuum potential is the reference. Once the materials are connected, charge will flow from the semiconductor to the metal until electronic equilibrium is established, and the Fermi levels line up. Relative to the Fermi level in the metal, the Fermi level in the semiconductor is lowered or raised by an amount equal to the difference between the two work functions. A contact potential ψ_c exists.

13.1.1 Space charge layers

Space charge layers are classified into three groups:

(1) **Accumulation layer:** the density of the majority carriers increases toward the surface.

(2) **Depletion layer:** the concentration of both types of charge carriers (electrons and holes) in the space charge region is smaller than the majority carrier density in the bulk.

(3) **Inversion layer:** the band bending is so large that the minority carrier density exceeds the majority carrier bulk density.

The energy diagrams for n-type as well as p-type semiconductors are depicted in Fig. 2, p. 228.

The maximum width, w_D, of the surface depletion region is limited by the breakdown field [69s]. It is approximately given by:

$$w_D \approx \sqrt{\frac{4\varepsilon_s k_B T \ln(N/n_i)}{e^2 N}} \tag{1}$$

(see also Fig. 3, p. 228).

Eisele

13.1.2 Concentration of charge carriers in space charge layers

The total concentration, n_{tot}, of the charge carriers is capacitively induced at the interface by the surface electric field ($E_z = V_G/d_i$ for the MIS structure):

$$n_{tot} = \frac{\varepsilon_i}{e\, d_i}(V_G - V_T). \tag{2}$$

V_T is the threshold or turn-on voltage and V_G the gate voltage. For depletion and accumulation layers V_T is the bias voltage to observe the flat band voltage, for which the surface potential is zero. For inversion layers, the threshold or turn-on voltage is given by:

$$V_T = 2\psi_B + \frac{d_i\sqrt{2\varepsilon_s\, e\, N\psi_B}}{\varepsilon_i} + \psi_{MS} + \frac{e\cdot N_{it}\cdot d_i}{\varepsilon_i} - \frac{e\cdot N_{ox}\cdot d_i}{\varepsilon_i}. \tag{3}$$

The total carrier concentration is partially trapped in surface states, N_{it}, and partially free to move along the surface:

$$n_{tot} = n_s + N_{it}. \tag{4}$$

For state of the art SiO_2—Si interfaces, one can assume that $N_{it} \ll n_s$.

13.1.3 Energy band structure and charge carrier distribution

(Description of electrical surface quantization)

Whereas the depletion layer extends quite far into the bulk semiconductor, the mobile charge carriers in inversion or accumulation layers are confined in a narrow potential well near the semiconductor interface. The allowed energy states in this potential well are quantized. This quantization effect has to be taken into account in narrow channels (MIS-structures) or in thin semiconductor films (heterojunctions). The quantization occurs in the direction perpendicular to the surface and the interface, respectively. Within the surface plane, the charge carriers are freely mobile. The electrons (or holes) confined in a potential well are described by a quasi-two-dimensional electron (hole) gas.

The source for the band bending at the semiconductor interface is the electric field $E_z = -\operatorname{grad}\psi$ present inside the semiconductor. The electrostatic potential ψ is given by the solution of Poisson's equation:

$$\frac{d^2\psi}{dz^2} = \frac{\varrho(z)}{\varepsilon_s}. \tag{5}$$

The charge density can be treated either by a classical continuum model or by quantum mechanics.

For the continuum model $\varrho(z)$ is given by:

$$\varrho(z) = e\,[p(z) - n(z) + N_d^+ - N_a^-]. \tag{6}$$

For the quantum mechanical treatment the wave function φ has to be known. The charge density in this case is given by:

$$\varrho(z) = e(N_d^+ - N_a^-) - e\sum_i n_i|\varphi_i(z)|^2. \tag{7}$$

In the effective mass approximation φ is determined by the Schrödinger equation. Because φ is only a function of z, φ can be separated and the Schrödinger equation has two parts. For constant energy surfaces with rotational symmetry one obtains:

$$\left[-\frac{\hbar^2}{2m_x^*}\frac{\partial^2}{\partial x^2} - \frac{\hbar^2}{2m_y^*}\frac{\partial^2}{\partial y^2}\right] e^{ik_x x + ik_y y} = E_{x,y}\, e^{ik_x x + ik_y y} \tag{8}$$

for the quasi-two-dimensional free motion, and

$$\left[-\frac{\hbar^2}{2m_z^*}\frac{\partial^2}{\partial z^2} - e\,\varphi_s(z)\right]\xi_i(z) = E_i\,\xi_i(z) \tag{9}$$

describing the one-dimensional bound motion. In the case of warped surfaces more complicated relations are obtained. Each eigenvalue E_i relates to the bottom of the i-th subband with an energy level given by:

$$E_i(\vec{k}) = \frac{\hbar^2}{2m_x^*}k_x^2 + \frac{\hbar^2}{2m_y^*}k_y^2 + E_i(m_z^*, F_z), \qquad i-0, 1, 2\ldots. \tag{10}$$

m_x^* and m_y^* are the principal effective masses for the free carrier motion parallel to the surface. They are only in exceptional cases identical to the bulk masses. The effective mass perpendicular to the surface, m_z^*, is responsible for the subband splitting. Assuming a triangular potential well and a constant electric field, i.e. $\psi(z) = -E_z \cdot z$, one obtains E_i for energy surfaces with rotational symmetry:

$$E_i(m_z^*, E_z) = \frac{(\hbar e E_z)^{2/3}}{(2m_z^*)^{1/3}} \left[\tfrac{2}{3}\pi(i+\tfrac{3}{4})\right]^{2/3}, \quad i = 0, 1, 2 \dots. \tag{11}$$

Consequently the valleys with highest m_z^* have the lowest energy levels in the quantum mechanical treatment. Contrary to the bulk, different valley degeneracies, g_v, are obtained in many cases for different surface orientations. For more accurate values, self-consistent calculations have to be performed [78a].

The two-dimensional density of states for one subband is given by:

$$D(E) = g_s g_v \frac{\sqrt{m_x^* m_y^*}}{2\pi \hbar^2}. \tag{12}$$

In contrast to the bulk material, $D(E)$, is energy independent. $\sqrt{m_x^* m_y^*}$ is defined as the two-dimensional density of states effective mass.

13.1.4 Effect of high magnetic fields

Applying a magnetic field perpendicular to the surface forces an electron (hole) into a one-dimensional bound state with discrete energy [72B1]:

$$E = E_i + (n + \tfrac{1}{2})\hbar\omega_c + s g^* \mu_B B \quad (n = 0, 1, 2 \dots). \tag{13}$$

The cyclotron resonance frequency, ω_c, is given by:

$$\omega_c = \frac{eB}{\sqrt{m_x^* m_y^*}}. \tag{14}$$

The Hall resistance of a two-dimensional electron gas depends exclusively on fundamental constants. This occurs because the thickness parameter of the sample is incorporated in the two-dimensional carrier density n_s. For strong magnetic fields the ratio n_s/B becomes quantized in integer multiples of e/h. This leads to a quantization of the Hall resistance (Quantum Hall Effect) with resistance values of [80K1]:

$$R_H = \frac{h}{e^2 i} = \frac{\mu_0 c}{2\alpha} \cdot \frac{1}{i}, \quad i = 1, 2, 3 \dots \tag{15}$$

where μ_0 is the permeability of vacuum which is exactly equal to $4\pi \cdot 10^{-7} \mathrm{H\,m^{-1}}$, c is the velocity of light in vacuum and α is the fine-structure constant.

The accuracy for the determination of h/e^2 (and therefore α) is basically limited by the uncertainty in the realization of the unit Ohm. On the other hand if α is assumed to be known from other experiments, Eq. (15) may be used to derive a standard resistance. The following value of α can be evaluated:

$$\alpha^{-1} = 137.03592\,(18). \tag{16}$$

Recently for two dimensional electron systems with mobilities of about $10^6\ \mathrm{cm^2/Vs}$ fractions of the integer i have been observed [84S1]. This was only possible with modulation doped GaAs—(AlGa)As heterostructures which are discussed elsewhere. The "Fractional Quantum Hall Effect" is explained by a crystallization of the electrons. (See Fig. 38, p. 238).

13.1.5 Electronic transport properties

The transport without magnetic field is governed by:

$$\sigma = e\,n_s\,\mu_c = e^2\,n_s\,\frac{\tau}{m^*}. \tag{17}$$

The surface (interface) influences the transport parameters by the following effects:

(a) The free carrier concentration depends on the surface field and trapping of charge carriers in surface states.

(b) The effective mass in the direction of the transport, m^*, depends on quantization effects, surface orientation, electric surface field, and on mechanical stress.

(c) The scattering time, τ, depends on surface roughness, and the quasi-two-dimensional motion.

The dependencies of the surface mobility on temperature and electric field for various scattering contributions are listed in Table 1. The average mobility is given by $1/\mu = \sum_i (1/\mu_i)$ (see Fig. 91, p. 251).

Table 1. Dependencies on temperature T and surface electric field E_z of the channel mobilities for various scattering mechanisms as indicated.

Phonon scattering	Coulomb scattering	Surface roughness scattering	Diffuse scattering
$T^{-1} E_z^{-1/3}$ [72K1]	$T^{3/2} N_{\text{ion}}^{-1}$ [66K1]	$E_z^{-4/3}$ [69G1]	$T^{1/2} E_z^{-1}$ [57S1]
$T^{-\alpha}$ $\alpha = 1$ acoustic $\alpha > 1.5$ acoustic + optic [72S1]	$T^1 \cdot N_{\text{ion}}^{-1}$ [72S1]		
	$E_z^\alpha N_{\text{ion}}^\beta$ $\alpha < 0,\ \beta > 0$ [67S1]	$E_z^{-\alpha}$ $0.5 < \alpha < 2$ [73C1]	

13.2 Experimental methods

Space charge layers are characterized by specifically designed measurement methods.

13.2.1 Capacitance and conductance measurements

Surface charges are characterized by the insulator capacitance, the semiconductor space charge capacitance, and the capacitance and conductance as a function of the electric field. A change of charge in surface states occurs whenever levels cross the Fermi level. This charge change results in a frequency-dependent capacitance in parallel with a frequency-dependent conductance [67N1, 73G1].

The concentration of surface states is determined by measuring the voltage shift between the ideal (no surface states) and the measured capacitance-voltage (CV) curves.

13.2.2 Transport measurements

Transport measurements are performed for surface channels e.g. on FETs. The field effect mobility, μ_{FE}, is determined by differentiating the channel conductivity with respect to the carrier concentration, which is dependent on the electrical surface field:

$$\mu_{\text{FE}} = \frac{\varepsilon_i}{d_i} \frac{L}{w} \frac{1}{V_D} \frac{\partial I_D}{\partial V_G}. \tag{18}$$

This field effect mobility differs from the common mobility definitions:

conductivity mobility (transconductance):
$$\mu_c = \frac{1}{e\, n_s} \frac{V_D}{I_D} \frac{w}{L}, \tag{19}$$

Hall mobility:
$$\mu_H = \frac{V_{\text{Hall}}}{V_D} \cdot \frac{L}{w} \cdot \frac{1}{B_z} \tag{20}$$

13.3 Surface data for various semiconductor layers

The representative properties are depicted in figures as listed.

13.3.1 Silicon (Si)

Data on silicon surfaces are presented for three different space charge layer systems.
— Electron inversion and accumulation layers: **Figs. 4 ··· 38,** p. 229 ff.
— Hole inversion and accumulation layers: **Figs. 39 ··· 54,** p. 238 ff.
— Space charge layers in epitaxially grown silicon (silicon on insulators): **Figs. 55 ··· 73,** p. 242 ff.

Technological data such as insulator thickness, bulk doping, and threshold voltage are indicated as taken from the references.

For all the silicon data, the insulator material is SiO_2 (relative permittivity $\varepsilon_r = 3.8$). Also confer to volume 17c for silicon SiO_2—Si interface data.

13.3.1.1 Electron inversion and accumulation layers on Si

Flatband voltage, threshold voltage and **workfunction differences** are listed in Table 2. Properties are depicted in **Figs. 5···14**, p. 229 ff.

Valley degeneracies and various **effective masses** are listed in Tables 3···5. Properties are depicted in **Figs. 15···17**, p. 231 ff.

Table 2a. Flatband voltage V vs. insulator thickness d for Al—SiO$_2$—Si structures (Si: $\varrho = 10\,\Omega\,$cm, n-type).

d [nm]	V [V]
50	−0.8
100	−1.3
150	−1.8
200	−2.3

Table 2b. Threshold voltages V_T in [V] for inversion of Al—SiO$_2$—Si structures ($N_{it} = 2 \cdot 10^{11}\,$cm^{-2}).

Oxide thickness [nm]	50	100	150	200
bulk doping [cm^{-3}]	V_T [V]			
$N_d = 10^{13}$	−1.1	−1.6	−2.2	−2.7
$N_a = 10^{13}$	−1.1	−1.6	−2.0	−2.5
$N_d = 10^{14}$	−1.3	−1.8	−2.4	−3.0
$N_a = 10^{14}$	−0.9	−1.4	−1.8	−2.2
$N_d = 10^{15}$	−1.5	−2.2	−2.9	−3.6
$N_a = 10^{15}$	−0.7	−1.0	−1.3	−1.6
$N_d = 10^{16}$	−2.1	−3.1	−4.5	−5.7
$N_a = 10^{16}$	−0.1	−0.1	+0.3	+0.5

Table 2c. Workfunction difference of metal—SiO$_2$—Si structures

Metal	$\Delta\Phi_{MS}$ [eV]	
	Si: $10\,\Omega\,$cm p-type	Si: $10\,\Omega\,$cm n-type
Ag	+0.4	+1.1
Au	+0.3	+1.0
Ni	−0.3	+0.4
Al	−1.0	−0.3
p$^+$ Si	+0.3	+0.9
n$^+$ Si	−0.9	−0.3

Table 3. Valley degeneracy g_v for electron inversion layers on silicon surfaces.

Surface orientation	g_v (Theory)	g_v (Experiment)	Ref.
(100)	2	2.0(2)	68F1
(110)	4	2.0(2)	75N1
(111)	6	2.0(2)	75N1 76D1
		6	79T1

Table 4. Si. Calculated relative effective masses m^*/m_0 along the principal axes of the ellipsoids in Fig. 9, p. 230, for electron inversion layers on various Si surface orientations.

Surface orientation	Effective mass along the surface		Effective mass perpendicular to surface
	m_x^*/m_0	m_y^*/m_0	m_z^*/m_0
(100)			
lower valleys (1, 2)	0.190	0.190	0.916
higher valleys (3–6)	0.190	0.916	0.190
(110)			
lower valleys (1–4)	0.190	0.553	0.315
higher valleys (5, 6)	0.190	0.916	0.190
(111)	0.190	0.674	0.258

Table 5. Si. Relative density of states effective mass m^*/m_0 for electron inversion layers on different surface orientations, theoretical and experimental values. SdH: Evaluation from Shubnikov-de Haas oscillations CR: Evaluation from cyclotron resonance. (Experiments for $n_s = 3 \cdot 10^{12}\,$cm^{-2}).

Surface orientation	m^*/m_0 Theory	m^*/m_0 Experiment	Method	Ref.
(100) lower valleys	0.19	0.197(5)	CR	76A2
		0.21(2)	SdH	78E1
(100) higher valleys	0.43	0.43(2)	SdH	77E1
		0.42	CR	78K1
(110)	0.324	0.37(3)	SdH	75L1
		0.34(3)	CR	76A2
(111)	0.368	0.38(3)	CR	76A2
		0.40(3)	SdH	75L1

Mobilities of electrons in surface channels and various dependencies are depicted in **Figs. 18···30,** p. 232 ff. The effect of **mechanical stress** on the valley shift is explained in **Table 6.** Various dependencies are depicted in **Figs. 31···38,** p. 236 ff.

Table 6. Shift of energy valleys by means of mechanical stress on electron inversion layers on (100), (110), and (111) surface orientations for various current- and electric surface field directions. Drain current I_D and electric surface field E_z perpendicular to the surface as indicated.
L: longitudinal, T: transverse.

surface orientation		(100)	(110)	(111)
directions (current and surface field)		$I_D \parallel [010]$ $E_z \parallel [100]$	$I_D \parallel [001]$ $E_z \parallel [110]$	$I_D \parallel [\bar{1}10]$ $E_z \parallel [111]$
constant energy surfaces				
$E(k)$ valleys		I II,III	I,II III	I,II,III
shift of valleys	tension	$\parallel[010]$L I↓ II↑ III↓ $\parallel[001]$T I↓ II↓ III↑	$\parallel[001]$L I↓ II↓ III↑	$\parallel[\bar{1}10]$L I↑ II↑ III↓
	compression	$\parallel[010]$L I↑ II↓ III↑ $\parallel[001]$T I↑ II↑ III↓	$\parallel[001]$L I↑ II↑ III↓	$\parallel[\bar{1}10]$L I↓ II↓ III↑

13.3.1.2 Hole inversion and accumulation layers on Si

Properties of the energy band structure and effective masses for holes are depicted in **Figs. 39···41,** p. 238. The effect of mechanical stress is depicted in **Figs. 42, 51, and 52,** p. 239 ff.

Mobilities of holes in surface channels of Si and various dependencies are depicted in **Figs. 43···50,** p. 239 ff. The **magnetoresistance** of holes is depicted in **Figs. 53 and 54,** p. 242.

13.3.1.3 Space charge layers on epitaxially grown silicon

The basic properties of silicon films deposited on insulating substrates are listed in Table 7. Properties observed are depicted in **Figs. 55···58,** p. 242 ff.

Mobilities and various dependencies are depicted in **Figs. 59···70,** p. 243 ff. The **magnetoresistance** is shown in **Figs. 71···73,** p. 246 ff.

Table 7a, b. Basic properties of epitaxial layers of silicon on insulating sapphire and spinel substrates:

a) Substrate material: Al_2O_3 (sapphire)

Epitaxial growth of (100) silicon layers
on ($\bar{1}012$) Al_2O_3:

Residual stress due to lattice mismatch:

anisotropic compression in the silicon film:
‖[100] direction: 950 N/mm^2,
‖[010] direction: 895 N/mm^2 [73H1];

Valley degeneracy:

n-type silicon:

valleys 3–6 are lifted above valleys 1, 2
(cf. Fig. 9 and Fig. 55, p. 230 ff.);

valleys 3, 5 are splitted from valleys 4, 6
by about 6.5 meV;

calculated value: $g_v=2$ for low n_s and
$g_v=4$ for high n_s;

experimental value: $g_v=1.85\,(20)$ [77K2];

p-type silicon:

heavy and light hole bands are splitted

Density of states effective mass:

n-type silicon:

theory: $m^*=0.43\,m_0$;

experiment: $m^*=0.50\,(5)\,m_0$ [77K2]
(from Shubnikov-de Haas measurements).

b) Substrate material: $Al_2O_3:MgO=2:1$
to 3.5:1 (spinel):

Epitaxial growth of (100) and (111) silicon
layers on (100) and (111) surfaces:

Residual stress due to lattice mismatch:

isotropic compression in the silicon film:
(100) orientation: 900 N/mm^2,
(111) orientation: 1150 N/mm^2 [68S1];

Valley degeneracy:

n-type silicon:

calculated value: $g_v=4$ (cf. table 7a)

experimental value: $g_v=1.8\,(2)$ [78E1];

p-type silicon:
(cf. table 7a)

Density of states effective mass:

n-type silicon:

theory: $m^*=0.43\,m_0$;

experiment: $m^*=0.45\,m_0$ [78E1]
(from Shubnikov-de Haas measurements).

13.3.2 Germanium (Ge)

The (110) and (111) surfaces have been investigated.

Valley degeneracy for electron inversion layers:
calculated value: $g_v=1$;
experimental value: $g_v=1$ (from Shubnikov-de Haas) [76W2].

Density of states effective mass for the (111) surface:
experimental value: $m^*=0.082\,m_0$ (from cyclotron resonance) [76W2].
The **surface mobility** and **inversion layer carrier density** are depicted in **Figs. 74···76**, p. 247.

13.3.3 Gallium arsenide (GaAs)

The interface state density of GaAs is high. Various results are depicted in **Fig. 77,** p. 248, and listed in
Table 8. Results of the **surface mobility** are depicted in **Figs. 78, 79,** p. 248.

Table 8. GaAs. Interface state density near midgap of GaAs for various insulators [74M1].

Type	Orientation	Insulator film	Minimum density ($cm^{-2} eV^{-1}$)
n	(100)	Si_3N_4	$4.2 \cdot 10^{12}$
	$(\bar{1}\bar{1}\bar{1})$	$(SiH_4 + NH_3, 700\,°C)$	$2.8 \cdot 10^{12}$
p	(100)		$1.3 \cdot 10^{12}$
	$(\bar{1}\bar{1}\bar{1})$		$1.5 \cdot 10^{12}$
p	(100)	Al_2O_3	$4.2 \cdot 10^{11}$
	$(\bar{1}\bar{1}\bar{1})$	$(Al(C_4H_9)_3 + O_2, 450\,°C)$	$5.0 \cdot 10^{11}$
p	(100)	SiO_2	$6.8 \cdot 10^{11}$
n		$(Si(C_2H_5O)_4, 700\,°C)$	$1.0 \cdot 10^{12}$
p	(100)	SiO_2 (glow discharge)	$1.8 \cdot 10^{11}$
		SiO_2—Si_3N_4 (double layer)	$9.5 \cdot 10^{10}$

13.3.4 Indium antimonide (InSb)

Subband energies, effective masses and **conductances** are depicted in **Figs. 80$\cdots$89,** p. 248 ff.

13.3.5 Indium arsenide (InAs)

Density of states effective mass for electron inversion layers:
$m^* = 0.024\, m_0$, bulk value;
$m^* = 0.04\, m_0$, surface value from cyclotron resonance measurements [78W1].
Fraction of **electrons in subbands** and **mobilities** are depicted in **Figs. 90$\cdots$93,** p. 251 ff.

13.3.6 Zink oxide (ZnO)

Accumulation layers occur on free ZnO surfaces [74M2, 79G1] after:
(a) illumination by bandgap light (photon energy ≈ 3.2 eV),
(b) exposure to atomic hydrogen,
(c) exposure to thermalized He^+ ions,
(d) heating in vacuum to about 600 K,
(e) electron bombardment.

Workfunctions, mobilities, and surface **density of electrons** are depicted in **Figs. 94$\cdots$97,** p. 252 ff.

13.3.7 Tellurium (Te)

Effective masses and **magnetoresistance** are depicted in **Figs. 98 and 99,** p. 253. More information is given in [75O2, 78O2].

13.3.8 Other semiconductors

PbTe: [74T1, 77S1, 78S1]. **$Hg_{1-x}Cd_xTe$:** [76K3].

Figures for chapter 13

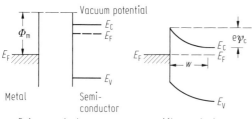

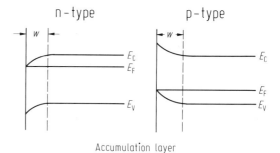

Accumulation layer

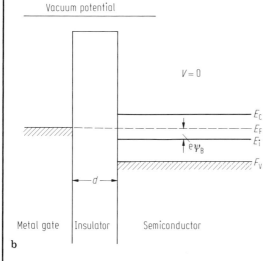

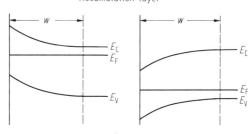

Depletion layer

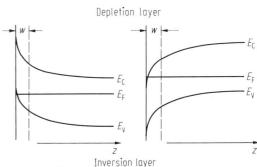

Inversion layer

Fig. 2. Energy band diagrams for an n- and a p-type semi-conductor, respectively, in the presence of a surface field for the three groups of space charge layers. The width w of the space charge for the accumulation and the depletion layers and the width w of the free electron gas for the inversion layer are indicated.

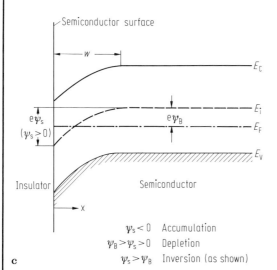

$\psi_s < 0$ Accumulation
$\psi_B > \psi_s > 0$ Depletion
$\psi_s > \psi_B$ Inversion (as shown)

Fig. 1 a–c. Energy band diagrams for a metal-semiconductor before and after contacting a), an ideal metal-insulator-semiconductor structure b), and an ideal semiconductor surface c). w is the width of the space charge region, ψ_s is the surface potential, ψ_c is the contact potential, ψ_B is the bulk potential, Φ_m is the metal work function, E_i is the intrinsic energy level.

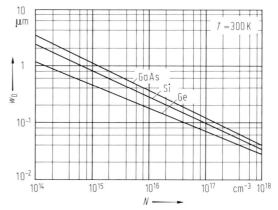

Fig. 3. Ge, Si, GaAs. Maximum depletion layer width w_D as a function of bulk doping concentration N for Ge, Si, and GaAs under heavy inversion condition [69s].

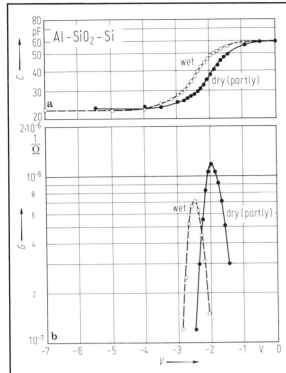

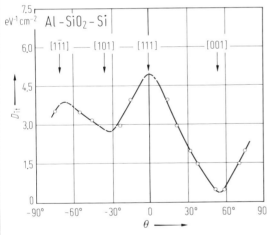

Fig. 4a, b. Al–SiO$_2$–Si. Capacitance C (a) and conductance G (b) for a Si(111) surface measured vs. bias voltage V before annealing (wet) and after heating in dry nitrogen for 17 hours at $T = 350\,°C$ (partly dried). Substrate: 0.75 Ωcm, p-type; oxide thickness: 60 nm [67N1].

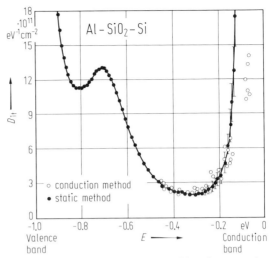

Fig. 5. Al–SiO$_2$–Si. Distribution of interface state density D_{it} vs. energy E across the bandgap of Si [72D1, 73G1].

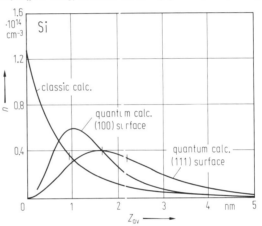

Fig. 7. Si. Continuum and quantum calculations for the electron concentration n in inversion layers on Si. The electron concentration n is plotted vs. the average penetration depth z_{av} at 4.2 K. $N_s = 10^{13}$ cm^{-2} is the total surface concentration, $N_a - N_d = 10^{15}$ cm^{-3} is the net doping concentration of the substrate [70S1].

Fig. 6. Al–SiO$_2$–Si. Interface state density D_{it} for Si vs. the orientation of the oxidized Si surface; positive angles θ are measured by rotation about the [1$\bar{1}$0] axis, negative angles θ by rotation about the [10$\bar{1}$] axis [68A1].

Figs. 8 and 9, see next page.

Fig. 10. Si. Electronic g^* factor for inversion layer electrons on Si (100), (110), and (111) surfaces as a function of the total surface electron density n_s (the surface orientations are indicated in the figure) [68F1, 75L1]. ▶

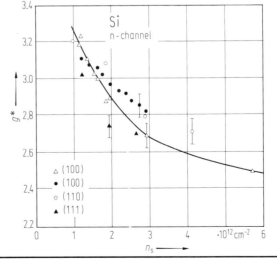

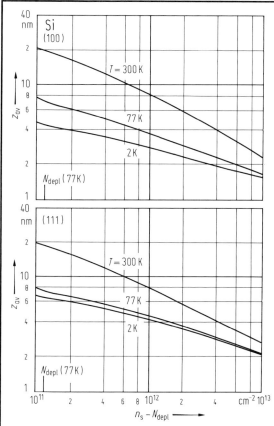

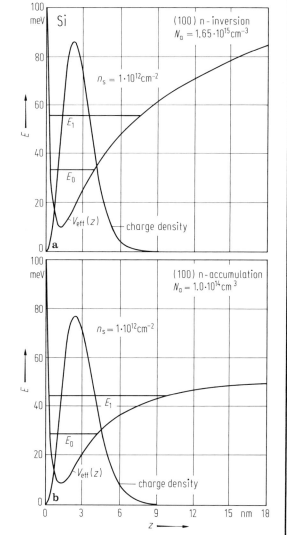

Fig. 8. Si. Average penetration depth z_{av} of inversion layer electrons as a function of the total density of charges in the inversion and depletion layers for (100), and (111) surfaces of p-type Si with $N_a - N_d = 10^{15}$ cm^{-3} at various temperatures. The values for the (110) surface are close to those for a (111) surface [70S1].

Fig. 11. Si. Calculated subband structure in electron space charge layers on the Si (100) surface as a function of depth z. $V_{eff}(z)$ is the effective surface potential perpendicular to the surface.
a) inversion layer with $N_{depl} = 1.6 \cdot 10^{11}$ cm^{-2},
b) accumulation layer.
(E_0 and E_1 are the ground and first excited states, respectively).

Fig. 9. Si. Constant energy ellipses and Brillouin zones of the two-dimensional electron gas in Si inversion layers on (100), (110), and (111) surfaces. (The numbers correspond to Table 4 in the text).

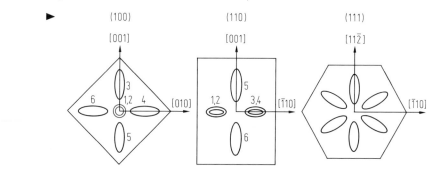

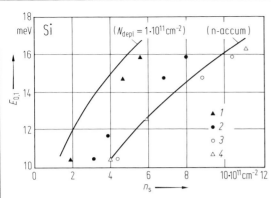

Fig. 12. Si. Energy separation $E_{0,1}$ of the ground and the first excited subband in an electron inversion layer on the Si (100) surface. The full lines represent the theory [76A1]. Experimental results for various methods:
1: $N_{depl} = 1 \cdot 10^{11}$ cm, optical absorption [76K1],
2: $N_{depl} = 6 \cdot 10^{10}$ cm^{-2}, photoconductivity [77N1],
3: $N_{depl} = 1 \cdot 10^{11}$ cm^{-2}, photoconductivity [76W1],
4: $N_{depl} = 1.2 \cdot 10^{11}$ cm^{-2}, optical emission [76G1].

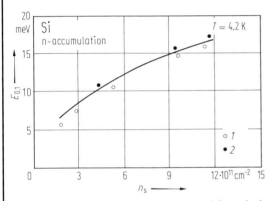

Fig. 13. Si. Subband energies $E_{0,1}$ measured from the bottom of the ground subband as a function of electron concentration n_s in an electron accumulation layer on (100) Si. The full line represents the theory [76A1]. Experimental results:
1: optical absorption [76K1], 2: photoconductivity [77N1].

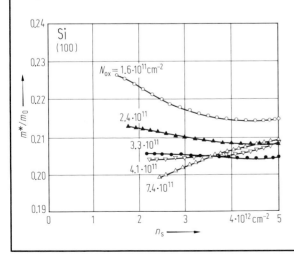

Fig. 16. Si. Relative density of states effective mass m^*/m_0 as a function of electron concentration n_s for an inversion layer on a Si (100) surface. Parameter is the charge in the oxide N_{ox}. Shubnikov-de Haas experiments at $B = 3.88$ T [77F1].

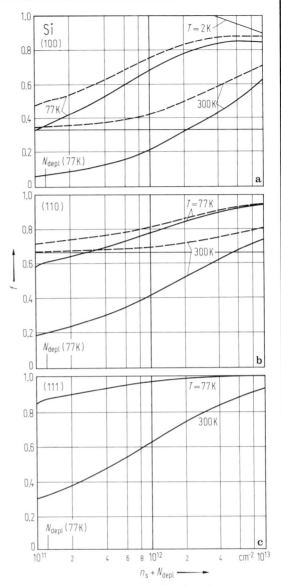

Fig. 14a–c. Si. Calculated fraction f of electrons in the lowest subband (full curves), and in all subbands associated with the valleys having the largest value of m_z^* (dashed curves) ($N_a - N_d = 10^{15}$ cm^{-3}).
a) Si (100) surface,
b) Si (110) surface,
c) Si (111) surface [72S2].

For Fig. 15, see next page

◄

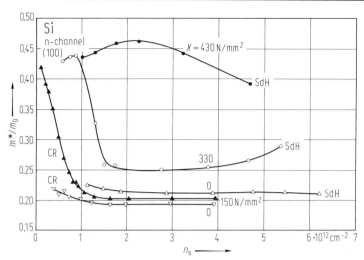

Fig. 15. Si. Relative density of states effective mass m^*/m_0 for an electron inversion layer on a Si (100) surface as a function of electron concentration n_s for uniaxial mechanical stress X. Measurements are obtained from cyclotron resonance CR [76S1] and Shubnikov-de Haas (SdH) measurements [78E1].

Fig. 17. Si. Relative density of states effective mass m^*/m_0 as a function of electron concentration n_s for an inversion layer on a Si (100) surface. Parameter is the substrate bias voltage V_{sub}. Shubnikov-de Haas experiments at $B = 2.5$ T; substrate: 0.5 Ωcm, p-type; oxide thickness: 120 nm [77F1].

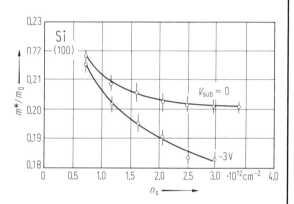

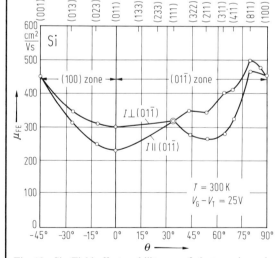

Fig. 18. Si. Field effect mobility μ_{FE} of electrons in an inversion Si layer as a function of surface orientation indicated at the upper scale (θ at the abscissa denotes the angle between the surface normal and the normal of the (011) surface) for two current directions indicated at the curves. Oxide thickness: 200 nm, substrate doping and threshold voltage are listed in the table [71S1].

Table for Fig. 18. Si.

Surface orientation	current direction	substrate [Ω cm]	threshold voltage [V]
(100)		23.7	−1.4
(811)	∥ [01$\bar{1}$]	24.0	−1.9
(811)	⊥ [01$\bar{1}$]	24.6	−3.2
(411)	∥ [01$\bar{1}$]	8.0	−1.7
(411)	⊥ [01$\bar{1}$]	8.3	−1.6
(311)	∥ [01$\bar{1}$]	23.1	−3.0
(311)	⊥ [01$\bar{1}$]	24.7	−2.9
(211)	∥ [01$\bar{1}$]	24.3	−2.3
(211)	⊥ [01$\bar{1}$]	24.0	−2.6
(322)	∥ [01$\bar{1}$]	8.7	−2.1
(322)	⊥ [01$\bar{1}$]	8.5	−1.7
(111)		8.3	−4.0
(011)	⊥ [01$\bar{1}$]	27.5	−2.5
(011)	∥ [01$\bar{1}$]	27.3	−2.2
(023)	∥ [100]	15.8	−2.3
(023)	⊥ [100]	15.7	−2.5
(013)	∥ [100]	8.7	−1.3
(013)	⊥ [100]	8.5	−1.6

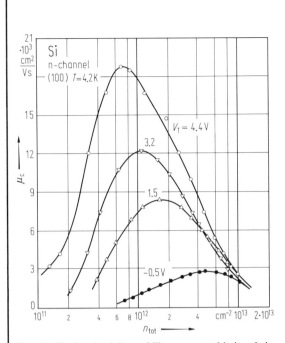

Fig. 19. Si. Conductivity mobility μ_c vs. total induced electron concentration n_{tot} at 4.2 K for an inversion layer on a Si (100) surface. Parameter is the threshold voltage V_T which reflects the influence of charges in the insulator. Substrate: $4\cdots7\,\Omega$cm, p-type; oxide thickness: 120 nm [84E1].

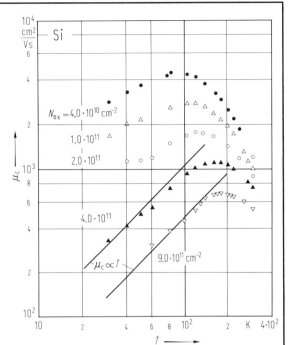

Fig. 21. Si. Conductivity mobility μ_c of electrons for an inversion layer on a Si (100) surface vs. temperature T. Parameter is the oxide charge density N_{ox}. Acceptor doping $N_a = 2.2 \cdot 10^{14}\,\mathrm{cm}^{-3}$; surface field $E_z = 2.5 \cdot 10^4\,\mathrm{V/cm}$; substrate: $70\,\Omega$cm; oxide thickness: 500 nm [72S3].

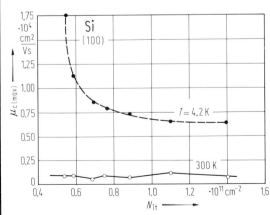

Fig. 20 Si Maximum conductivity mobility $\mu_{c\,(max)}$ vs. surface trap density N_{it} at 4.2 and 300 K for an electron inversion layer on a Si (100) surface; oxide thickness: 100 nm [74K1].

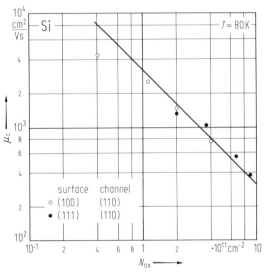

Fig. 22. Si. Conductivity mobility μ_c of electrons for an inversion layer on a Si (100) and (111) surface vs. total oxide charge density N_{ox} at $T = 80$ K. Full line:

$$\mu^{-1} \propto N_{ox} = \frac{1}{e}(Q_{ox} + Q_{ot} + Q_{it})$$

with Q_{ox}: fixed oxide charge, Q_{ot}: charge in oxide traps, Q_{it}: charge in interface traps. Acceptor doping $N_a = 2.2 \cdot 10^{14}\,\mathrm{cm}^{-3}$; substrate: $70\,\Omega$cm; oxide thickness 500 nm [72S3].

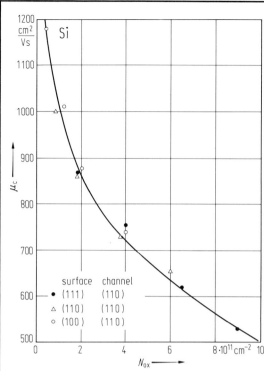

Fig. 23. Si. Conductivity mobility μ_c in [110] direction of electrons for inversion layers on a Si (100), (110) and (111) surface vs. total oxide charge density N_{ox} at 300 K. Acceptor doping $N_a = 2.2 \cdot 10^{14}\,cm^{-3}$; substrate: $70\,\Omega cm$; oxide thickness 500 nm [72S3].

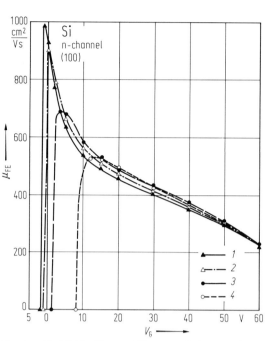

Fig. 25. Si. Field effect mobility μ_{FE} of electrons for an inversion layer on a Si(100) surface vs. gate voltage V_G for various acceptor concentrations N_a at $T = 295$ K.
1: $N_a = 1.2 \cdot 10^{14}\,cm^{-3}$,
2: $N_a = 3.3 \cdot 10^{15}\,cm^{-3}$,
3: $N_a = 3.2 \cdot 10^{16}\,cm^{-3}$,
4: $N_a = 1.3 \cdot 10^{17}\,cm^{-3}$.
Oxide thickness: 200 nm [71S1].

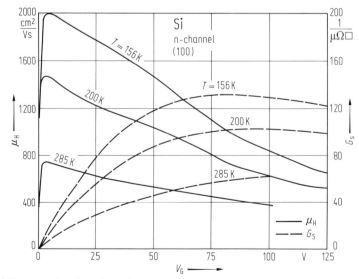

Fig. 24. Si. Hall mobility μ_H and surface channel conductance G_s of electrons for an inversion layer on a Si(100) surface vs. gate voltage V_G at various temperatures. Substrate: $5 \cdots 20\,\Omega cm$; oxide thickness: $200 \cdots 300$ nm [69M1].

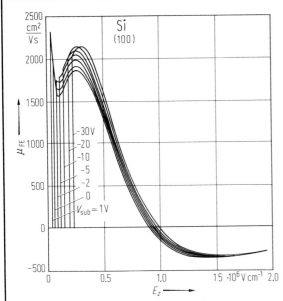

Fig. 26. Si. Field effect mobility μ_{FE} of electrons in an inversion layer on a Si (100) surface as a function of electric surface field E_z perpendicular to the surface at $T = 77$ K. Parameter is the substrate bias voltage V_{sub}. Substrate: $2\,\Omega$cm [68F2].

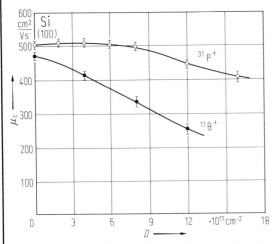

Fig. 27. Si. Conductivity mobility μ_c of inversion layer electrons at a Si (100) surface vs. total implantation dose D of 50 keV ^{11}B$^+$ and 120 keV ^{31}P$^+$ ions. $T = 300$ K. Mobility measurements taken at a drain voltage $V_D = 5$ V and an effective gate voltage $V_G - V_T = 5$ V. Implantation through gate oxide. Substrate: $4\,\Omega$cm; oxide thickness: 100 nm [74K2].

▶

Fig. 30. Si. Drift velocity v_d of electrons in an inversion layer (measured at source and drain of an MOS structure) vs. average electric field V_D/L along the Si (100) surface for two gate voltages V_G. Surface doping concentration (by implantation): $N = 1.5 \cdot 10^{16}$ cm^{-3}; oxide thickness: 50 nm; channel length $L = 0.5$ µm; $V_T = 0$ V [80M1].

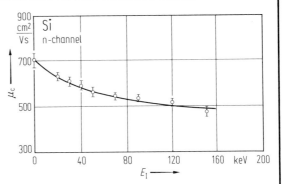

Fig. 28. Si. Conductivity mobility μ_c of electrons for an inversion layer on a Si (100) surface vs. implantation energy E_I for an implantation of ^{11}B$^+$ ions with a dose of $D = 5 \cdot 10^{11}$ cm^{-2}; $T = 300$ K. Measurements taken at an effective gate voltage $V_G - V_T = 5$ V. Implantation through gate oxide. Substrate: $4\,\Omega$cm; oxide thickness: 100 nm [74K2].

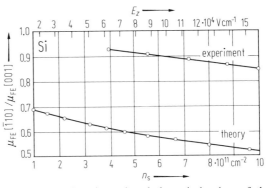

Fig. 29. Si. Experimental and theoretical values of the ratio of electron field effect mobility in two directions $(\mu_{FE}[\bar{1}10]/\mu_{FE}[001])$ on a Si (110) surface vs. electric surface field E_z and free electron concentration n_s at $T = 77$ K. Oxide thickness: 200 nm [72S4].

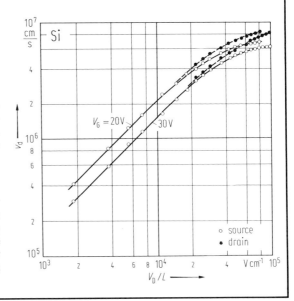

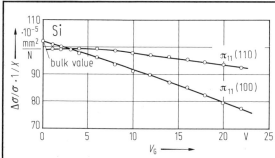

Fig. 31. Si. Longitudinal piezoresistance coefficients

$$\pi = \frac{\Delta\sigma}{\sigma} \cdot \frac{1}{X}$$

for electron inversion layers on Si (100) and (110) surfaces vs. gate voltage V_G at $T = 300$ K. Substrate: $10\,\Omega$cm; oxide thickness: 120 nm [71D1].

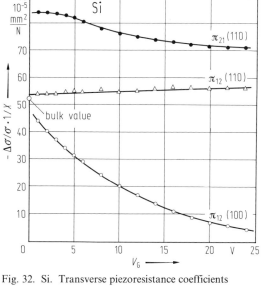

Fig. 32. Si. Transverse piezoresistance coefficients

$$\pi = \frac{\Delta\sigma}{\sigma} \cdot \frac{1}{X}$$

for electron inversion layers on Si (100) and (110) surfaces vs. gate voltage V_G at $T = 300$ K. Substrate: $10\,\Omega$cm; oxide thickness: 120 nm [71D1].

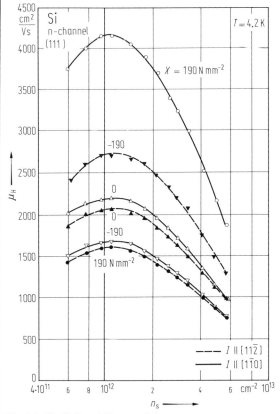

Fig. 34. Si. Hall mobility μ_H of electrons vs. concentration of free electrons n_s in an inversion layer on a Si (111) surface, for various stresses X in [11$\bar{2}$] direction. $T = 4.2$ K, $B = 1$ T. Substrate: $10\,\Omega$cm; oxide thickness: 120 nm [78D1].

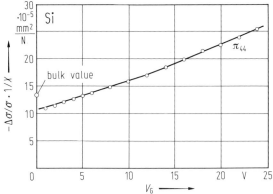

Fig. 33. Si. Shear piezoresistance coefficient

$$\pi = \frac{\Delta\sigma}{\sigma} \cdot \frac{1}{X}$$

for electron inversion layers on a Si (100) surface vs. gate voltage V_G. $T = 300$ K; substrate: $10\,\Omega$cm; oxide thickness: 120 nm [71D1].

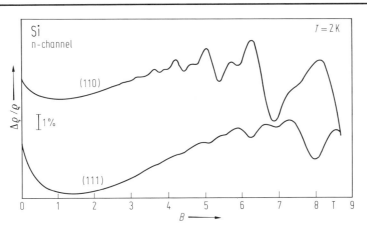

Fig. 35. Si. Oscillatory magnetoresistivity $\Delta\varrho/\varrho$ of electrons vs. magnetic induction B perpendicular to the Si (110) and (111) surfaces (Shubnikov-de Haas effect). $T=2$ K; substrate: $7\,\Omega$cm; oxide thickness: 120 nm; free electron concentration $n_s = 2.3 \cdot 10^{12}$ cm^{-2} [71S2].

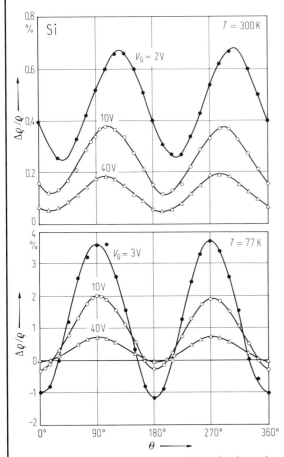

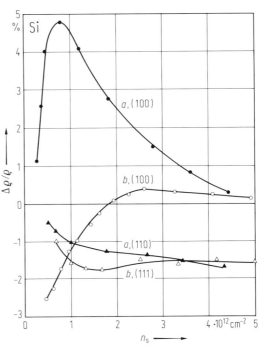

Fig. 36. Si. Magnetoresistivity $\Delta\varrho/\varrho$ of inversion layer electrons as a function of the angle θ between the direction of the magnetic field ($B=1$ T) and the (111) surface for various gate voltages V_G at $T=300$ K, and $T=77$ K. Substrate: $10\,\Omega$cm; oxide thickness: 300 nm [71S2].

Fig. 37. Si. Magnetoresistivity $\Delta\varrho/\varrho$ vs. electron concentration n_s for inversion layers on Si (100), (110), and (111) surfaces. $T=4.2$ K, $B=1$ T. (a) $N_{it}=1\cdot 10^{11}$ cm^{-2}, (b) $N_{it}=2\cdot 10^{12}$ cm^{-2}. Substrate: $7\,\Omega$cm; oxide thickness: 120 nm [74D1].

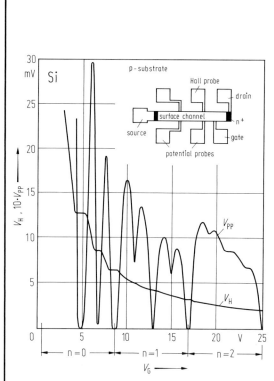

Fig. 38. Si. Hall voltage V_H, and voltage drop between potential probes V_{pp} as a function of gate voltage V_G for a Si (100) surface. The source drain current is 1 μA. The insert shows a top view of the device with length $L = 400$ μm, and a distance between the potential probes $L_{pp} = 130$ μm. $T = 1.5$ K, $B = 18$ T. Substrate: 10 Ωcm; oxide thickness: 100 nm [80K1]. (n: number of subband).

Fig. 40. Si. Energy spacing between the lowest two subbands $E_{0,1}$ as a function of hole concentration p_s for inversion and accumulation layers. The solid lines represent the theoretical calculations [75O1]. The circles are experimental results obtained by optical absorption measurements at $T = 4.2$ K. Substrate: 6···10 Ωcm; oxide thickness: 230 nm [76K1].

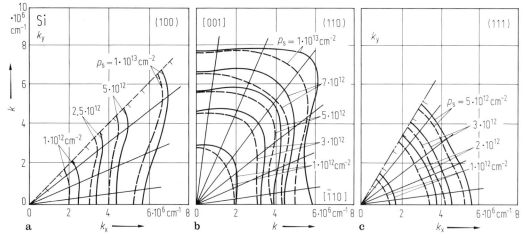

Fig. 39a–c. Si. Section of constant energy contours as a function of wave vector k for the valence band in Si inversion layers. $N_d - N_a = 10^{15}$ cm^{-3}. Parameter is the concentration of holes p_s in the surface. The full and the dashed curves correspond to a splitting of the spin degeneracy. a) Si (100) surface, b) Si (110) surface, c) Si (111) surface. It should be noted that the warped energy surfaces are caused by the coupling of the so-called "heavy hole band" and the "light hole band" (see Fig. 51) [75O1].

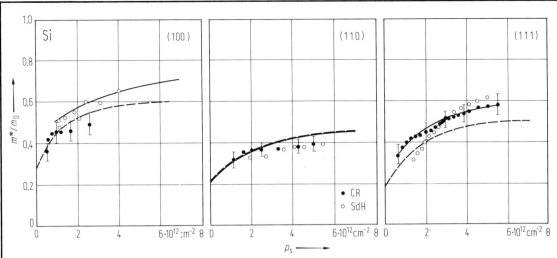

Fig. 41. Si. Density of states effective mass m^*/m_0 vs. hole concentration p_s in an inversion layer on Si (100), (110), and (111) surfaces.
SdH: Evaluation from Shubnikov-de-Haas oscillations; substrate: 7 Ωcm; oxide thickness: 120 nm [74K3, 74K4].
CR: Evaluation from cyclotron resonance; substrate: ≈ 10 Ωcm; oxide thickness: 200 nm [77K1].
Theoretical results are included: solid curves from [75O1], dashed curves from [74B1].

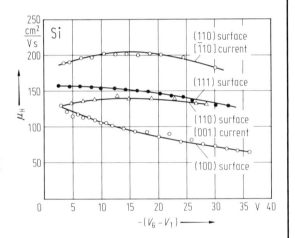

Fig. 43. Si. Hall mobility μ_H of holes vs. effective gate voltage $(V_G - V_T)$ for inversion layers on Si (100), (110), and (111) surfaces at $T = 300$ K. The current direction along the surface is indicated. Substrate: 2 Ωcm; oxide thickness: 100 nm [68C1].

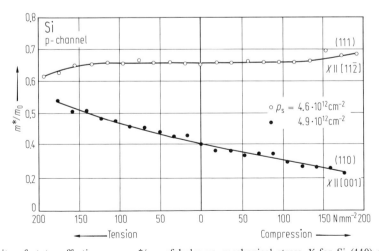

Fig. 42. Si. Density of states effective mass m^*/m_0 of holes vs. mechanical stress X for Si (110) and (111) surfaces. Evaluation from Shubnikov-de-Haas measurements. p_s is the hole concentration. Substrate: 10 Ωcm; oxide thickness: 120 nm [78E1].

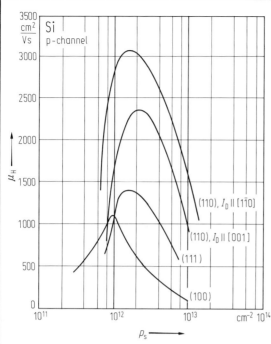

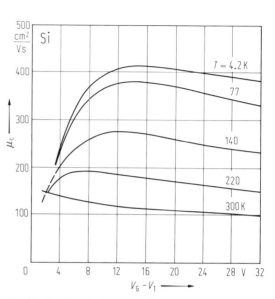

Fig. 44. Si. Hall mobility μ_H vs. hole concentration p_s for inversion layers on Si (100), (110), and (111) surfaces at $T = 4.2$ K. Results for two current directions I_D along the surface are plotted for the (110) surface. Substrate: 7 Ωcm, oxide thickness: 120 nm [84E1].

Fig. 45. Si. Conductivity mobility μ_c of holes vs. effective gate voltage $(V_G - V_T)$ for inversion layers on a Si (100) surface at various temperatures. Surface state density $N_{it} = 9.5 \cdot 10^{12}$ cm^{-2}; substrate: 10 Ωcm; oxide thickness: 110 nm [73C2].

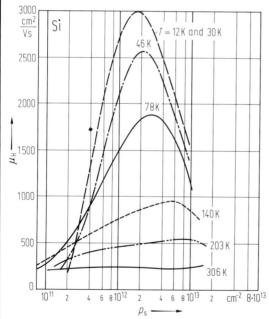

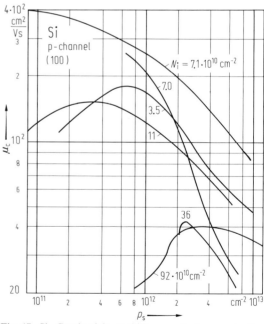

Fig. 46. Si. Hall mobility μ_H vs. hole concentration p_s for inversion layers on a Si (110) surface at various temperatures. Current flow along the surface in [001] direction. Substrate: 7 Ωcm; oxide thickness: 120 nm [84E1].

Fig. 47. Si. Conductivity mobility μ_c vs. hole concentration p_s for inversion layers on a Si (100) surface for various impurity concentrations N_i at the surface. $T = 4.2$ K; substrate: 10 Ωcm; oxide thickness: 110 nm [73C1].

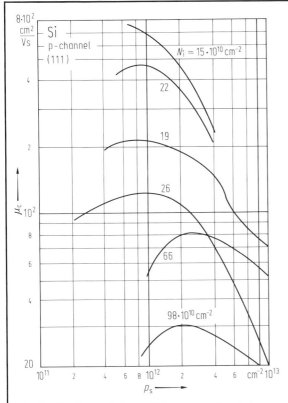

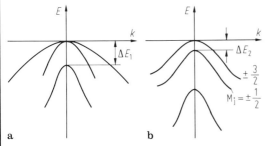

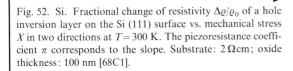

Fig. 48. Si. Conductivity mobility μ_c vs. surface hole concentration p_s for inversion layers on a Si (111) surface for various impurity concentrations N_i at the surface. $T = 4.2$ K; substrate: $10\,\Omega$cm; oxide thickness: 110 nm [73C1].

Fig. 51a, b. Si. Schematic diagrams of the valence band structure of silicon. a) unstrained, ΔE_1 due to spin-orbit coupling, b) mechanical compression causing ΔE_2.

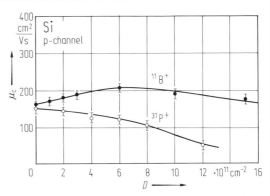

Fig. 49. Si. Conductivity mobility μ_c of inversion layer holes vs. implantation dose D of 50 keV ^{11}B$^+$ and 129 keV ^{31}P$^+$ ions at $T = 300$ K. Implantation through gate oxide. Mobility measurements taken at a drain voltage $V_D = 5$ V and an effective gate voltage $V_G - V_T = 5$ V. Substrate: $4\,\Omega$cm; oxide thickness: 100 nm [74K2].

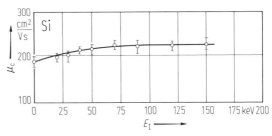

Fig. 50. Si. Conductivity mobility μ_c of inversion layer holes as a function of the implantation energy E_I of ^{11}B$^+$ ions. Implantation dose $D = 5 \cdot 10^{11}$ cm^{-2}, $T = 300$ K. Implantation through gate oxide. Substrate: $4\,\Omega$cm; oxide thickness: 100 nm [74K2].

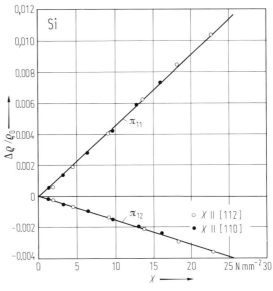

Fig. 52. Si. Fractional change of resistivity $\Delta\varrho/\varrho_0$ of a hole inversion layer on the Si (111) surface vs. mechanical stress X in two directions at $T = 300$ K. The piezoresistance coefficient π corresponds to the slope. Substrate: $2\,\Omega$cm; oxide thickness: 100 nm [68C1].

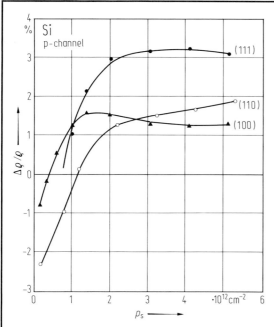

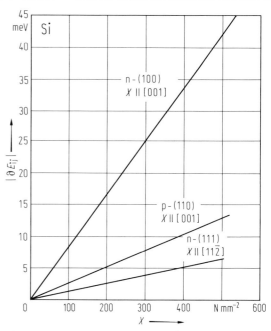

Fig. 53. Si. Longitudinal magnetoresistivity $\Delta\varrho/\varrho$ vs. hole concentration p_s for an inversion layer on Si (100), (110), and (111) surfaces. $T = 4.2$ K, $B = 1$ T in z-direction. Substrate: $7\,\Omega$cm; oxide thickness: 120 nm [74D1].

Fig. 55. Si. Energy shift ∂E_{ij} of $E(k)$ valleys for n- and p-type surface space charge layers vs. stress X. Si n-(100) surface: lowering of valleys 3, 5 with respect to 1, 2, 4, 6 (cf. Fig. 9). Si p-(110) surface: for all p-type surfaces, the heavy and light hole bands are splitted (cf. Fig. 51).

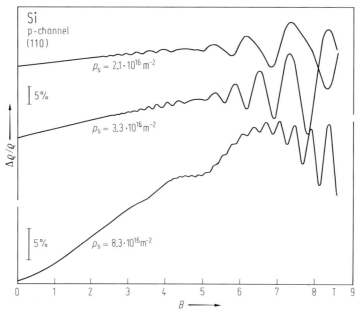

Fig. 54. Si. Magnetoresistivity $\Delta\varrho/\varrho$ of holes vs. magnetic induction B perpendicular to the (110) surface for three hole concentrations p_s. $T = 1.64$ K; substrate: $4\,\Omega$cm; oxide thickness: 120 nm [74D1].

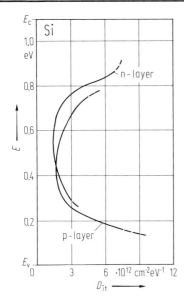

Fig. 56. Si. Energy distribution of the interface state density D_{it} in the forbidden energy gap between valence band E_V and conduction band E_C of Si at the Si-sapphire interface [75G1].

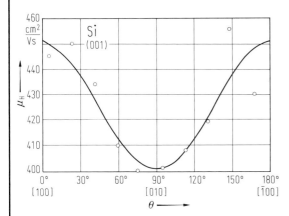

Fig. 59. Si. Room temperature Hall mobility μ_H of electrons for (001) Si on (01$\bar{1}$2) Al$_2$O$_3$. Silicon film thickness: 1.5⋯1.8 μm; doping concentration: $N = (1⋯6) \cdot 10^{16}$ cm^{-3} [73H1].

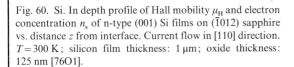

Fig. 60. Si. In depth profile of Hall mobility μ_H and electron concentration n_s of n-type (001) Si films on ($\bar{1}$012) sapphire vs. distance z from interface. Current flow in [110] direction. $T = 300$ K; silicon film thickness: 1 μm; oxide thickness: 125 nm [76O1].

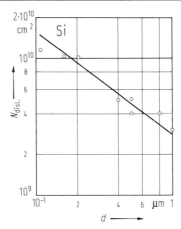

Fig. 57. Si. Surface dislocation density N_{disl} vs. film thickness d in (100) Si films on sapphire [76O1].

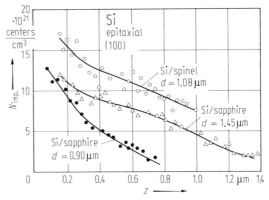

Fig. 58. Si. In depth profile of imperfections in (100) epitactic Si films of various thickness d, observed by ion channeling technique. Density of imperfections N_{imp} vs. distance z from epitaxial interface [73P1].

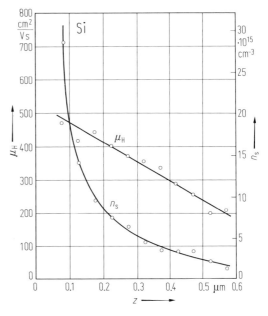

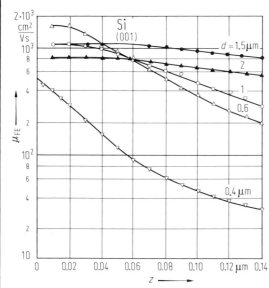

Fig. 61. In depth profile of the field effect mobility μ_{FE} vs. distance z from interface for deep depletion in n-type (001) Si on sapphire for various values of the Si film thickness d [75H1].

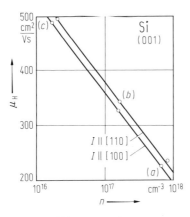

Fig. 62. Si. Hall mobility μ_H vs. electron concentration n in (001) Si films on ($\bar{1}$012) sapphire. The two current flow directions are indicated. $T = 300$ K; silicon film thickness: 1 μm; doping concentrations of the various samples are characterized by their resistivities: for $I \parallel [100]$: (a) $4.11 \cdot 10^{-2}$ Ωcm, (b) $1.19 \cdot 10^{-1}$ Ωcm, (c) $7.47 \cdot 10^{-1}$ Ωcm; for $I \parallel [110]$: (a) $3.18 \cdot 10^{-2}$ Ωcm, (b) $1.10 \cdot 10^{-1}$ Ωcm, (c) $5.96 \cdot 10^{-1}$ Ωcm [76O1].

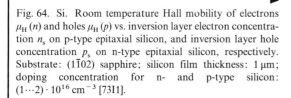

Fig. 64. Si. Room temperature Hall mobility of electrons μ_H (n) and holes μ_H (p) vs. inversion layer electron concentration n_s on p-type epitaxial silicon, and inversion layer hole concentration p_s on n-type epitaxial silicon, respectively. Substrate: ($1\bar{1}$02) sapphire; silicon film thickness: 1 μm; doping concentration for n- and p-type silicon: $(1 \cdots 2) \cdot 10^{16}$ cm^{-3} [73I1].

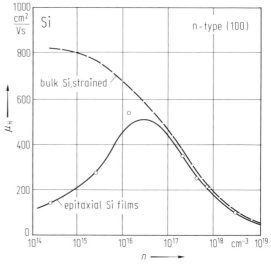

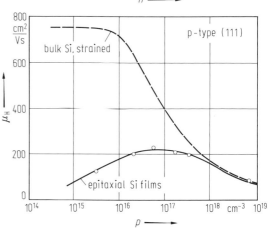

Fig. 63. Si. Room temperature Hall mobility μ_H vs. concentration of electrons n in (100) epitaxial Si and holes p in (111) epitaxial Si on spinel at $T = 300$ K. The dotted line has been calculated for the compressive strain listed in Table 7b. Si film thickness: 20 μm [68S1].

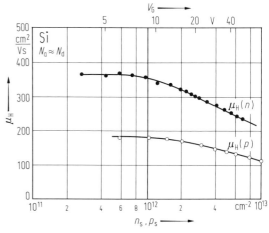

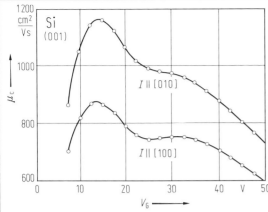

Fig. 65. Si. Conductivity mobility μ_c vs. gate voltage V_G for an electron inversion layer on (001) Si films. The two different current directions are indicated. $T = 1.4$ K; substrate: ($\bar{1}012$) sapphire [76K2].

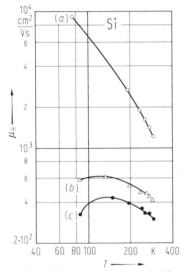

Fig. 66. Si. Hall mobility μ_H vs. temperature T for electrons in epitaxial Si films. Electron concentration $n \approx 10^{16}$ cm^{-3}. (a) epitaxial Si on silicon: Low crystal defect density, (b) epitaxial Si on silicon: high crystal defect density, (c) epitaxial silicon on spinel [68S1].

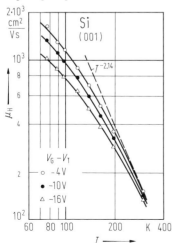

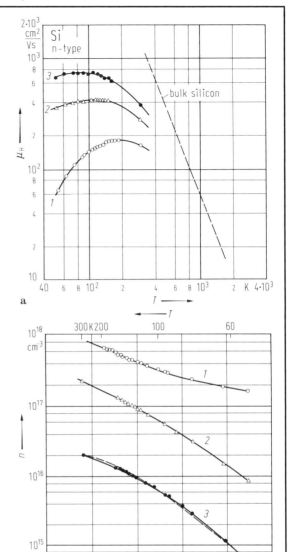

a

b

Fig. 67a, b. Si. a) Hall mobility μ_H for n-type (001) Si films on ($\bar{1}012$) sapphire vs. temperature T. b) Corresponding temperature dependence of the electron concentration n. Silicon film thickness: 1 μm. Parameter is the doping concentration between $2.0 \cdot 10^{16}$ cm^{-3} and $1.0 \cdot 10^{18}$ cm^{-3} [76O1].

For Fig. 68a, b, see next page.

◄

Fig. 69. Si. Hall mobility μ_H for holes in space charge layers vs. temperature T. Current in [110] direction. Parameter is the effective gate voltage $V_G - V_T$. Substrate: (001) Si on ($\bar{1}012$) sapphire; silicon film thickness: 0.5 μm; doping concentration: $1 \cdot 10^{15}$ cm^{-3}; oxide thickness: 120 nm [78O1].

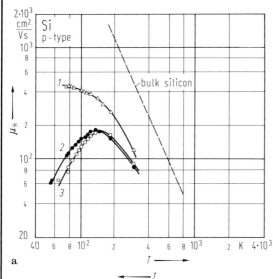

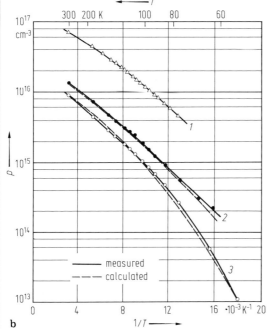

a

b

Fig. 68a, b. Si. a) Hall mobility μ_H for p-type (001) Si films on ($\bar{1}$012) sapphire vs. temperature T. b) Corresponding temperature dependence of the hole concentration p. Silicon film thickness: 1 μm. Dislocation density: $N = (2.05 \cdots 3.34) \cdot 10^9$ cm^{-2}. Parameter is the doping concentration: *1*: $N_a = 7.0 \cdot 10^{16}$ cm^{-3}, *2*: $N_a = 1.80 \cdot 10^{16}$ cm^{-3}, $N_d = 3.6 \cdot 10^{14}$ cm^{-3}, *3*: $N_a = 1.34 \cdot 10^{16}$ cm^{-3}, $N_d = 1.09 \cdot 10^{15}$ cm^{-3} [78O1].

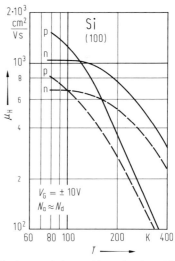

Fig. 70. Si. Accumulation and inversion layer Hall mobility μ_H vs. temperature T for space charge layer concentrations (n_s, p_s) of the order of $(1 \cdots 3) \cdot 10^{12}$ cm^{-3}. The Si films have a (100) surface and are grown on ($\bar{1}$02) sapphire. Silicon film thickness: 1 μm; n- and p-type doping: $(1 \cdots 2) \cdot 10^{16}$ cm^{-3}. The solid curves represent the accumulation layer mobilities (p = holes, n = electrons) and the dashed curves represent the inversion layer mobilities (p = holes, n = electrons) [73I1].

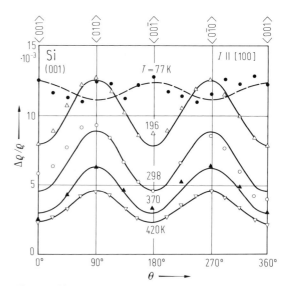

Fig. 71. Si. Transverse magnetoresistivity $\Delta\varrho/\varrho$ in [100] current direction for a n-type (001) Si film on ($\bar{1}$012) sapphire as a function of the angle θ between magnetic induction B and the surface at various temperatures. $B = 1.46$ T [76O2].

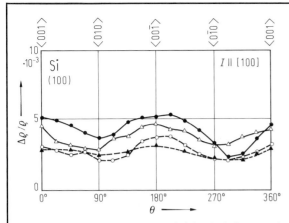

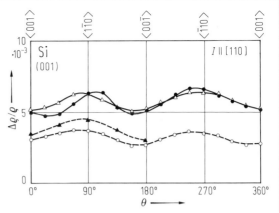

Fig. 72. Si. Transverse magnetoresistivity $\Delta\varrho/\varrho$ for several samples in [100] current direction for a p-type (100) Si film on ($\bar{1}012$) sapphire as a function of the angle θ between magnetic induction B and the surface. $T=298$ K; $B=1.62$ T; silicon film thickness: $1\cdots2$ µm [76O3].

Fig. 73. Si. Transverse magnetoresistivity $\Delta\varrho/\varrho$ for several samples in [110] current direction for a p-type (001) Si film on ($\bar{1}012$) sapphire as a function of the angle θ between magnetic induction B and the surface. $T=298$ K; $B=1.62$ T; silicon film thickness: $1-2$ µm [76O3].

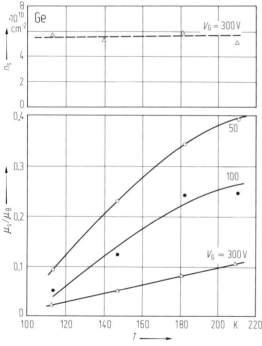

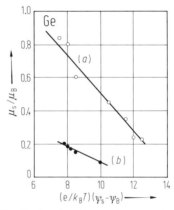

Fig. 74. Ge. Ratio of surface to bulk mobility μ_s/μ_B, as a function of the energy band bending $(e/k_B T)(\psi_s-\psi_B)$ at $T=80$ K. (a) electron inversion layer [77D1], (b) hole inversion layer [76D2]. Substrate: $N_d-N_a\approx5\cdot10^{10}$ cm^{-3}; insulator: 0.01 mm polyester.

Fig. 75. Ge. Concentration of inversion layer electrons n_s and ratio of surface to bulk mobility μ_s/μ_B as a function of temperature T. Parameter is the gate voltage V_G. Substrate: $N_d-N_a\approx5\cdot10^{10}$ cm^{-3}; insulator: 0.01 mm polyester [77D1].

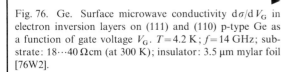

Fig. 76. Ge. Surface microwave conductivity $d\sigma/dV_G$ in electron inversion layers on (111) and (110) p-type Ge as a function of gate voltage V_G. $T=4.2$ K; $f=14$ GHz; substrate: $18\cdots40$ Ωcm (at 300 K); insulator: 3.5 µm mylar foil [76W2].

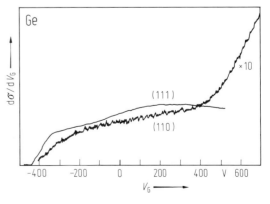

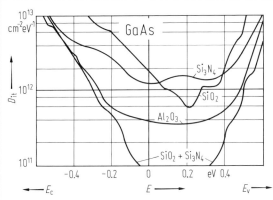

Fig. 77. GaAs. Interface state density D_{it} across the band-gap of GaAs for different insulators. Substrate: (100) surface of p-type GaAs (doping concentration $(1\cdots2)\cdot10^{17}$ cm^{-3}) [74M1].

For Fig. 80, see next page.

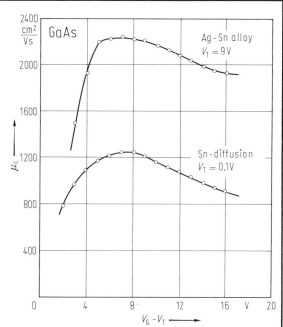

Fig. 78. GaAs. Conductivity mobility μ_c of inversion layer electrons vs. effective gate voltage $V_G - V_T$. Parameter is the diffusion process for source and drain contacts. $T = 300$ K; substrate: (111) surface on p-type, Zn-doped $(N_a = 7\cdots9\cdot10^{16}$ cm$^{-3})$ GaAs; insulator: Al$_2$O$_3$ $(50\cdots150$ nm) and SiO$_2$ $(100\cdots200$ nm) [74I1].

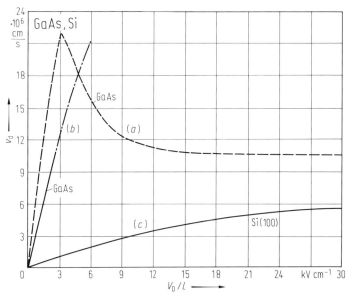

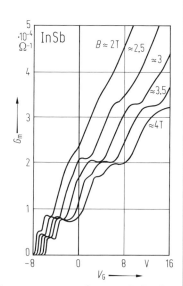

Fig. 79. GaAs, Si. Steady state drift velocity v_d as a function of electric field $E = V_D/L$ for electrons in Si n-type inversion layers and an electron channel in a GaAs metal-semiconductor structure [79E1]. (a) GaAs $> 10^6$ Ωcm; substrate: semiinsulating ($\approx10^8$ Ωcm) GaAs, double implantation for the channel [67R1]. (b) GaAs substrate: $N_d = 2\cdot10^{17}$ cm^{-3}. (c) Si (100); $n_s = 6.6\cdot10^{12}$ cm^{-2}.

Fig. 81. InSb. Oscillatory magnetoconductance G_m for electron inversion layer on (111) p-type InSb as a function of gate voltage V_G. Parameter is the magnetic induction B. $T = 4.2$ K; substrate: $N_a \approx 2\cdot10^{13}$ cm^{-3}; insulator: SiO$_2$ with a thickness of 550 nm; $w/L = 1.5$ [72K2].

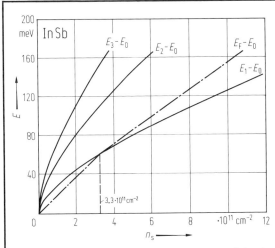

Fig. 80. InSb. Calculated energy of the bottom of the electric subbands $E_n - E_o$ and the Fermi energy $E_F - E_o$ as a function of surface electron concentration n_s for n-type inversion layers on (111) surfaces. $T = 0$ K; E_o is the energy of the lowest subband. Assumptions: $m^* = 0.013\,m_o$, $\varepsilon_s = 16$ [72K2].

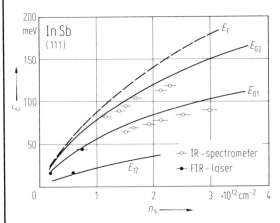

Fig. 82. InSb. Subband spacings for electron inversion layers on (111) surfaces as a function of electron concentration n_s. Substrate: p-type, doped with $0.7 \cdots 1.4 \cdot 10^{13}$ cm^{-3} Ge (at 77 K); insulator: 2 μm lacquer [77B1]. The lines represent calculations of [77A1].

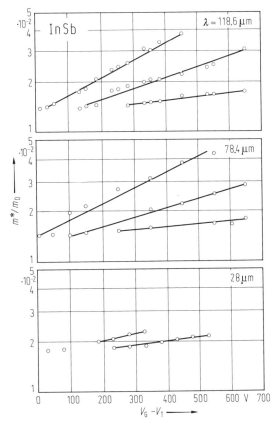

Fig. 83. InSb. Density of states effective mass m^*/m_o determined by cyclotron resonance vs. effective gate voltage $V_G - V_T$ for electron inversion layers on the (111) InSb surface. Each line indicates the mass variation in a given subband [75D1]. λ is the wavelength of the incident infrared radiation. For comparison: density of states effective mass in the bulk: $m^* = 0.0155\,m_o$; substrate: p-type InSb ($N_a \approx 1.2 \cdot 10^{13}$ cm^{-3}) at 77 K; insulator: 3.5 μm mylar foil.

Fig. 84. InSb. Electron concentration n_s for various subbands as a function of total electron concentration n_{tot} for the InSb (111) surface at $T = 77$ K. Substrate: p-type InSb ($N_a \approx 1.2 \cdot 10^{13}$ cm^{-3}); insulator: 3.5 μm mylar foil [78D2]. The dotted lines correspond to theory [77A1].

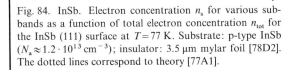
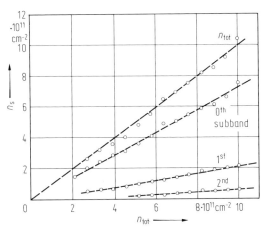

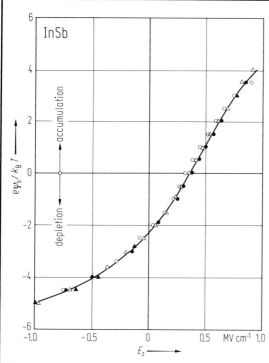

Fig. 85. InSb. Normalized surface potential $(e\psi_s/k_B T)$ as a function of electric field strength E_z at $T=174$ K. Substrate: $N_d \approx 10^{14}$ cm^{-3}; insulator: 4 μm mylar foil [70P1].

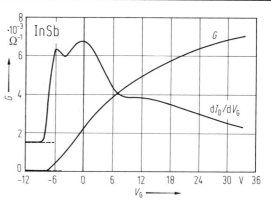

Fig. 87. InSb. Conductance G and transconductance (dI_D/dV_g) for an electron inversion layer on (111) p-type InSb as a function of gate voltage V_G. $T=4.2$ K; substrate: $N_a \approx 2 \cdot 10^{13}$ cm^{-3}; insulator: SiO$_2$ with a thickness of 550 nm [72K2].

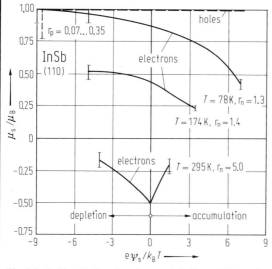

Fig. 86. InSb. Ratio of surface and bulk mobility μ_s/μ_B of electrons and holes as a function of normalized surface potential $(e\psi_s/k_B T)$ for InSb (110) surface. Parameter is the temperature T. r_n, r_p is the scattering parameter defined as the ratio of the mean free path and the Debye length. Substrate: $N_d \approx 10^{14}$ cm^{-3}; insulator: 4 μm mylar foil [70P1].

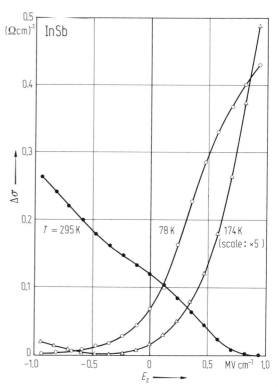

Fig. 88. InSb. Conductivity change $\Delta\sigma$ as a function of electric field strength E_z for various temperatures T. (110) surface of InSb; substrate: $N_d \approx 10^{14}$ cm^{-3}, insulator: 4 μm mylar foil [70P1].

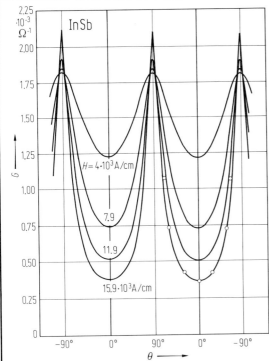

Fig. 89. InSb. Surface conductance G in a magnetic field
vs. the angle θ between the surface normal and the magnetic
field direction for various field strength values H. The points
are calculated: $G(\theta) = G(\theta = 0°)/H \cos \theta$. $T = 4.2$ K; substrate: $N_a \approx 2 \cdot 10^{13}$ cm^{-3}; insulator: SiO$_2$ with a thickness
of 550 nm [72K2].

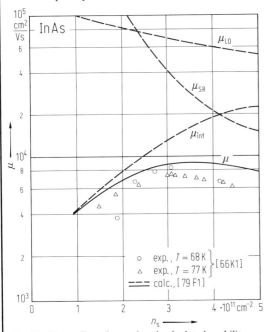

Fig. 91. InAs. Experimental and calculated mobility μ and
the calculated contributions for carriers in an electron inversion layer on (110) InAs as a function of electron concentration n_s. μ_{LO}: phonon scattering, μ_{SR}: surface roughness, μ_{int}: Coulomb scattering, substrate: p-type InAs
($N_a = 7 \cdot 10^{16}$ cm^{-3}) [79F1].

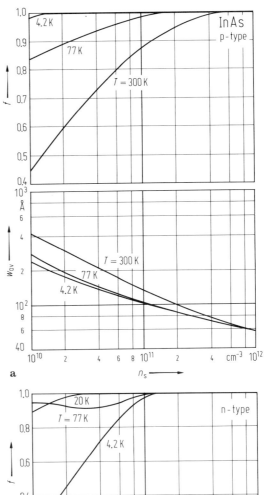

a

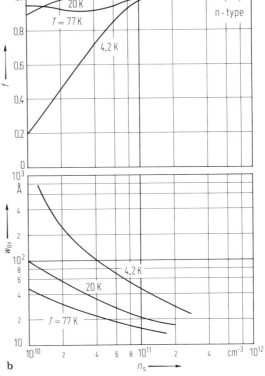

b

Fig. 90a, b. InAs. Calculated fraction f of electrons in the
lowest subband and the average width w_{av} of the quantized
layer as a function of the electron concentration n_s. a) Inversion layer on lightly doped p-type InAs, b) accumulation
layer on moderately doped n-type InAs [79F1].

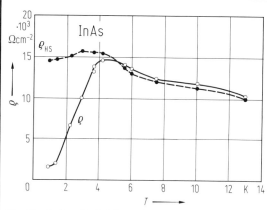

Fig. 92. InAs. Resistivity ϱ (solid curve) and saturation value of the longitudinal magnetoresistivity ϱ_{HS} (dashed curve) vs. temperature T for an electron inversion layer on (110) InAs. Substrate: p-type InAs ($N_a = 7 \cdot 10^{16} \, \text{cm}^{-3}$) [66K1].

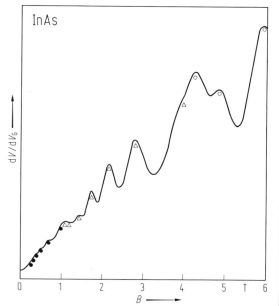

Fig. 93. InAs. Oscillatory magnetoresistance dV/dV_G for a strongly accumulated surface on n-type InAs. $V_G = 20 \, \text{V}$; substrate: $N_d = 2 \cdot 10^{15} \, \text{cm}^{-3}$; insulator: 150 nm SiO$_2$ [78W1]. The peaks corresponding to the three subbands are identified by: open circles: ground state subband, open triangles: first excited subband, full circles: second excited subband.

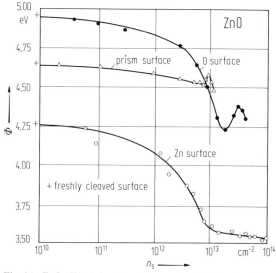

Fig. 94. ZnO. Work function Φ vs. surface electron density n_s (generated by hydrogen treatment) for different surfaces at $T = 100$ K [78K2].

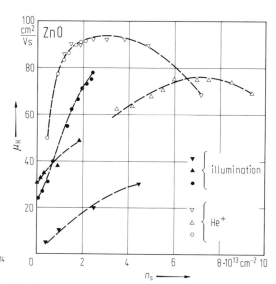

Fig. 95. ZnO. Hall mobility μ_H on the ZnO (000$\bar{1}$) surface as a function of surface electron density n_s at room temperature for various samples. Generation of the accumulation layer by exposure to He$^+$ ions in vacuum or illumination with light. Substrate: 10 Ωcm [79G1].

Eisele

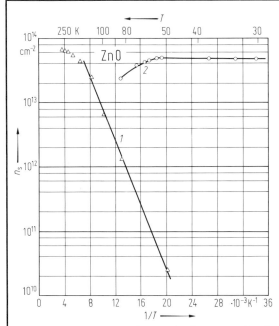

Fig. 96. ZnO. Apparent electron density n_s as a function of $1/T$ on a conducting sample. *1*: Bulk values (before surface treatment), activition energy $\Delta E = 0.1$ eV, $\Delta n_s = 0$; *2*: strong electron accumulation layer on the $(000\bar{1})$ surface, $\Delta n_s = 5.2 \cdot 10^{13}$ cm^{-2}, substrate: 10 Ωcm [79G1].

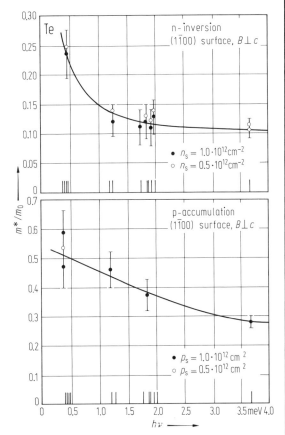

Fig. 98. Te. Density of states effective mass m^*/m_0 for electrons and holes on $(1\bar{1}00)$ Te surfaces as a function of photon energy $h\nu$ measured by cyclotron resonance [78O2].

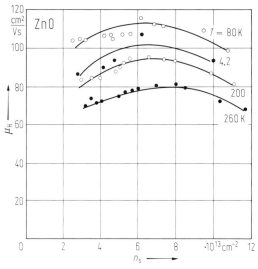

Fig. 97. ZnO. Hall mobility μ_H as a function of the electron concentration n_s on a ZnO $(000\bar{1})$ surface. Parameter is the temperature T. Preparation of the accumulation layer by exposure to He$^+$ ions, substrate: 10 Ωcm [79G1].

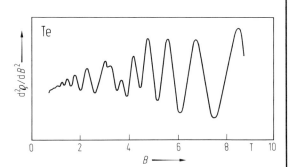

Fig. 99. Te. Second derivative $d^2\varrho/dB^2$ of the magnetoresistivity vs. magnetic induction B for strong p-type accumulation. $T = 4.2$ K; substrate: p-type ($N_a \approx 10^{14}$ cm^{-3}) [71K1].

References for 13
Books and review papers

69s   Sze, S.M.: Physics of Semiconductor Devices, New York: Wiley-Interscience **1969**.
73d   Dorda, G.: Festkörperprobleme XIII, Braunschweig: F. Vieweg **1973**, p. 215.
73g   Goetzberger, A., Schulz, M.: Festkörperprobleme XIII, Braunschweig: F. Vieweg **1973**, p. 309.
75l   Landwehr, G.: Festkörperprobleme XV, Braunschweig: F. Vieweg **1975**, p. 49.
78a   Ando, T.: Surf. Sci. **73** (1978) 1.
82a   Ando, T., Fowler, A.B., Stern, F.: Reviews of Modern Physics, Vol. 54, **1982**, p. 437.
82n   Nicollian, E.H., Brews, J.R.: MOS (Metal Oxide Semiconductor) Physics and Technology, New York: Wiley, **1982**.
82p   Paul, W.: Handbook on Semiconductors, Vol. 1, Amsterdam: North Holland Pub. Co., **1982**.

Bibliography

57S1   Schrieffer, J.R.: Semiconductor Surface Physics, Kingston, R.H., (ed.), Philadelphia: Univ. of Pennsylvania Press **1957**, p. 55.
66K1   Kawaji, S., Kawaguchi, Y.: J. Phys. Soc. Jpn. Suppl. **21** (1966) 336.
67N1   Nicollian, E.H., Goetzberger, A.: Bell Syst. Techn. Journal **46** (1967) 1055.
67R1   Ruch, J.G., Kino, G.S.: Appl. Phys. Lett. **10** (1967) 40.
67S1   Stern, F., Howard, W.E.: Phys. Rev. **163** (1967) 816.
68A1   Arnold, E., Ladell, J., Abowitz, G.: Appl. Phys. Lett. **13** (1968) 413.
68C1   Colman, D., Bate, R.T., Mize, J.P.: J. Appl. Phys. **39** (1968) 1923.
68F1   Fang, F.F., Stiles, P.J.: Phys. Rev. **174** (1968) 823.
68F2   Fang, F.F., Fowler, A.B.: Phys. Rev. **169** (1968) 619.
68S1   Schlötterer, H.: Solid State Electron. **11** (1968) 947.
69G1   Greene, R.F.: Molecular Process on Solid Surfaces, Drauglis, E., Gretz, R.D., Jaffce, R.J., (eds.), New York: Mac Graw-Hill Book C. **1969**.
69M1   Murphy, N.S.J., Berz, F., Flinn, J.: Solid State Electron. **12** (1969) 775.
70P1   Preuss, E.: Phys. Rev. B **1** (1970) 3392.
70S1   Stern, F.: 10th Int. Conf. Physics of Semiconductors, Cambridge **1970**, p. 451.
71D1   Dorda, G.: J. Appl. Phys. **42** (1971) 2053.
71K1   Klitzing, K.v., Landwehr, G.: Solid State Commun. **9** (1971) 2201.
71S1   Sato, T., Takeishi, Y., Hara, H., Okamoto, Y.: Phys. Rev. B **4** (1971) 1950.
71S2   Sakaki, H., Sugano, T.: Jpn. J. Appl. Phys. **10** (1971) 1016.
72B1   Baraff, G.A., Appelbaum, J.A.: Phys. Rev. B **5** (1972) 475.
72D1   Deuling, H., Klausmann, E., Goetzberger, A.: Solid State Electron. **15** (1972) 559.
72K1   Kawaji, S., Ezawa, H., Nakamura, K.: J. Vac. Sci. Technol. **9** (1972) 762.
72K2   Kotera, N., Katayama, Y., Komatsubara, K.F.: Phys. Rev. B **5** (1972) 3065.
72S1   Sah, C.T., Ning, T.H., Tschopp, L.L.: Surf. Sci. **30** (1972) 131.
72S2   Stern, F.: Phys. Rev. B **5** (1972) 4891.
72S3   Sah, C.T., Ning, T.H., Tschopp, L.L.: Surf. Sci. **32** (1972) 561.
72S4   Sah, C.T., Edwards, J.R., Ning, T.H.: Phys. Status Solidi (a) **10** (1972) 153.
73C1   Cheng, Y.C., Sullivan, E.A.: Surf. Sci. **34** (1973) 717.
73C2   Cheng, Y.C., Sullivan, E.A.: J. Appl. Phys. **44** (1973) 44.
73G1   Goetzberger, A., Schulz, M.: Festkörperprobleme XIII, Braunschweig: F. Vieweg **1973**, 309.
73H1   Hughes, A.J., Thorsen, A.C.: J. Appl. Phys. **44** (1973) 2304.
73I1   Ipri, A.C.: Appl. Phys. Lett. **22** (1973) 16.
73P1   Picraux, S.T.: J. Appl. Phys. **44** (1973) 594.
74B1   Bangert, E., Klitzing, K.v., Landwehr, G.: Proc. 12th Int. Conf. on Physics of Semiconductors, Stuttgart: B.G. Teubner **1974**, 714.
74D1   Dorda, G., Eisele, I.: Proc. 12th. Int. Conf. on Physics of Semiconductor, Stuttgart: B.G. Teubner (1974) 704.
74I1   Ito, T., Sakai, Y.: Solid State Electron. **17** (1974) 751.
74K1   Komatsubara, K.F., Narita, K., Katayama, Y., Kotera, N., Kobayashi, M.: J. Phys. Chem. Solids **35** (1974) 723.
74K2   Kudoh, O., Nakamura, K., Kamoshida, M.: Appl. Phys. **45** (1974) 4514.
74K3   Klitzing, K.v., Landwehr, G., Dorda, G.: Solid State Commun. **14** (1974) 387.
74K4   Klitzing, K.v., Landwehr, G., Dorda, G.: Solid State Commun. **15** (1974) 489.

74M1	Miyazaki, T., Nakamura, N., Doi, A., Tokuyama, T.: Jpn. J. Appl. Phys. Suppl. **2** (1974) 441.
74M2	Many, A.: Crit. Rev. Solid State Sci. **4** (1974) 515.
74T1	Tsui, D.C., Kaminsky, G., Schmidt, P.H.: Phys. Rev. B **9** (1974) 3524.
75D1	Daerr, A., Kotthaus, J.P., Koch, J.F.: Solid State Commun. **17** (1975) 455.
75G1	Goodman, A.M.: IEEE Trans. Electron. Devices ED-22 (1975) 63.
75H1	Hsu, S.T., Scott, J.H.: RCA Rev. **36** (1975) 240.
75L1	Landwehr, G.: Festkörperprobleme XV, Braunschweig: F. Vieweg **1975**, 49.
75N1	Neugebauer, T., Klitzing, K.v., Landwehr, G., Dorda, G.: Solid State Commun. **17** (1975) 295.
75O1	Ohkawa, F.J., Uemura, Y.: Suppl. Progress of Theoretical Physics **57** (1975) 164.
75O2	Ortenberg, M.v., Silbermann, R.: Solid State Commun. **17** (1975) 617.
76A1	Ando, T.: Phys. Rev. B **13** (1976) 3468.
76A2	Abstreiter, G., Kotthaus, J.P., Koch, J.F., Dorda, G.: Phys. Rev. B **14** (1976) 2480.
76D1	Dorda, G., Gesch, H., Eisele, I.: Solid State Commun. **20** (1976) 429.
76D2	Dinger, R.: J. Electrochem. Soc. **123** (1976) 1398.
76G1	Gornik, E., Tsui, D.C.: Phys. Rev. Lett. **37** (1976) 1425.
76K1	Kneschaurek, P., Kamgar, A., Koch, J.F.: Phys. Rev. B **14** (1976) 1610.
76K2	Kawaji, S., Hatanaka, K., Nakamura, K., Onga, S.: J. Phys. Soc. Jpn. **41** (1976) 1073.
76K3	Kuroda, T., Narita, S.: J. Phys. Soc. Jpn. **41** (1976) 709.
76O1	Onga, S., Yoshii, T., Hatanaka, K., Yasuda, Y.: Jpn. J. Appl. Phys. Suppl. **15** (1976) 225.
76O2	Ohmura, Y.: Jpn. J. Appl. Phys. Suppl. **15** (1976) 233.
76O3	Ohmura, Y., Yoshii, T., Yasuda, Y.: Solid State Commun. **20** (1976) 203.
76S1	Stallhofer, P., Kotthaus, J.P., Koch, J.F.: Solid State. Commun. **20** (1976) 519.
76W1	Wheeler, R.G., Goldberg, H.S.: IEEE Trans. Electron. Devices ED-**22** (1976) 1001.
76W2	Weber, W., Abstreiter, G., Koch, J.F.: Solid State Commun. **18** (1976) 1397.
77A1	Arai, K., Uemura, Y.: unpublished.
77B1	Beinvogl, W., Koch, J.F.: Solid State Commun. **24** (1977) 687.
77D1	Dinger, R.: Thin Solid Films **43** (1977) 311.
77E1	Eisele, I., Gesch, H., Dorda, G.: Solid State Commun. **22** (1977) 185.
77F1	Fang, F.F., Fowler, A.B., Hartstein, A.: Phys. Rev. B **16** (1977) 4446.
77K1	Kotthaus, J.P., Ranvaud, R.: Phys. Rev. B **15** (1977) 5758.
77K2	Klitzing, K.v., Englert, T., Landwehr, G., Dorda, G.: Solid State Commun. **24** (1977) 703.
77N1	Neppl, F., Kotthaus, J.P., Koch, J.F.: Phys. Rev. B **16** (1977) 1519.
77S1	Schaber, H., Doezema, R.E., Stiles, P.J., Lopez-Otero, A.: Solid State Commun. **23** (1977) 405.
78D1	Dorda, G., Eisele, I., Gesch, H.: Phys. Rev. **17** (1978) 1785.
78D2	Daerr, A., Kotthaus, J.P.: Surf. Sci. **73** (1978) 549.
78E1	Eisele, I.: Proc. Int. Conf. Application of High Magnetic Fields in Semiconductor Physics, Oxford **1978**, 302.
78K1	Kotthaus, J.P.: Surf. Sci. **73** (1978) 472.
78K2	Kohl, D., Moorman, H., Heiland, G.: Surf. Sci. **73** (1978) 160.
78O1	Onga, S., Hatanaka, K., Kawaji, S., Nishi, Y., Yasuda, Y.: Jpn. J. Appl. Phys. **17** (1978) 1587.
78O2	Ortenberg, M.v., Tuchendler, J., Silbermann, R., Tuilier, J.C.: Surf. Sci. **73** (1978) 496.
78S1	Schaber, H., Doezema, R.E., Koch, J.F., Lozez-Otero, A.: Surf. Sci. **73** (1978) 503.
78W1	Washburn, H.A., Sites, J.R.: Surf. Sci. **73** (1978) 537.
79E1	Eden, R.C., Welch, B.M., Zucca, R., Long, S.I.: IEEE J. Solid State Circuits Sc-**14** (1979) 221.
79F1	Ferry, D.K.: Thin Solid Films **56** (1979) 243.
79G1	Grinshpan, Y., Nitzan, M., Goldstein, Y.: Phys. Rev. B **19** (1979) 1098.
79T1	Tsui, D.C., Kaminsky, G.: Phys. Rev. Lett. **42** (1979) 595.
80K1	Klitzing, K.v., Dorda, G., Pepper, M.: Phys. Rev. Lett. **45** (1980) 494.
80M1	Müller, W., Eisele, I.: Solid State Commun. **34** (1980) 447.
84E1	Eisele, I.: unpublished data.
84S1	Störmer, H.L.: Festkörperprobleme XXIV, Braunschweig: F. Vieweg **1984**, 25.

14 Hot electrons

14.0 Introduction

14.0.1 General remarks

Hot electrons or, more generally, hot carriers are created by the application of an electric field to a semiconductor crystal. The electric field feeds energy into the electrons or the holes and raises their temperature above that of the crystal lattice. The temperature rise of the carriers depends on the strength of the field. For sufficiently small fields, the temperature rise is negligible and Ohm's law is valid. At high fields the temperature rise is large, Ohm's law breaks down, and the hot carriers produce a variety of effects, which are represented in the graphs of this chapter. The electric field therefore appears as the abscissa scale on most of the graphs.

Hot carriers cannot only be created by the application of electric fields to a semiconductor, but also by the absorption of intense optical radiation (for recent reviews of this field, see [78s1] and [80f1]). Experimental results of this kind are not included in the present article; this article is restricted to the application of electric fields.

14.0.2 Explanations for the graphs

The figures of this section present a variety of semiconductor parameters as a function of the electric field for various semiconductors. In some cases, the dependence of these parameters on the lattice temperature, impurity concentration, or magnetic induction is also provided.

The materials are characterized in the figure captions either by their donor and/or acceptor concentration, or by the free carrier density, or by their resistivity, where such information is available.

The data points were transferred directly from the original volumes to the presented graphs. In many graphs there appear curves only; they do represent experimental data. In some cases lines or curves are provided in addition to the experimental data points. They should be understood always as aids for easier interpolation but not as results of theoretical calculations. Theoretical results are not included in this article.

14.0.3 List of symbols

A_I	$cm^3 s^{-1}$	impact ionization coefficient
A_T	s^{-1}	thermal ionization coefficient
$\boldsymbol{B}$	T	magnetic induction ($\|\boldsymbol{B}\| = B$)
B_I	$cm^6 s^{-1}$	Auger recombination coefficient
B_T	$cm^3 s^{-1}$	phonon-assisted recombination coefficient
D	$cm^2 s^{-1}$	diffusion coefficient
D_0	$cm^2 s^{-1}$	diffusion coefficient in the limit of zero electric field
$D_\parallel$	$cm^2 s^{-1}$	diffusion coefficient parallel to the electric field
$D_\perp$	$cm^2 s^{-1}$	diffusion coefficient perpendicular to the electric field
D_n	$cm^2 s^{-1}$	noise diffusion coefficient
e	C	elementary charge
$\boldsymbol{E}$	$V cm^{-1}$	electric field strength ($\|\boldsymbol{E}\| = E$)
E_{br}	$V cm^{-1}$	breakdown electric field strength
E_l	$V cm^{-1}$	longitudinal electric field strength
E_t	$V cm^{-1}$	transverse electric field strength
g_e	s^{-1}	generation rate of electrons
g_h	s^{-1}	generation rate of holes
h_{ijkl}	$cm^2 V^{-2}$	components of the warm carrier transport-coefficient tensor
I	A	electric current
I_n	A	electron current
I_p	A	hole current
I_d	A	drain-source current of a MOS transistor
$\boldsymbol{j}$	$A cm^{-2}$	electric current density ($\|\boldsymbol{j}\| = j$)
j_{100}	$A cm^{-2}$	current density in the $\langle 100 \rangle$ direction
j_{111}	$A cm^{-2}$	current density in the $\langle 111 \rangle$ direction

j_θ	A cm^{-2}	current density in a certain cristallographic direction characterized by the angle θ
k_B	eV K^{-1}	Boltzmann constant
n	cm^{-3}	electron concentration
N_I	cm^{-3}	ionized impurity concentration
N_a	cm^{-3}	acceptor concentration
N_d	cm^{-3}	donor concentration
n_0	cm^{-3}	electron concentration for $E=0$
n_s	cm^{-2}	surface electron concentration
p	cm^{-3}	hole concentration
p_s	cm^{-2}	surface hole concentration
R_H	cm^3 C^{-1}	Hall coefficient
R_{H0}	cm^3 C^{-1}	Hall coefficient in the limit of zero electric field
t	s	time
T	K	lattice temperature
T_e	K	electron temperature
T_h	K	hole temperature
T_n	K	noise temperature
$T_{n\parallel}$	K	noise temperature parallel to the electric field
$T_{n\perp}$	K	noise temperature perpendicular to the electric field
V	V	voltage
V_{th}	V	thermoelectric voltage
v_{dr}	cm s^{-1}	drift velocity
α	cm^{-1}	optical absorption coefficient
α_0	cm^{-1}	optical absorption coefficient in the limit of zero electric field
$\Delta\alpha$	cm^{-1}	electric field induced change of the absorption coefficient
α_n	cm^{-1}	electron ionization coefficient
α_p	cm^{-1}	hole ionization coefficient
β	cm^2 V^{-2}	warm carrier coefficient
β_0	cm^2 V^{-2}	isotropic part of the warm carrier coefficient
β_{111}	cm^2 V^{-2}	warm carrier coefficient with the electric field parallel to $\langle 111\rangle$
γ_0	cm^2 V^{-2}	warm carrier anisotropy coefficient
$\langle\varepsilon\rangle$	eV	mean carrier energy with applied electric field
$\langle\varepsilon_0\rangle$	eV	equilibrium mean carrier energy
ε_h	eV	hole energy
θ	°	angle between the current density and a particular cristallographic direction
$\Delta\vartheta$	°, rad	electric field induced change of the Faraday rotation angle
λ	μm	wavelength of optical radiation
λ_i		cosine of the angle between the electric field and the coordinate axes
μ	cm^2 V^{-1} s^{-1}	drift mobility
μ_0	cm^2 V^{-1} s^{-1}	drift mobility in the limit of zero electric field
μ_H	cm^2 V^{-1} s^{-1}	Hall mobility
v	s^{-1}	frequency of ac electric fields
ϱ	Ω cm	resistivity
ϱ_0	Ω cm	resistivity in the limit of zero electric field
$\varrho_\parallel$	Ω cm	resistivity parallel to the magnetic induction
$\varrho_\perp$	Ω cm	resistivity perpendicular to the magnetic induction
$\Delta\varrho$	Ω cm	electric-field induced change of the resistivity
σ	Ω^{-1} cm^{-1}	conductivity
σ_0	Ω^{-1} cm^{-1}	conductivity in the limit of zero electric field
$\Delta\sigma$	Ω^{-1} cm^{-1}	electric-field induced change of the conductivity
τ	s	scattering time
τ_ε	s	energy relaxation time
ψ	°	arctan (E_t/E_1)
$\hbar\omega$	eV	photon energy

14.1 Carrier heating by electric fields

14.1.1 Conductivity in the warm carrier range

The field dependence of the conductivity can in general be mathematically approximated by an expansion in even powers of the electric field E. When the electric field is sufficiently small, only the first non-trivial term is taken into account. One obtains

$$\sigma = \sigma_0 (1 + \beta E^2)$$

The range of fields where such a relation describes the data is called the warm carrier range. The warm carrier coefficient β can be measured by dc methods using short voltage pulses in order to avoid lattice heating or by applying microwave electric fields [80B1]. In practice, β-values in the order of $10^{-4} \cdots 10^{-8}$ cm^2 V^{-2} have to be measured.

The coefficient β usually depends on the doping and on the temperature. It can be positive or negative, depending on the physical origin of the field dependence of the conductivity. Measurements of the coefficient β for Si, Ge, InSb, and HgCdTe are presented in

Figs. 1···10, p. 264f.

14.1.2 Conductivity and current density in the hot carrier range

For arbitrary electric fields it is not possible to give an analytic expression for the dependence of the current density j on the field E or for the ratio of these two quantities, which is the conductivity $\sigma(E) = j(E)/E$. Measurements are usually performed by applying short voltage pulses to the sample. The current is monitored by a resistor connected in series to the sample. The field strength E is determined from the voltage drop across potential probes along the sample to avoid contact effects. The current density is measured by the voltage drop across the series resistor by dividing it through the resistance value of the series resistor and the cross sectional area of the sample [80B1]. Results for Si, Ge, InAs, CdS, HgCdTe, and Te are shown in

Figs. 11···19, p. 266f.

14.1.3 Conductivity anisotropy

In cubic crystals with nonspherical energy surfaces the conductivity is a scalar only for weak fields. At increasing fields, however, the directions of the current density and the electric field strength do not necessarily coincide, and the values of the current density depend on the orientation [54S1, 55S1]. These effects are caused by differing effective masses in the valleys for different orientations, which imply a differing heating power by the field to the carriers accelerated in various directions.

In many-valley semiconductors this anisotropy effect causes a deviation of the distribution function of carriers in different valleys. The sum of the contribution of all the valleys is not longer isotropic, as it is in the ohmic case.

In the region of weak fields, where the deviations from Ohm's law can be characterized by the coefficient β, the anisotropy can be described by a fourth rank tensor for the conductivity. The tensor allows a phenomenological description of the effects [61S1]. The current density vector can be expanded with respect to the components of the electric field E, where λ_i is the cosine of the angle between E and the coordinate axis i.

$$j_i = \sigma_0 E (\delta_{ij} \lambda_j + E^2 h_{ijkl} \lambda_j \lambda_k \lambda_l + \cdots)$$

The effects of warm electrons can be described by h_{ijkl}. For the point groups to which Si and Ge belong, this tensor reduces to two independent constants because of symmetry

$$h_{iiii} = \beta_0 \quad \text{and} \quad h_{iiii} - 3 h_{iijj} = \gamma_0$$

with $h_{iijj} = h_{ijij} = h_{ijji}$, all the other $h_{ijkl} = 0$.

The departure of the current direction from the field direction causes the appearance of a transverse electric field E_t; the ratio of the transverse field E_t to the longitudinal field E_1 is called $\tan \psi = E_t / E_1$. Results on the anisotropy for Si and Ge are presented in

Figs. 20···28, p. 267ff.

14.1.4 Frequency dependence of warm carrier parameters

The deviations from Ohm's law are determined by the rates at which carriers gain energy from the field and lose energy to the crystal lattice. The energy loss rate depends on the nature of the scattering processes and is phenomenologically described by a time constant τ_ε called energy relaxation time, whereas the momentum relaxation time controls the returning of a disturbed carrier distribution function to its equilibrium value. If the measuring frequency for deviations from Ohm's law is close to the reciprocal of one of these time constants, strong frequency dependences of the warm carrier parameters occur. Because of the low magnitude of the relevant time constants, these frequencies usually are in the microwave or submillimeter region. Various techniques have been developed to determine β at microwave frequencies [80B1]. Microwave measurements have the advantage compared to dc experiments that contact problems can be avoided. On the other hand, it is difficult to determine the accurate value of the microwave electric field strength inside the sample. Typical results on the frequency dependence of the anisotropy in Ge are presented in

Figs. 29···32, p. 269.

14.1.5 Electric field dependence of the carrier mobility

An important reason for the electric field dependence of the conductivity of many semiconductors is the field dependence of the mobility μ. It can be obtained from a measurement of the conductivity σ by using the relation $\sigma(E) = e\,n(E)\,\mu(E)$. If an independent measurement of $n(E)$ is performed, $\mu(E)$ is immediately accessible. In many cases it suffices to assume a field-independent carrier concentration. This is particularly correct, if all impurities in an extrinsic semiconductor are already thermally ionized, and if the electric field is small compared to values, where impact ionization across the energy gap occurs. However, experimental precautions have to be taken to avoid injection or extraction of carriers through the end contacts. Results on mobility measurements for Si, Ge, and InAs are presented in

Figs. 33···37, p. 270f.

14.1.6 Drift velocity

The most widely applicable technique for obtaining the drift velocity as a function of electric field is based on a measurement of the current density which is present in a sample of an extrinsic semiconductor material at a given applied electric field E, and a calculation of v_{dr} from the relation $j(E) = n\,e\,v_{dr}(E)$. Provided the carrier concentration n has been determined by an independent measurement, the drift velocity can be determined when the carrier concentration n does not depend on the applied electric field. When the carrier concentration n is obtained from a measurement of the Hall effect, it is difficult to obtain exact values for the carrier concentration because of the uncertainty in the Hall scattering factor r_H. The scattering factor must be known for the calculation of n from the measurement of the Hall coefficient R_H.

Injection of carriers from the contacts also affects conductivity measurement techniques which cannot be applied when the resistivity is high, particularly in pure materials and at low temperatures. Only the drift velocity of majority carriers can be determined by conductivity measurements. The generally accepted estimate of the uncertainty of v_{dr} obtained is within 5% [79j1].

The determination of the drift velocity by the time-of-flight technique is based on an analysis of the waveform of the current signal induced by a sheet of carriers crossing the sample in the presence of an applied electric field. Charge pairs are created by a suitable ionizing radiation in a narrow region of the sample close to one contact. One type of carriers is collected at the contact, while the other is swept across the sample toward the opposite contact. The drift velocity of the latter type of carriers is obtained by measurement of the time duration of the current pulse, which is the time elapsed when the carriers have travelled across the specimen of known thickness. This technique yields a more direct measurement of v_{dr}. The drift velocity of both types of carriers can be measured within the same sample. The method is not always applicable, however, since it requires both the lifetime of the carriers and the dielectric relaxation time of the material to be longer than the transit time. These two conditions are satisfied in samples of sufficiently high purity and resistivity. The generally obtained uncertainty of the values obtained is within 5%. The conductivity technique and the time-of-flight technique are complementary since they are applicable in low- and high-resistivity materials, respectively [79j1].

The derivative of the v_{dr} vs. E curves yields a differential mobility $\mu = d\,v_{dr}/d\,E$, which in certain cases becomes negative, when carriers are transferred into valleys with large values of the effective mass. This effect is called Gunn-effect; it can be used to generate or amplify microwave radiation. For a review, see [71a1].

Drift velocities of electrons and holes as a function of electric field and temperature for diamond, Si, Ge, GaAs, InP, InSb, CdTe, AgCl, PbSe, and PbTe are shown in

Figs. 38···57, p. 271ff.

14.1.7 Noise and diffusion of hot carriers

The noise of semiconductors in the hot carrier regime is due to fluctuations of the velocity of the carriers. Noise and diffusion of hot carriers are closely related properties. When the noise power ΔP within a bandwidth Δf is measured under impedance-matched conditions, one can deduce a noise temperature T_n by using the relation [80N2]:

$$\Delta P = k_B T_n \Delta f.$$

The noise temperature measured parallel to the applied electric field is called longitudinal noise temperature $T_{n\parallel}$; the temperature measured transverse to the applied field is called transverse noise temperature $T_{n\perp}$. Related to these quantities are noise diffusion coefficients D_n defined by the relation

$$D_n(E) = (k_B/e) T_n(E) \, dv_{dr}/dE.$$

The difference between the longitudinal and transverse diffusion coefficients $D_\parallel$ and $D_\perp$, respectively, is related to the energy dependence of the microscopic scattering probabilities. The diffusion coefficient D may depend on the orientation of the applied electric field with respect to the cristallographic axes, reflecting the peculiarities in the shape of the energy band structure.

The time of flight technique for measuring $D_\parallel$ makes use of the diffusive broadening of the carrier layer which drifts across the sample in the applied electric field [68R2]. The uncertainty observed strongly depends on the experimental conditions but is rarely below 20%. $D_\perp$ is experimentally accessible by a beam spreading technique [71P1], whereas both quantities $D_\parallel$ and $D_\perp$ can be measured by detection of the noise [80N2]. Results on noise and diffusion of hot carriers for Si, Ge, and GaAs are presented in

Figs. 58···70, p. 274ff.

14.1.8 Thermoelectric power of hot carriers

The ordinary thermoelectric or Seebeck voltage arises from spatial variations of the carrier distribution due to a gradient in lattice temperature. A spatial variation of the distribution due to a gradient in electrical field intensity causes an open-circuit voltage. Because of the analogy, the voltage is frequently called "thermoelectric voltage" of hot carriers; a more accurate title would be "field gradient voltage" [67c1]. The experimental methods are described in [71d1]. Results on the thermoelectric power of hot carriers for Si and Ge are presented in

Figs. 71 and 72, p. 277.

14.1.9 Hot carrier Hall effect

The measurement of the Hall effect under hot carrier conditions yields important information about the field dependence of the carrier concentration. Changes in R_H also indicate changes in the Hall scattering factor r_H, since

$$R_H(E) = -r_H(E)/(e\, n(E))$$

for an n-type extrinsic semiconductor. A determination of the Hall coefficient R_H permits a calculation of the Hall mobility μ_H defined by the relation

$$\mu_H(E) = R_H(E) \cdot \sigma(E)$$

when the electric field dependence of the conductivity is additionally measured. The experimental techniques are described in [80B1]. Results on the Hall effect in Si, Ge, and PbTe are presented in

Figs. 73···79, p. 277f.

14.1.10 Hot carrier magnetoresistance

A magnetic field has always a cooling influence on warm and hot carriers:

Due to the Lorentz force, the carriers are deflected from the drift direction and thus gain less energy from the accelerating field than without a magnetic field [73s1]. The electric field dependence of the magneto-resistance is caused by the change in the carrier distribution function. In many-valley semiconductors, repopulation effects have to be additionally considered. Results on properties induced by magnetic fields for Si, GaSb, and InAs are presented in

Figs. 80···83, p. 278f.

14.1.11 Hot carrier effects in quantizing magnetic fields

The influence of a magnetic field on the electrons and holes in semiconductors is to quantize their energies into a set of Landau levels. This quantization can lead to several quantum effects in the electrical transport properties at high magnetic fields. When the semiconductor is degenerately doped, one observes an oscillatory variation of the conductivity with magnetic field: the Shubnikov-de Haas effect. The temperature dependence of the oscillation amplitudes may be used to investigate the temperature increase of the electron gas, when it is heated by an electric field [74b1].

In nondegenerate semiconductors, a resonant cooling of carriers can occur when the separation between one or more Landau levels is equal to the energy of a monoenergetic relaxation process, e.g. when LO phonons are emitted. This resonant cooling may be visible in one of the transport coefficients. The most widely investigated example of this effect is the "hot-electron magnetophonon effect" where oscillations are found to occur in the magnetoresistance of many different semiconductors due to the emission of both single and pairs of phonons. Another example is the 'magneto-impurity effect', where carriers are found to lose energy by exciting electrons or holes from shallow impurity levels into higher lying levels or into the conduction or valence band. Because of the complexity and vast amount of experimental transport data in quantizing magnetic fields, the reader is referred to the original literature, which is reviewed in [80N1].

Other consequences of the quantization of the energy levels are the appearance of cyclotron resonance absorption, which can be modified by the application of electric fields (for a review, see [80B1]), and the possibility of emission of submillimeter radiation from hot carriers in quantizing magnetic fields [78M1, 80G1].

14.1.12 Hot carrier effects in quantizing electric field

Due to the small spatial separation of the source and drain contacts of small MOS transistor structures, at moderate applied voltages, the electric fields are even in a range, where Ohm's law is no longer valid and hot carrier effects are important. The effects observed differ from the effects in bulk crystals: Since the carriers are confined to a narrow layer at the semiconductor surface, electrical quantization occurs due to the high surface electric field strength. The energy bands are split into a series of electrical subbands. Additional scattering mechanisms are important for the carrier transport close to the semiconductor-oxide interface, which do not occur in the bulk [78F1, 78H1]. Results of surface/interface effects on hot carriers for Si-structures are presented in

Figs. 84···87, p. 279.

14.1.13 Impurity breakdown

Conductivity changes in high electric fields are not only caused by mobility variations. Especially at low temperatures, where the carriers are frozen out and the donors or acceptors are neutral, impact ionization by hot carriers in the conduction or valence band may occur. The change of the carrier density can be described by the equation

$$\frac{dn}{dt} = A_T(N_d - N_a - n) - B_T(N_a + n)\,n + A_I(N_d - N_a - n)\,n - B_I(N_a + n)\,n^2$$

where A_T denotes the thermal ionization coefficient, A_I the impact ionization coefficient, B_T the phonon assisted recombination coefficient, and B_I the Auger recombination coefficient [79A1]. All these quantities depend on the electric field strength. The threshold electric field, where a rapid increase of the conductivity sets in, is called breakdown field E_{br}. It depends on the impurity concentration, on the mobility, and on the applied magnetic field. Results for impurity breakdown in Si, Ge, and InSb are presented in

Figs. 88···106, p. 280 ff.

14.1.14 Avalanche breakdown across the energy gap

In small gap semiconductors, the energy of hot carriers may be sufficient to cause impact ionization of electrons from the valence band to the conduction band. Electron-hole pairs are created. The avalanche breakdown process is visible in a dramatic increase of the current density. The electron-hole pair generation rate can be determined from the observation of the time-dependent increase of the current under constant voltage conditions. When holes do not contribute to the generation and to the current (a situation valid in the case of a high mobility ratio between electrons and holes), the generation rate g_e of electrons may be approximated by

$$g_e(E) = \frac{1}{n_0}\frac{dn}{dt} \approx \frac{1}{I_0}\frac{dI}{dt}\Big|_{t\to 0}$$

where n_0 is the thermal equilibrium density of electrons, I is the total current, and I_0 the current corresponding to the carrier concentration n_0. The ionization coefficients α_n for electrons and α_p for holes are defined by the relations

$$\frac{\partial I_n}{\partial x} = \alpha_n I_n \quad \text{and} \quad \frac{\partial I_p}{\partial x} = \alpha_p I_p$$

where I_n is the electron current and I_p is the hole current. Results on avalanche breakdown are presented in

Figs. 107···117, p. 284 ff.

14.1.15 Carrier temperatures

An immediate consequence of the application of electric fields to a semiconductor is a change in the carrier distribution function, since the carriers gain energy from the electric field. In general, this disturbed distribution function is not of the Maxwell-Boltzmann type in nondegenerately doped samples and not of the Fermi-Dirac type in degenerately doped samples. However, if carrier-carrier scattering is strong and dominates over carrier-phonon scattering and carrier-ionized impurity scattering, it is reasonable to assume that the distribution function will be of the Maxwell-Boltzmann or Fermi-Dirac type. Under these circumstances, which crucially depend on a sufficiently high carrier concentration, an electron temperature T_e or hole temperature T_h may be defined as a parameter of the distribution function [74b1]. A variety of experimental methods have been developed to determine carrier temperatures and their variation with electric field [74b1]. The method used for the measurement of each particular set of data and relevant results for Ge, GaSb, InAs, InSb, and PbTe, are presented in

Figs. 119···130, p. 287 ff.

14.1.16 Energy relaxation times

In the warm electron region, the introduction of an energy relaxation time τ_ε [61G1] has been found convenient. This quantity, which should describe in a phenomenological way the time-dependent return of the isotropic part of the disturbed distribution function towards its equilibrium value, is in general a function of the electric field and the lattice temperature. It is defined by

$$\tau_\varepsilon(E, T) = \frac{\langle \varepsilon \rangle - \langle \varepsilon_0 \rangle}{e\, v_{dr}\, E}$$

where $\langle \varepsilon_0 \rangle$ is the thermal-equilibrium mean energy of the carriers, $\langle \varepsilon \rangle$ is the mean energy in presence of the field, and v_{dr} is the drift velocity. This definition of τ_ε is often also used outside the warm carrier region. The energy relaxation time has been found to be strictly correlated to the warm carrier coefficient β, as discussed in [73C2] and [73s1]. The measurements are usually performed by applying microwave electric fields. The various techniques are described in [80B1]. Energy relaxation times for Ge, GaAs, InSb, and Te are shown in

Figs. 131···141, p. 289 f.

14.1.17 Electric field-induced changes of the optical absorption

When charge carriers are heated by an electric field, not only the transport parameters but also the optical properties change due to the disturbance of the distribution function. Three types of optical absorption processes have been studied under application of electric fields:

1) free carrier absorption in the infrared region [79S1],

2) intervalence-band absorption in p-type materials, and

3) changes in the fundamental absorption of degenerately doped semiconductors, where the Burstein-Moss shift of the fundamental absorption edge is partially removed by the application of electric fields.

The latter two methods are reviewed in [74b1]. Changes of the optical absorption constant α observed are related to changes in the carrier distribution function. The methods (2) und (3) offer the possibility for a direct determination of the shape of the disturbed distribution function. Results on the optical absorption in high electric fields for Ge and GaAs are presented in

Figs. 142···149, p. 291f.

14.1.18 Hot carrier Faraday effect

The Faraday effect is the rotation of the plane of polarization of linearly polarized light for a propagation direction parallel to the applied magnetic field. The contribution of free carriers to the Faraday effect is of interest for hot-electron investigations. The Faraday rotation angle ϑ describes the different velocities for left-hand and right-hand circularly polarized light. Hot electron effects appear in the Faraday rotation by a change of the effective mass with electric field (e.g. caused by a band nonparabolicity or by carrier intervalley transfer) or by the energy dependence of the scattering time τ. If the angular frequency ω of the radiation satisfies the condition $\omega\tau \gg 1$, changes in the Faraday rotation are only created by a changing effective mass. Under this condition, the Faraday rotation can be measured to obtain information on the amount of carrier transfer of electrons from low-lying conduction-band valleys to higher bandminima. For a review of the experimental methods see [80B1]. Faraday rotation measurements for GaAs and GaSb are shown in

Figs. 150 and 151, p. 292.

14.1.19 Hot electrons in insulators

Hot carriers are not only important for the transport and optical properties of doped semiconductors, but under certain circumstances also have been studied in insulators. Prominent examples are (i) the emission of hot carriers from Si into SiO_2 in silicon transistor structures [78N1] and (ii) the role of hot electron processes for dielectric aging and dielectric breakdown in polymeric insulators [84P1]. The injection of hot carriers from Si into SiO_2 can result in device degradation and instability, but it has also found application in electrically programmable memory devices. Typical emission probabilities studied by optically induced hot electron injection and corresponding effective electron temperatures are shown in

Figs. 152···154, p. 292f.

Figures for chapter 14

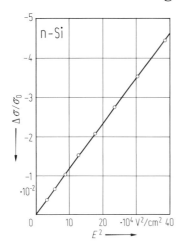

Fig. 1. n-Si. Normalized conductivity change $\Delta\sigma/\sigma_0$ vs. square of the electric field E^2 at $T = 273$ K; $n = 1.1 \cdot 10^{15}$ cm^{-3}; [65K1].

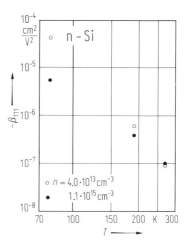

Fig. 2. n-Si. Warm electron coefficient β_{111} vs. temperature T; [65K1].

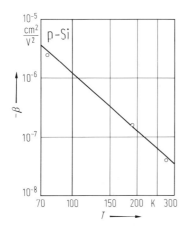

Fig. 3. p-Si. Warm hole coefficient β vs. temperature T; $j \parallel \langle 100 \rangle$; $\varrho = 3.5 \, \Omega$ cm at 300 K; [68R1].

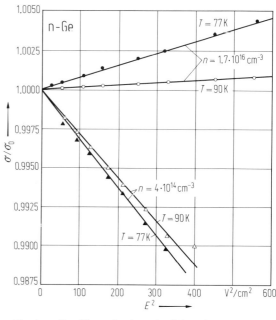

Fig. 4. n-Ge. Normalized conductivity σ/σ_0 vs. square of the electric field E^2; [57G1].

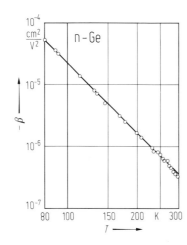

◄

Fig. 5. n-Ge. Warm electron coefficient β vs. temperature T; $\varrho = 21.5 \, \Omega$ cm at 300 K; [59M1].

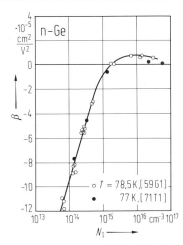

Fig. 6. n-Ge. Warm electron coefficient β vs. ionized impurity concentration N_I.

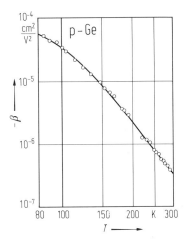

Fig. 7. p-Ge. Warm hole coefficient β vs. temperature T. $\varrho = 23.5\ \Omega$ cm at 300 K; [59S1].

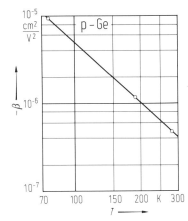

Fig. 8. p-Ge. Warm hole coefficient β vs. temperature T; $\varrho = 0.9\ \Omega$ cm at 300 K, $E \parallel \langle 100 \rangle$; [68R1].

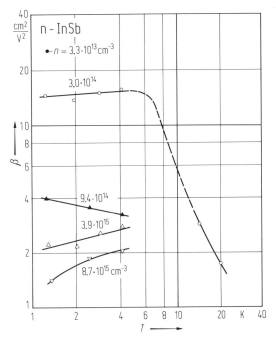

Fig. 9. n-InSb. Warm electron coefficient β vs. temperature T; [60S1].

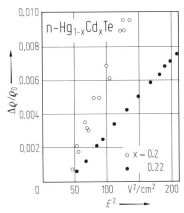

Fig. 10. n-Hg$_{1-x}$Cd$_x$Te. Normalized resistivity change $\Delta\varrho/\varrho_0$ vs. square of the electric field E^2 at $T = 295$ K; [74E1].

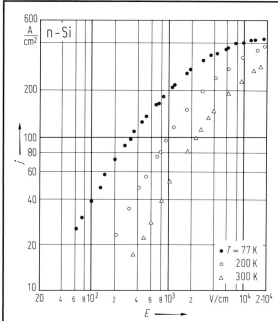

Fig. 11. n-Si. Current density j vs. electric field E. $\varrho = 20\,\Omega\,\mathrm{cm}$ at 300 K; $j \parallel \langle 111 \rangle$; [65V1].

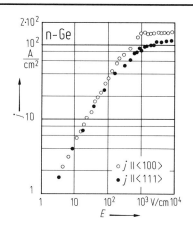

Fig. 12. n-Ge. Current density j vs. electric field E. $n = 1 \cdot 10^{14}\,\mathrm{cm}^{-3}$; $T = 80\,\mathrm{K}$; [67M1].

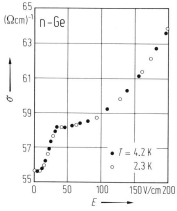

Fig. 13. n-Ge. Conductivity σ vs. electric field E. $n = 3.4 \cdot 10^{17}\,\mathrm{cm}^{-3}$; [69T1].

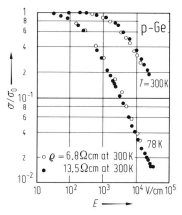

Fig. 14. p-Ge. Normalized conductivity σ/σ_0 vs. electric field E. [60Z1].

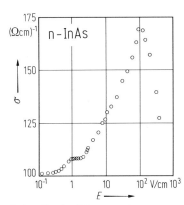

Fig. 15. n-InAs. Conductivity σ vs. electric field E at $T = 4.2\,\mathrm{K}$. $n = 2.5 \cdot 10^{16}\,\mathrm{cm}^{-3}$; [70B1].

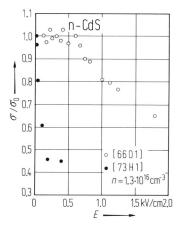

Fig. 16. n-CdS. Normalized conductivity σ/σ_0 vs. electric field E at $T = 77\,\mathrm{K}$.

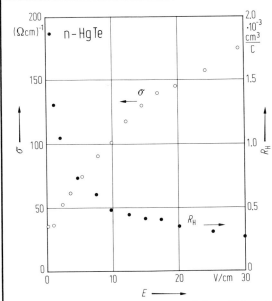

Fig. 17. n-HgTe. Conductivity σ and Hall coefficient R_H vs. electric field E at $T=4.2$ K. $n=3\cdot10^{15}$ cm^{-3}; [68I1, 74I1].

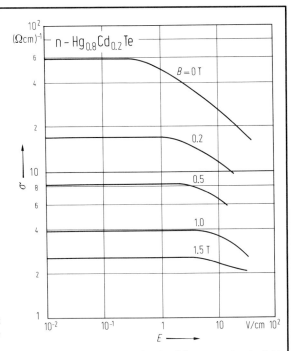

Fig. 18. n-Hg$_{0.8}$Cd$_{0.2}$Te. Conductivity σ vs. electric field E at $T=4.2$ K; $n=9\cdot10^{14}$ cm^{-3} at 77 K; [74D1].

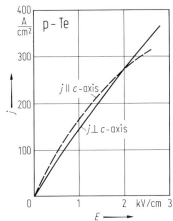

Fig. 19. p-Te. Current density j vs. electric field E at $T=77$ K. $p=4\cdot10^{14}$ cm^{-3} at 77 K; [69N1].

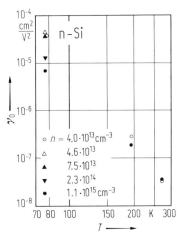

Fig. 20. n-Si. Warm electron anisotropy coefficient γ_0 vs. temperature T; [65K1].

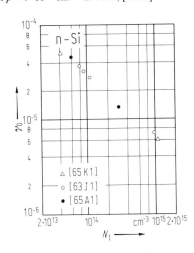

Fig. 21. n-Si. Warm electron anisotropy coefficient γ_0 vs. ionized impurity concentration N_I at $T=77$ K.

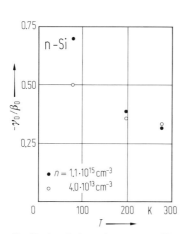

Fig. 22. n-Si. Ratio of the anisotropy coefficient to the warm electron coefficient γ_0/β_0 vs. temperature T; [65K1].

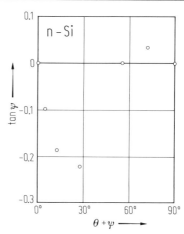

Fig. 23. n-Si. Maximum of the ratio of the transverse to the longitudinal field $E_t/E_1 = \tan \psi$ vs. field direction $\theta + \psi$ in the {110} plane at $T = 77$ K. $n = 3 \cdot 10^{14}$ cm^{-3} at 300 K. [65A1].

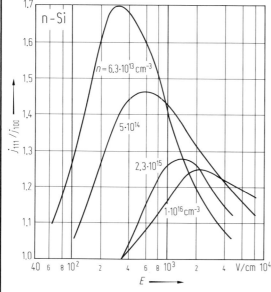

Fig. 24. n-Si. Ratio of the current density in $\langle 111 \rangle$ direction to the current density in $\langle 100 \rangle$ direction j_{111}/j_{100} vs. electric field E at $T = 77$ K; [69a1].

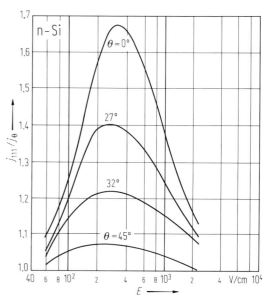

Fig. 25. n-Si. Ratio of the current density in [111] direction j_{111} to the current density j_θ vs. electric field E at $T = 77$ K. θ is the angle between j_θ and the [001] axis in the (100) plane. $n = 6.3 \cdot 10^{13}$ cm^{-3}; [65A1].

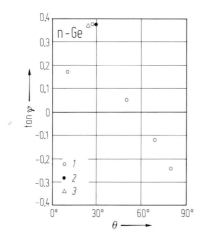

◄

Fig. 26. n-Ge. Ratio of the transverse to the longitudinal field $E_t/E_1 = \tan \psi$ vs. sample orientation θ. θ is the angle between j and the [001] direction in the (110) plane.
1: $T = 90$ K, $E_1 = 750$ V/cm; [58S1, 59S2];
2: $T = 80$ K, $E_1 = 4000$ V/cm; [60K1];
3: $T = 78$ K, $E_1 = 3000$ V/cm; [62r1].

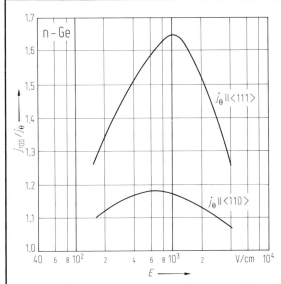

Fig. 27. n-Ge. Ratio of the current density in ⟨100⟩ direction j_{100} to the current density j_θ vs. electric field E at $T = 77$ K. $\varrho = 18\,\Omega$ cm at 297 K; [63N1].

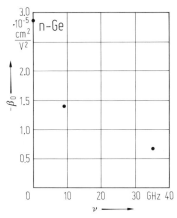

Fig. 29. n-Ge. Isotropic part of the warm electron coefficient β_0 vs. measuring frequency v at $T = 100$ K. $n = 5.6 \cdot 10^{13}$ cm^{-3}; [63S1].

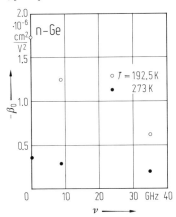

Fig. 31. n-Ge. Isotropic part of the warm electron coefficient β_0 vs. measuring frequency v. $n = 5.6 \cdot 10^{13}$ cm^{-3}; [63S1].

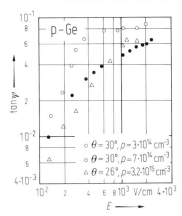

Fig. 28. p-Ge. Ratio of the transverse to the longitudinal electric field $E_t / E_1 = \tan\psi$ vs. electric field E at $T = 77$ K. θ is the angle between j and the [001] direction in the (110) plane; [62G1].

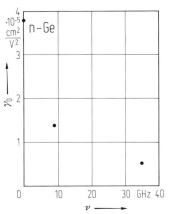

Fig. 30. n-Ge. Warm electron anisotropy coefficient γ_0 vs. measuring frequency v at $T = 100$ K. $n = 5.6 \cdot 10^{13}$ cm^{-3}; [63S1].

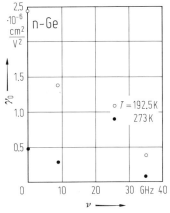

Fig. 32. n-Ge. Warm electron anisotropy coefficient γ_0 vs. measuring frequency v. $n = 5.6 \cdot 10^{13}$ cm^{-3}; [63S1].

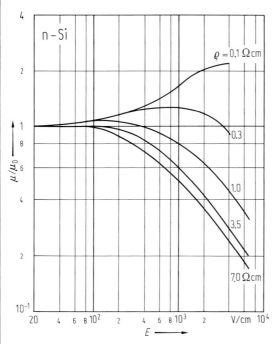

Fig. 33. n-Si. Normalized mobility μ/μ_0 vs. electric field $E \parallel \langle 111 \rangle$ at $T - 77$ K; all ϱ-values at 300 K; [69a1].

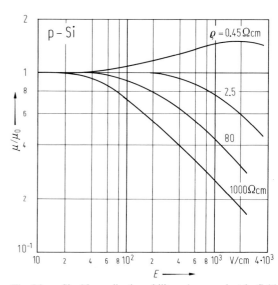

Fig. 34. p-Si. Normalized mobility μ/μ_0 vs. electric field E at $T = 77$ K; all ϱ-values at 300 K [69a1].

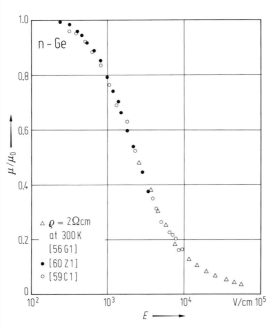

Fig. 35. n-Ge. Normalized mobility μ/μ_0 vs. electric field E at $T = 300$ K.

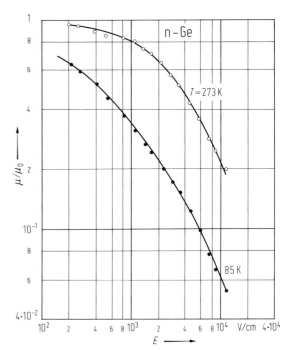

Fig. 36. n-Ge. Normalized mobility μ/μ_0 vs. electric field E for $j \parallel \langle 100 \rangle$. $n = 4.6 \cdot 10^{14}$ cm^{-3}; [65S1].

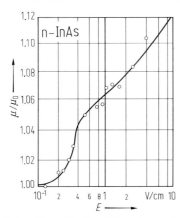

Fig. 37. n-InAs. Normalized mobility μ/μ_0 vs. electric field E at T 4.2 K. $n = 2.5 \cdot 10^{16}$ cm^{-3}; [72B1].

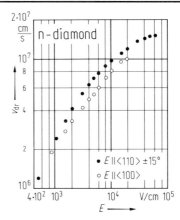

Fig. 38. n-Diamond. Electron drift velocity v_{dr} vs. electric field E at $T = 300$ K; [78C2].

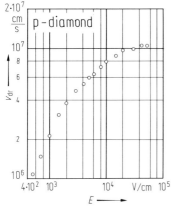

Fig. 39. p-Diamond. Hole drift velocity v_{dr} vs. electric field E at $T = 300$ K; $E \| \langle 110 \rangle \pm 15°$; [78C2].

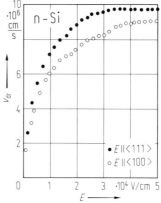

Fig. 41. n-Si. Electron drift velocity v_{dr} vs. electric field E at $T = 300$ K; [75J1].

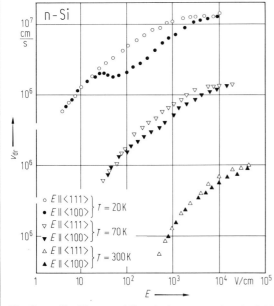

Fig. 40. n-Si. Electron drift velocity v_{dr} vs. electric field E at various temperatures; [75C1].

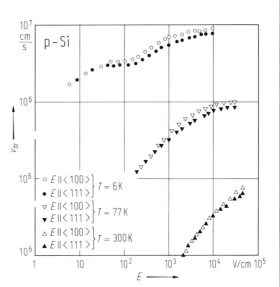

Fig. 42. p-Si. Hole drift velocity v_{dr} vs. electric field E at various temperatures; [77J1].

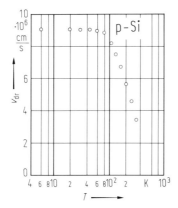

Fig. 43. p-Si. Hole drift velocity v_{dr} at $E=10^4$ V/m vs. temperature T; [75O1].

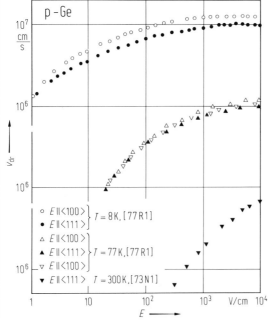

Fig. 45. p-Ge. Hole drift velocity v_{dr} vs. electric field E at various temperatures.

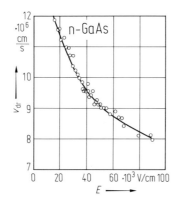

Fig. 48. n-GaAs. Electron drift velocity v_{dr} vs. electric field E at $T=300$ K; [71B1].

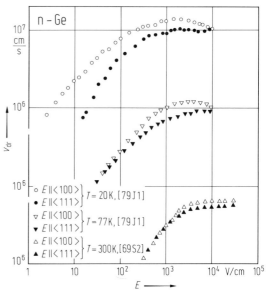

Fig. 44. n-Ge. Electron drift velocity v_{dr} vs. electric field E at various temperatures.

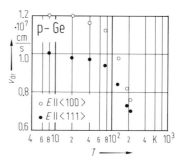

Fig. 46. p-Ge. Hole drift velocity v_{dr} at $E=10^4$ V/cm vs. temperature T; [77R1].

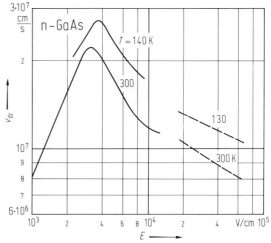

Fig. 47. n-GaAs. Electron drift velocity v_{dr} vs. electric field E. Full curves are from [68R2]. Dashed curves are from [77H1].

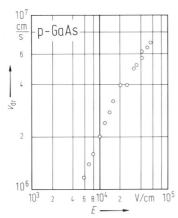

Fig. 49. p-GaAs. Hole drift velocity v_{dr} vs. electric field E at $T = 300$ K; [70D1].

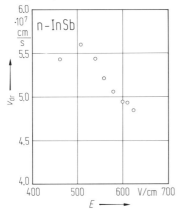

Fig. 51. n-InSb. Electron drift velocity v_{dr} vs. electric field E at $T = 300$ K; [70N1].

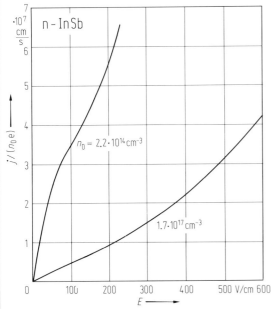

Fig. 53. n-InSb. Electron drift velocity $j/(n_0 e)$ vs. electric field E from current density measurements at $T = 77$ K; [70S1].

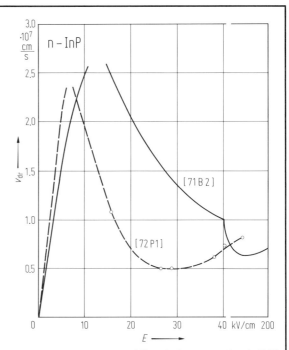

Fig. 50. n-InP. Electron drift velocity v_{dr} vs. electric field E at $T = 300$ K.

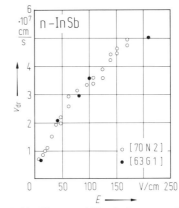

Fig. 52. n-InSb. Electron drift velocity v_{dr} vs. electric field E at $T = 77$ K.

For Fig. 54, see next page.

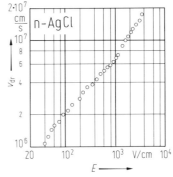

Fig. 55. n-AgCl. Electron drift velocity v_{dr} vs. electric field E at $T = 4.2$ K; [74K1].

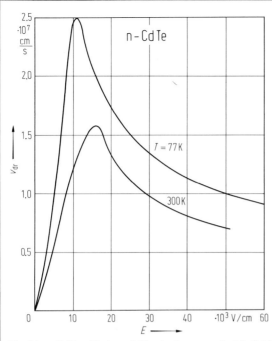

Fig. 54. n-CdTe. Electron drift velocity v_{dr} vs. electric field E; [71C1].

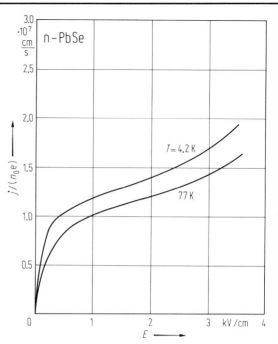

Fig. 56. n-PbSe. Electron drift velocity $j/(n_0 e)$ vs. electric field E from current density measurements. $n_0 = 5.3 \cdot 10^{17}$ cm^{-3}; [70S1].

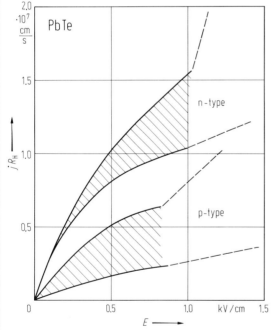

Fig. 57. PbTe. Drift velocity of electrons and holes $j \cdot R_H$ vs. electric field E from current density measurements at $T = 77$ K; $j \parallel \langle 112 \rangle$. Upper shaded region: Variation of measurements below threshold for an instability in n-type samples; lower shaded region: Variation of measurements below threshold for an instability in p-type samples; [78J1].

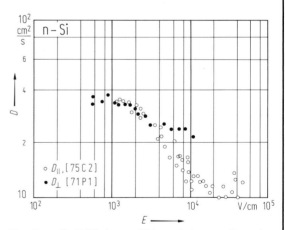

Fig. 58. n-Si. Diffusion coefficient of electrons D vs. electric field E at $T = 300$ K for $E \parallel \langle 111 \rangle$; [71P1].

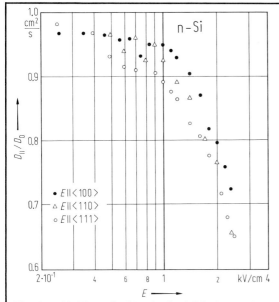

Fig. 59. n-Si. Normalized longitudinal diffusion coefficient $D_{\parallel}/D_0$ vs. electric field E at $T = 300$ K; [76N1].

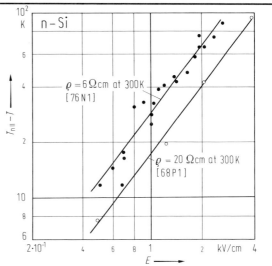

Fig. 60. n-Si. Longitudinal excess noise temperature $T_{n\parallel} - T$ vs. electric field E at $T = 300$ K for $E \parallel \langle 111 \rangle$.

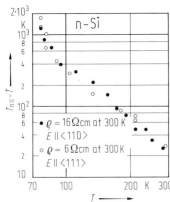

Fig. 61. n-Si. Longitudinal excess noise temperature $T_{n\parallel} - T$ vs. lattice temperature T at $E = 1$ kV/cm; [80N2].

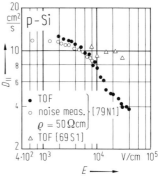

Fig. 62. p-Si. Longitudinal diffusion coefficient $D_{\parallel}$ vs. electric field E at $T = 300$ K for $E \parallel \langle 111 \rangle$; from time-of-flight (TOF) and noise measurements.

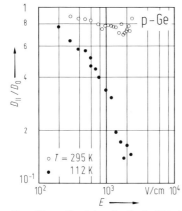

Fig. 64. p-Ge. Normalized longitudinal diffusion coefficient $D_{\parallel}/D_0$ vs. electric field E for $E \parallel \langle 110 \rangle$; $p = 7.5 \cdot 10^{13}$ cm^{-3}; [78R1].

Fig. 63. n-Ge. Longitudinal diffusion coefficient $D_{\parallel}$ vs. electric field E; [78C1]. ▶

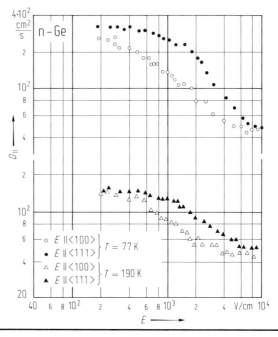

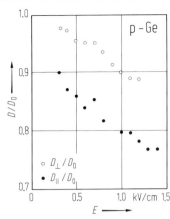

Fig. 65. p-Ge. Normalized diffusion coefficient D/D_0 vs. electric field E at $T=300$ K; $\varrho=40\,\Omega$ cm at $T=300$ K; $E\,\|\,\langle110\rangle$; $D_\perp/D_0$ measured $\|\,\langle111\rangle$; [73N1].

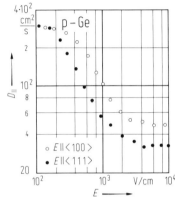

Fig. 67. p-Ge. Longitudinal diffusion coefficient $D_\|$ vs. electric field E at $T=77$ K; [78R1].

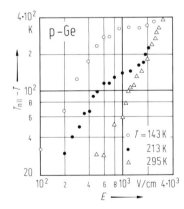

Fig. 69. p-Ge. Longitudinal excess noise temperature $T_{n\|}-T$ vs. electric field E; $\varrho=40\,\Omega$ cm at 300 K; $E\,\|\,[1\bar{1}0]$; [73N1].

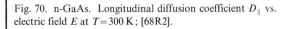

Fig. 70. n-GaAs. Longitudinal diffusion coefficient $D_\|$ vs. electric field E at $T=300$ K; [68R2].

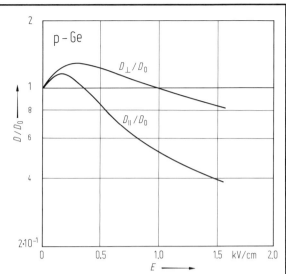

Fig. 66. p-Ge. Normalized diffusion coefficient D/D_0 vs. electric field E at $T=77$ K; $p=3.5\cdot10^{14}$ cm^{-3}; $E\,\|\,\langle100\rangle$. $D_\perp/D_0$, measured $\|\,\langle010\rangle$; [77B1].

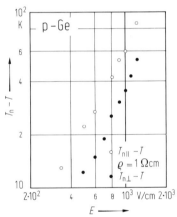

Fig. 68. p-Ge. Excess noise temperature T_n-T vs. electric field E at $T=300$ K; $E\,\|\,[1\bar{1}0]$; $T_{n\perp}-T$, measured $\|\,[111]$; [73N1].

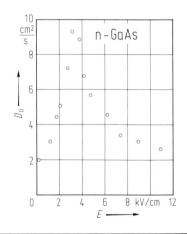

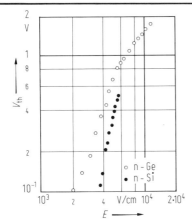

Fig. 71. n-Si, n-Ge. Thermoelectric voltage V_{th} of electrons vs. electric field E at $T = 300$ K; [71d1].

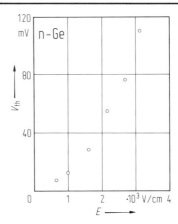

Fig. 72. n-Ge. Thermoelectric voltage V_{th} of electrons vs. electric field E at $T = 300$ K; [71d1].

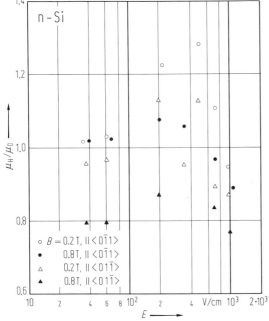

Fig. 73. n-Si. Normalized Hall mobility μ_{H}/μ_0 vs. electric field E at $T = 77$ K for $j \parallel \langle 111 \rangle$; [72K1],

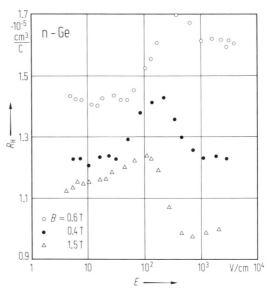

Fig. 74. n-Ge. Hall coefficient R_{H} vs. electric field E at $T = 80$ K for $j \parallel \langle 1\bar{1}0 \rangle$; [69M4].

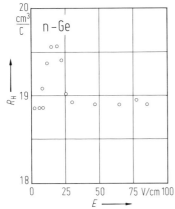

Fig. 76. n-Ge. Hall coefficient R_{H} vs. electric field E at $T = 4.2$ K; $n = 3.4 \cdot 10^{17}$ cm^{-3}; [69T1].

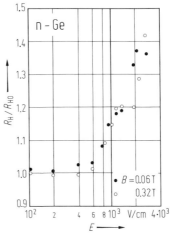

Fig. 75. n-Ge. Normalized Hall coefficient $R_{\text{H}}/R_{\text{H}0}$ vs. electric field E at $T = 200$ K for $j \parallel \langle 100 \rangle$; [70H3].

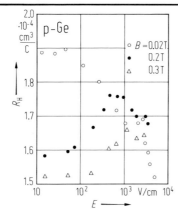

Fig. 77. p-Ge. Hall coefficient R_H vs. electric field E at $T = 78$ K; $\varrho = 6.4 \ \Omega$ cm at 300 K; [59Z1].

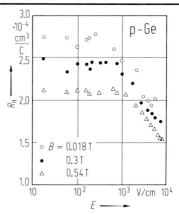

Fig. 78. p-Ge. Hall coefficient R_H vs. electric field E at $T = 300$ K; $\varrho = 6.4 \ \Omega$ cm at 300 K; [59Z1].

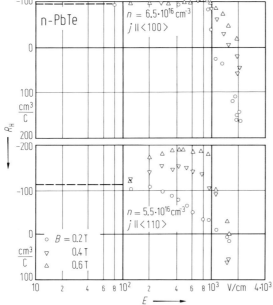

Fig. 79. n-PbTe. Hall coefficient R_H vs. electric field E at $T = 77$ K; [78J1].

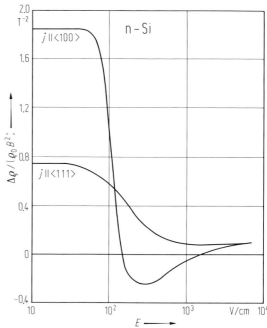

Fig. 80. n-Si. Weak-field transverse magnetoresistance $\Delta\varrho/(\varrho_0 B^2)$ vs. electric field E at $T = 77$ K, $B \perp j$; [70H2].

For Fig. 81, see next page.

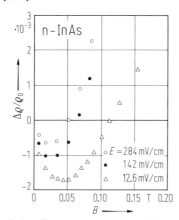

Fig. 83. n-InAs. Transverse magnetoresistance $\Delta\varrho/\varrho_0$ vs. magnetic induction B at $T = 4.2$ K; $n = 2.5 \cdot 10^{16}$ cm^{-3}; [70B1].

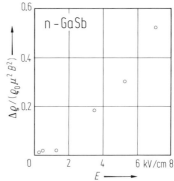

Fig. 82. n-GaSb. Longitudinal magnetoresistance $\Delta\varrho/(\varrho_0 \mu^2 B^2)$ vs. electric field E at $T = 300$ K. $B \| j$. $j \| \langle 111 \rangle$. $n = 6.8 \cdot 10^{16}$ cm^{-3}; [70H1].

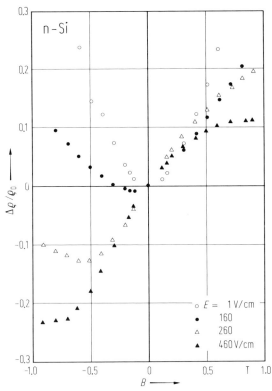

Fig. 81. n-Si. Transverse magnetoresistance $\Delta\varrho/\varrho_0$ vs. magnetic induction B. Angle between j and $\langle 100 \rangle = 24°$. $B \perp j$; [71A1].

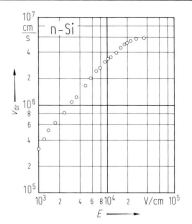

Fig. 84. n-Si. Drift velocity of electrons v_{dr} in an n-channel vs. electric field E at $T = 300$ K; {100} surface; $n_s = 6 \cdot 10^{12}$ cm^{-2}; [70F1].

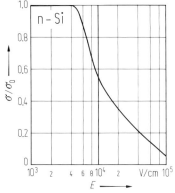

Fig. 85. n-Si. Normalized conductivity σ/σ_0 of an n-channel on a {100} surface vs. electric field E at $T = 300$ K; $n_s = 7 \cdot 10^{12}$ cm^{-2}; [74H1].

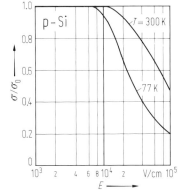

Fig. 86. p-Si. Normalized conductivity σ/σ_0 of a p-channel on a {100} surface vs. electric field E; $p_s = 7 \cdot 10^{12}$ cm^{-2}; [74H1].

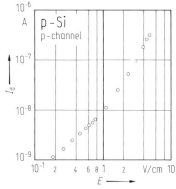

Fig. 87. p-Si. Source-drain current I_d of a p-channel on a {100} surface vs. electric field E at $T = 1.5$ K, $p_s = 1.2 \cdot 10^{11}$ cm^{-2}; [76E1].

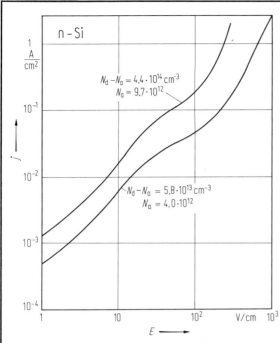

Fig. 88. n-Si. Current density j vs. electric field E at $T = 27.1$ K; $j \parallel \langle 111 \rangle$; [79A1].

Fig. 89. n-Si. Current density j vs. electric field E at $T = 20.36$ K; $j \parallel \langle 111 \rangle$; [79A1].

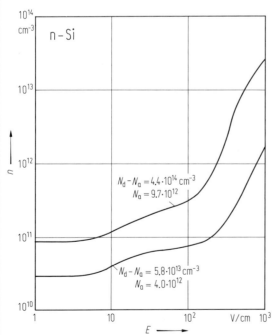

Fig. 90. n-Si. Stationary electron concentration n vs. electric field E at $T = 27.1$ K; [79A1].

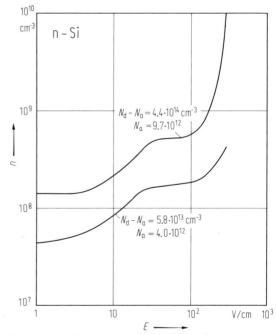

Fig. 91. n-Si. Stationary electron concentration n vs. electric field E at $T = 20.36$ K; [79A1].

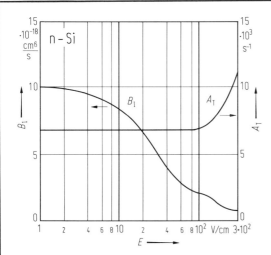

Fig. 92. n-Si. Thermal ionization coefficient A_T and Auger recombination coefficient B_I vs. electric field E at $T = 27.1$ K; [79A1].

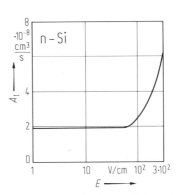

Fig. 93. n-Si. Impact ionization coefficient A_I vs. electric field E at $T = 27.1$ K; [79A1].

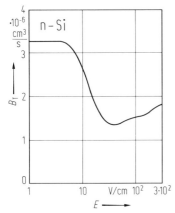

Fig. 94. n-Si. Phonon assisted recombination coefficient B_T vs. electric field E at $T = 27.1$ K; [79A1].

Fig. 95. n-Si. Thermal ionization coefficient A_T and phonon assisted recombination coefficient B_T vs. electric field E at $T = 20.36$ K; [79A1].

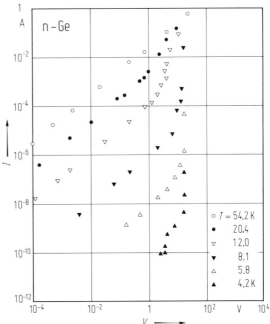

Fig. 96. n-Ge. Current I vs. voltage V. $n = 3 \cdot 10^{15}$ cm^{-3}; [61L1].

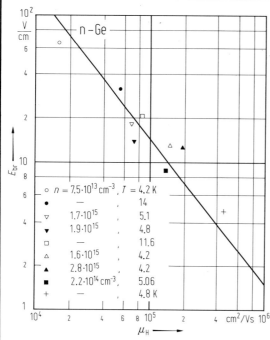

Fig. 97. n-Ge. Breakdown electric field E_{br} vs. Hall mobility μ_H at 4.2 K; [61L1].

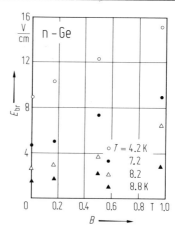

Fig. 98. n-Ge. Breakdown electric field E_{br} vs. magnetic induction B. $N_d = 5.6 \cdot 10^{14}$ cm^{-3}; $N_a = 2.1 \cdot 10^{14}$ cm^{-3}; [59F1].

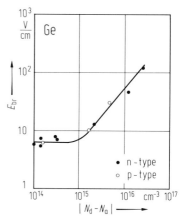

Fig. 99. Ge. Breakdown electric field E_{br} vs. carrier density $|N_d - N_a|$ at $T = 4.2$ K. n-type samples: $N_d > N_a$; p-type samples: $N_a > N_d$; [57S1].

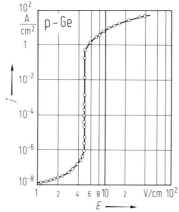

Fig. 100. p-Ge. Current density j vs. electric field E at $T = 4.2$ K; $p = 1 \cdot 10^{14}$ cm^{-3}; [68Z1].

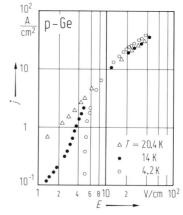

Fig. 101. p-Ge. Current density j vs. electric field E; $p = 1 \cdot 10^{14}$ cm^{-3}; [68Z1].

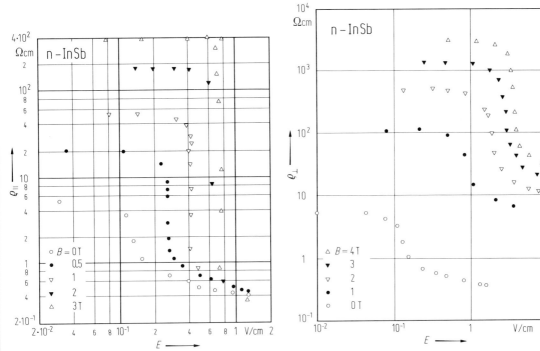

Fig. 102. n-InSb. Longitudinal resistivity $\varrho_\parallel$ vs. electric field E at $T = 4.2$ K; $j \parallel B$; $N_d - N_a = 4.7 \cdot 10^{13}$ cm^{-3}; $N_d = 2.5 \cdot 10^{14}$ cm^{-3}; [69M2].

Fig. 103. n-InSb. Transverse resistivity $\varrho_\perp$ vs. electric field E at $T = 4.2$ K; $j \perp B$; N_d and N_a as in Fig. 102; [69M2].

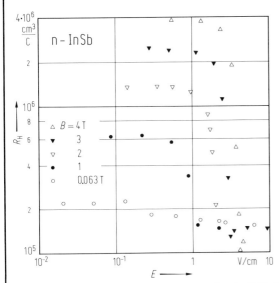

Fig. 104. n-InSb. Hall coefficient R_H vs. electric field E at $T = 4.2$ K. $j \perp B$. N_d and N_a as in Fig. 102; [69M2].

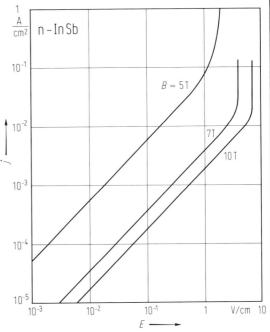

Fig. 105. n-InSb. Current density j vs. electric field E at $T = 1.8$ K. $N_d = 3 \cdot 10^{15}$ cm^{-3}; [80K1].

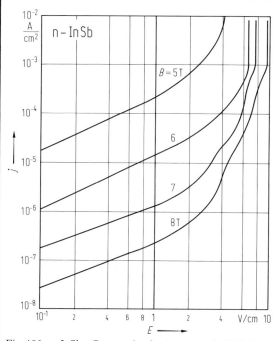

Fig. 106. n-InSb. Current density j vs. electric field E at $T = 0.37$ K. $N_d = 3 \cdot 10^{15}$ cm^{-3}; [80K1].

Fig. 107. n-GaSb. Current density j vs. electric field E at $T = 300$ K. $n = 6.8 \cdot 10^{16}$ cm^{-3}. Parameter: time after the rise of the voltage pulse, as indicated in the figure; [71J1].

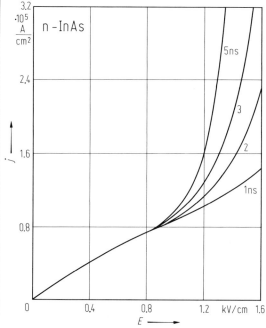

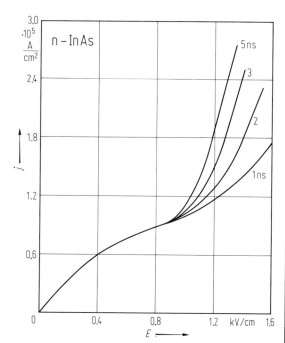

Fig. 108. n-InAs. Current density j vs. electric field E at $T = 300$ K. $n = 2.5 \cdot 10^{16}$ cm^{-3}; parameter: time after the rise of the voltage pulse as indicated in the figure; [72B2].

Fig. 109. n-InAs. Current density j vs. electric field E at $T = 77$ K. $n = 2.5 \cdot 10^{16}$ cm^{-3}; parameter: time after the rise of the voltage pulse as indicated in the figure; [72B2].

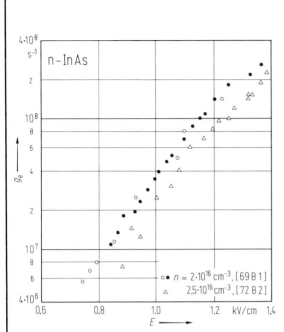

Fig. 110. n-InAs. Generation rate g_e of electrons vs. electric field E at $T=77$ K. The results from [69B1] are for two different samples.

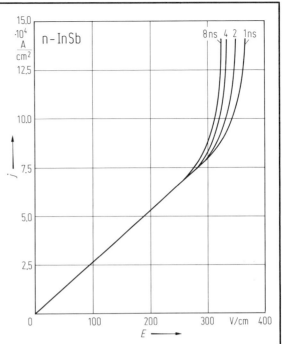

Fig. 111. n-InSb. Current density j vs. electric field E at $T=77$ K; $n=1.5 \cdot 10^{16}$ cm^{-3} at 77 K; parameter: time after the rise of the voltage pulse as indicated in the figure; [72B2].

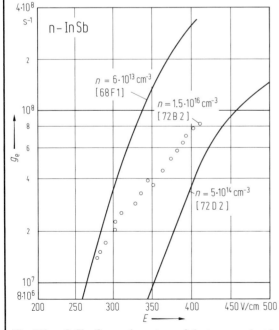

Fig. 112. n-InSb. Generation rate g_e of electrons vs. electric field E at $T=77$ K.

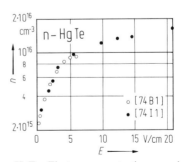

Fig. 113. n-HgTe. Electron concentration n vs. electric field E at $T=4.2$ K. $n=3 \cdot 10^{15}$ cm^{-3}.

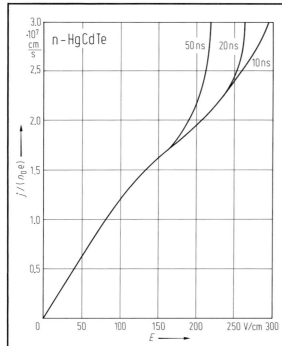

Fig. 114. n-Hg$_{0.8}$Cd$_{0.2}$Te. Current density j divided by $(n_0 e)$ vs. electric field E at $T = 77$ K. $n = 6 \cdot 10^{14}$ cm^{-3}; parameter is the time after the rise of the voltage pulse as indicated in the figure; [74N1].

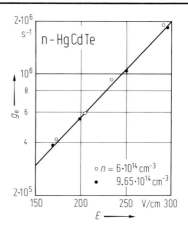

Fig. 115. n-Hg$_{0.8}$Cd$_{0.2}$Te. Generation rate g_e of electrons vs. electric field E at $T = 77$ K; [74N1].

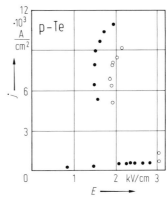

Fig. 116. p-Te. Current density j vs. electric field E at $T = 77$ K. $p = 4 \cdot 10^{14}$ cm^{-3} at 77 K. Open circles: measured 15 ns after the rise of the voltage pulse; closed circles: measured 36 ns after the rise of the voltage pulse; [69N1].

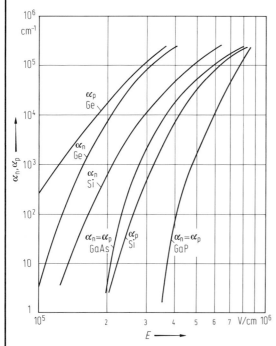

Fig. 118. Si, Ge, GaAs, GaP. Ionization coefficient α_n for electrons and α_p for holes in various semiconductors vs. electric field E at $T = 300$ K; [69S3].

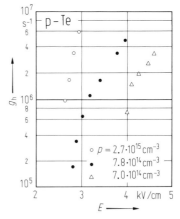

Fig. 117. p-Te. Generation rate g_h of holes vs. electric field E at $T = 77$ K. Densities indicated in the figure are given at 77 K; [70N3].

Kahlert

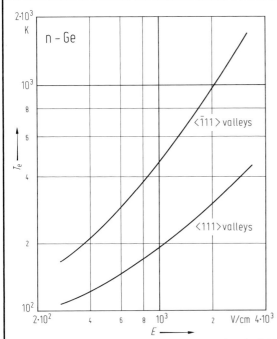

Fig. 119. n-Ge. Electron temperature T_e vs. electric field E from birefringence measurements at $T = 85$ K. $E \| \langle 111 \rangle$; $n = 5 \cdot 10^{14}$ cm^{-3}; [72V1].

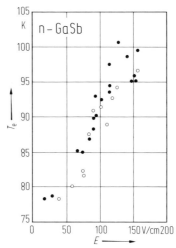

Fig. 120. n-GaSb. Electron temperature T_e vs. electric field E from optical absorption measurements at $T = 77$ K. $n = 3.5 \cdot 10^{17}$ cm^{-3}; data from two different samples; [71H2].

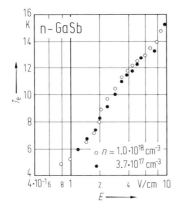

Fig. 121. n-GaSb. Electron temperature T_e vs. electric field E from Shubnikov-de Haas measurements at $T = 4.2$ K; [71K1].

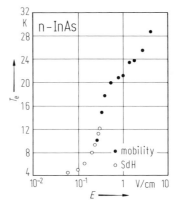

Fig. 122. n-InAs. Electron temperature T_e vs. electric field E at $T = 4.2$ K. $n = 2.5 \cdot 10^{16}$ cm^{-3}; from mobility measurements, and from Shubnikov-de Haas measurements, as indicated in the figure; [72B1].

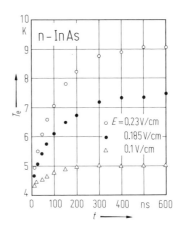

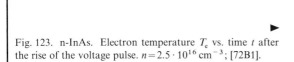

Fig. 123. n-InAs. Electron temperature T_e vs. time t after the rise of the voltage pulse. $n = 2.5 \cdot 10^{16}$ cm^{-3}; [72B1].

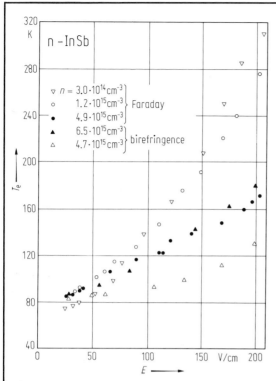

Fig. 124. n-InSb. Electron temperature T_e vs. electric field E at $T = 80$ K; from Faraday effect measurements, and from birefringence measurements as indicated in the figure [73V1, 72V2, 72V3].

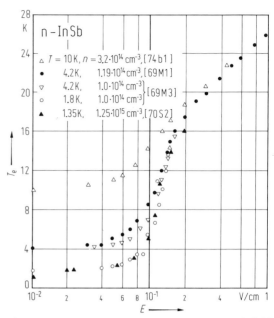

Fig. 125. n-InSb. Electron temperature T_e vs. electric field E from mobility measurements.

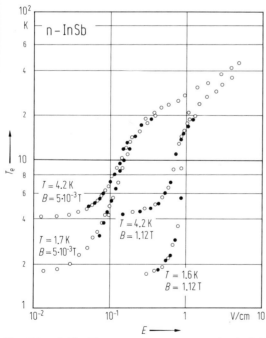

Fig. 126. n-InSb. Electron temperature T_e vs. electric field E from mobility (open circles) and Hall effect (closed circles) measurements; $n = 1 \cdot 10^{14}$ cm^{-3}; [69M3].

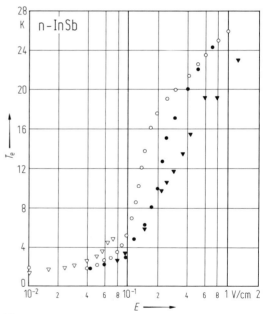

Fig. 127. n-InSb. Electron temperature T_e vs. electric field E. (open circles): $T = 1.8$ K; $n = 1 \cdot 10^{14}$ cm^{-3}, from mobility measurements; [69M3]; (closed circles): $T = 1.8$ K, $n = 1 \cdot 10^{14}$ cm^{-3}, from magnetoresistance measurements; [69M3]; (open triangles): $T = 1.3$ K, $n = 1.7 \cdot 10^{15}$ cm^{-3}, from Shubnikov-de Haas measurements; [66I1]; (closed triangles): $T = 1.7$ K, $n = 1.4 \cdot 10^{14}$ cm^{-3}, from electron spin resonance measurements; [64Z1].

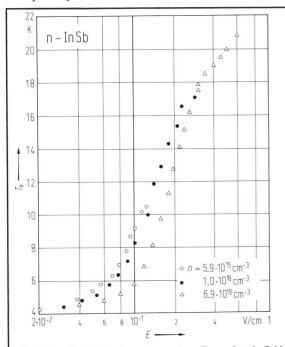

Fig. 128. n-InSb. Electron temperature T_e vs. electric field E at $T = 4.2$ K from Shubnikov-de Haas measurements; [73K1, 73B1].

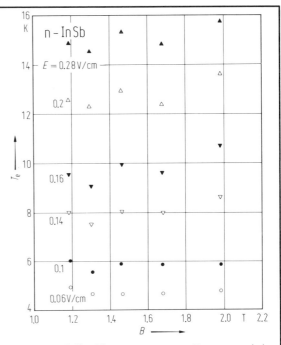

Fig. 129. n-InSb. Electron temperature T_e vs. magnetic induction B at $T = 4.2$ K; $B \parallel j$, from Shubnikov-de Haas effect measurements; $n = 6.9 \cdot 10^{16}$ cm^{-3}; [74b1].

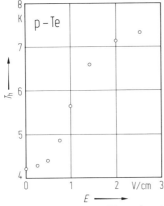

Fig. 130. p-Te. Hole temperature T_h vs. electric field E at $T = 4.2$ K. $p = 3.6 \cdot 10^{17}$ cm^{-3}; from Shubnikov-de Haas effect measurements; [75K1].

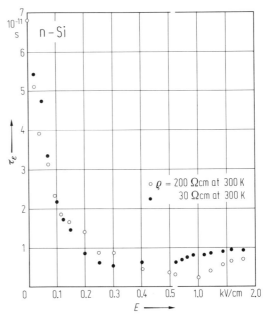

Fig. 132. n-Si. Energy relaxation time τ_ε vs. electric field E at $T = 77$ K, $E \parallel \langle 111 \rangle$; [72D1].

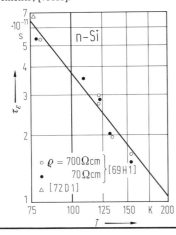

◀

Fig. 131. n-Si. Energy relaxation time τ_ε vs. lattice temperature T. The ϱ-values indicated in the figure are given at 300 K.

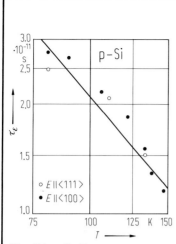

Fig. 133. p-Si. Energy relaxation time τ_ε vs. lattice temperature T. $p = 7.3 \cdot 10^{13}$ cm^{-3}; [69H1].

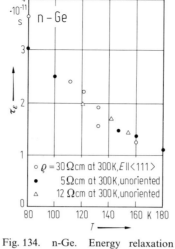

Fig. 134. n-Ge. Energy relaxation time τ_ε vs. lattice temperature T; [69H1].

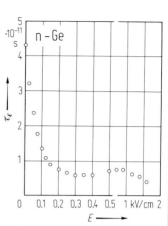

Fig. 135. n-Ge. Energy relaxation time τ_ε vs. electric field E at $T = 77$ K; $E \parallel \langle 100 \rangle$; $\varrho = 40 \,\Omega$ cm at 300 K; [72D1].

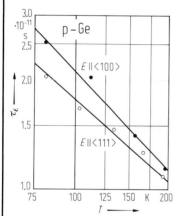

Fig. 136. p-Ge. Energy relaxation time τ_ε vs. lattice temperature T. $\varrho = 12 \,\Omega$ cm at 300 K; [69H1].

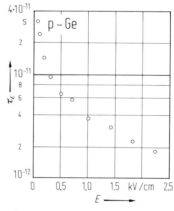

Fig. 137. p-Ge. Energy relaxation time τ_ε vs. electric field E at $T = 77$ K; $p = 3.5 \cdot 10^{14}$ cm^{-3}; $E \parallel \langle 100 \rangle$; [77B1].

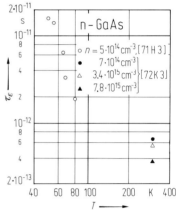

Fig. 138. n-GaAs. Energy relaxation time τ_ε vs. lattice temperature T.

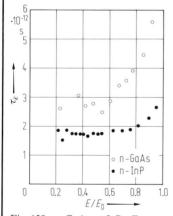

Fig. 139. n-GaAs, n-InP. Energy relaxation time τ_ε vs. normalized electric field E/E_0 at 300 K. E_0 is the threshold value for the onset of the negative differential mobility.
E_0 (GaAs) $= 3.36$ kV/cm;
E_0 (InP) $= 6.8$ kV/cm; [73G1].

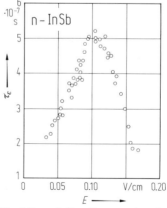

Fig. 140. n-InSb. Energy relaxation time τ_ε vs. electric field E at $T = 4.2$ K; $n = 1.19 \cdot 10^{15}$ cm^{-3}; [69M1].

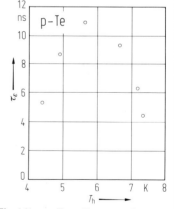

Fig. 141. p-Te. Energy relaxation time τ_ε vs. hole temperature T_h at $T = 4.2$ K; $p = 3.6 \cdot 10^{17}$ cm^{-3}; [75K1].

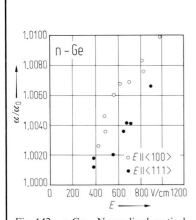

Fig. 142. n-Ge. Normalized optical absorption coefficient α/α_0 vs. electric field E at $T=300$ K; $n=5\cdot10^{15}$ cm^{-3}; $\lambda=10.6$ μm; light is polarized $\parallel E$; [79S1].

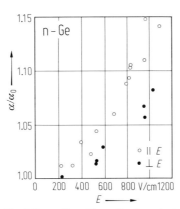

Fig. 143. n-Ge. Normalized optical absorption coefficient α/α_0 vs. electric field E at $T=150$ K; $n=5\cdot10^{15}$ cm^{-3}; $\lambda=10.6$ μm; $E\parallel\langle100\rangle$. The light is polarized $\parallel E$, and $\perp E$ as indicated in the figure; [79S1].

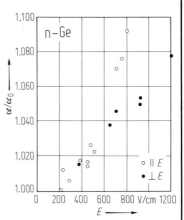

Fig. 144. n-Ge. Normalized optical absorption coefficient α/α_0 vs. electric field E at $T=150$ K; $n=5\cdot10^{15}$ cm^{-3}; $\lambda=10.6$ μm; $E\parallel\langle110\rangle$. The light is polarized $\parallel E$, and $\perp E$ as indicated in the figure; [79S1].

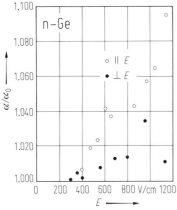

Fig. 145. n-Ge. Normalized optical absorption coefficient α/α_0 vs. electric field E at $T=150$ K; $n=5\cdot10^{15}$ cm^{-3}; $\lambda=10.6$ μm; $E\parallel\langle111\rangle$. The light is polarized $\parallel E$, and $\perp E$ as indicated in the figure; [79S1].

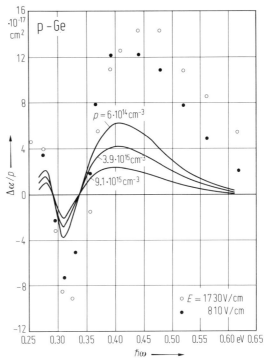

Fig. 146. p-Ge. Change of absorption cross section $\Delta\alpha/p$ vs. photon energy $\hbar\omega$, for various hole densities p, $T=93$ K, $E=300$ V/cm, [62B1]; and for two different electric field strengths; $T=77$ K, $p=1.9\cdot10^{15}$ cm^{-3}; [64P1].

◀

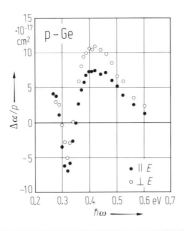

Fig. 147. p-Ge. Change of absorption cross section $\Delta\alpha/p$ vs. photon energy $\hbar\omega$ at $T=85$ K; $E=500$ V/cm; $p=2.1\cdot10^{15}$ cm^{-3}; $E\parallel\langle111\rangle$. The light is polarized $\perp E$, and $\parallel E$ as indicated in the figure; [73C1].

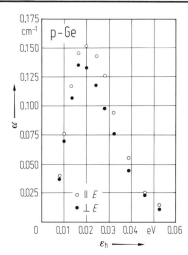

Fig. 148. p-Ge. Absorption coefficient α vs. heavy hole energy ε_h at $T = 77$ K; $p = 5.9 \cdot 10^{14}$ cm^{-3}; $E \parallel \langle 100 \rangle$; $E = 760$ V/cm. The light is propagating $\parallel E$, and $\perp E$ as indicated in the figure; [63B1].

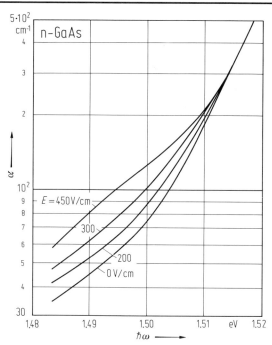

Fig. 149. n-GaAs. Absorption coefficient α in the vicinity of the fundamental absorption edge vs. photon energy $\hbar \omega$ at $T = 77$ K; $n = 1 \cdot 10^{18}$ cm^{-3}; [73J1].

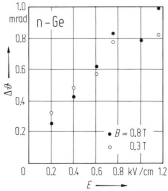

Fig. 150. n-Ge. Change of the Faraday rotation angle $\Delta \vartheta$ vs. electric field E at $T = 200$ K; $\lambda = 10.6$ μm; $n = 5.5 \cdot 10^{15}$ cm^{-3}; $j \parallel [1\bar{1}0]$; $B \parallel [111]$; [72K2].

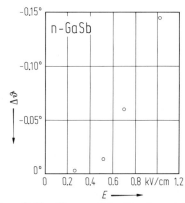

Fig. 151. n-GaSb. Change of the Faraday rotation angle $\Delta \vartheta$ vs. electric field E at $T = 300$ K; $B = 1.25$ T; $n = 6.8 \cdot 10^{16}$ cm^{-3}; [70H4, 71H1].

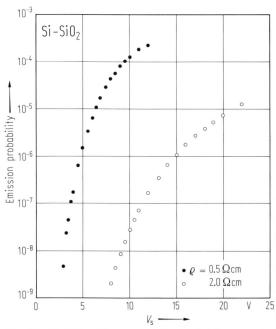

Fig. 152. Si–SiO$_2$. Electron emission probability vs. substrate voltage V_s at $T = 300$ K, electric field in the oxid layer: $E_{ox} = 3 \cdot 10^6$ V/cm; [78N1].

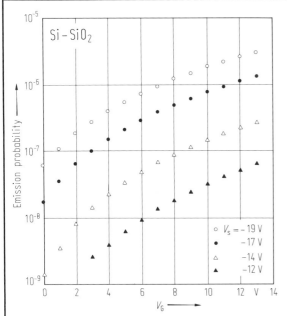

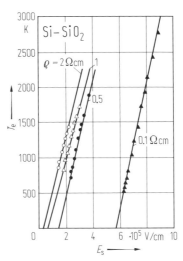

Fig. 153. Si–SiO$_2$. Electron emission probability vs. gate voltage V_G at various substrate voltages V_s, and at $T = 300$ K, $\varrho = 2\,\Omega$m, oxid layer thickness: $d_{ox} = 40.6$ nm; [78N1].

Fig. 154. Si–SiO$_2$. Effective electron temperature T_e vs. maximum surface electric field E_s at $T = 300$ K; [78N1].

References for 14

Textbook and reviews

62r1   Reik, H.G.: Festkörperprobleme I, Sauter, F., (ed.), Braunschweig: Vieweg **1962**, 89.

62s1   Schmidt-Tiedemann, K.J.: Festkörperprobleme I, Sauter, F., (ed.), Braunschweig: Vieweg **1962**, 122.

67c1   Conwell, E.M.: Solid State Phys. Suppl. **9**, Seitz, F., Turnbull, D., Ehrenreich, H., (eds.), New York: Academic Press, 1967.

69a1   Asche, M., Sarbei, O.G.: Phys. Status Solidi **33** (1969) 9.

69b1   Butcher, P.N., Hearn, C.J.: Sci. Prog. Oxf. **57** (1969) 229.

71a1   Alberigi-Quaranta, A., Jacoboni, C., Ottaviani, G.: Riv. Nuovo Cimento **1** (1971) 445.

71d1   Dienys, V., Pozhela, J.: Hot Electrons, Vilnius: Mintis Publishing House, **1971**.

73s1   Seeger, K.: Semiconductor Physics, Wien, New York: Springer, **1973**.

74b1   Bauer, G.: Springer Tracts in Modern Physics, Vol. **74**, p. 1; Berlin, Heidelberg, New York: Springer, 1974.

78s1   Seiler, D.G., Stephens, A.E. (eds.): Solid State Electron. **21** (1978).

79j1   Jacoboni, C., Reggiani, L.: Adv. Phys. **28** (1979) 493.

80f1   Ferry, D.D., Barker, J.R., Jacoboni, C. (eds.): Physics of Nonlinear Transport in Semiconductors, Nato Advanced Study Institute, Series B, Vol. **52**, New York, London: Plenum, 1980.

81n1   Nougier, J.P. (ed.): Third International Conference on Hot Carriers in Semiconductors, Montpellier France, Journal de Physique, Colloque C7, supplement au n° 10, Tome 42, 1981.

82f1   Ferry, D.K.: Handbook on Semiconductors, Moss, T.S. (ed.), Vol **1**; p. 563, Paul, W. (ed.), Amsterdam, Oxford, New York: North Holland Publishing Company, **1982**.

85r1   Reggiani, L. (ed.): Hot Electron Transport in Semiconductors; Topics in Applied Physics, Vol. **58**, Berlin, Heidelberg, New York, Tokyo: Springer, **1985**.

Bibliography

54S1   Shibuya, M.: Phys. Rev. **95** (1954) 1385.

55S1   Shibuya, M.: Phys. Rev. **99** (1955) 1189.

56G1   Gunn, J.B.: J. Electron. **2** (1956) 87.

57G1   Gunn, J.B.: Progress in Semiconductors, Gibson, A.F., Burgess, R.E., (eds.), Vol. **2**, London: Temple Press, **1957**, 213.

57S1   Sclar, N., Burstein, E.: J. Phys. Chem. Solids, **2** (1957) 1.

58S1   Sasaki, W., Shibuya, M., Mizuguchi, K.: J. Phys. Soc. Jpn. **13** (1958) 457.

59C1   Conwell, E.M.: Phys. Chem. Solids **8** (1959) 234.

59F1   Finke, G., Lautz, G.: Z. Naturforsch. A **14** (1959) 62.

59G1   Gunn, J.B.: Phys. Chem. Solids **8** (1959) 239.

59M1   Morgan, T.N.: Phys. Chem. Solids **8** (1959) 245.

59S1   Seeger, K.: Phys. Rev. **114** (1959) 476.

59S2   Sasaki, W., Shibuya, M., Mizuguchi, K., Hatoyama, G.M.: Phys. Chem. Solids **8** (1959) 250.

59Z1   Zucker, H.: Private communication in: Solid State Phys. Suppl. **9**, Seitz, F., Turnbull, D., Ehrenreich, H., (eds.), New York: Academic Press, 1967.

60K1   Koenig, S.H., Nathan, M.I., Paul, W., Smith, A.G.: Phys. Rev. **118** (1960) 1217.

60S1   Sladek, R.J.: Phys. Rev. **120** (1960) 1589.

60Z1   Zucker, J.: Phys. Chem. Solids **12** (1960) 350.

61G1   Gibson, A.F., Granville, J.W., Paige, E.G.S.: Phys. Chem. Solids **19** (1961) 198.

61L1   Lautz, G.: Halbleiterprobleme VI, Sauter, F., (ed.), Braunschweig: Vieweg **1961**, 21.

61S1   Schmidt-Tiedemann, K.J.: Phys. Rev. **123** (1961) 1999.

62B1   Brown, M.A.C.S., Paige, E.G.S., Simcox, L.N.: Proc. Int. Conf. Physics of Semiconductors, Exeter, Stickland, A.C., (ed.), London: Institute of Physics and Physical Society **1962**, 111.

62G1   Gibbs, W.E.K.: J. Appl. Phys. **33** (1962) 3369.

63B1   Bray, R., Pinson, W.E.: Phys. Rev. Lett. **11** (1963) 502.

63G1   Glicksman, M., Hicinbothem, W.A.: Phys. Rev. **129** (1963) 1572.

63J1   Jørgensen, M.H., Meyer, N.I., Schmidt-Tiedemann, K.J.: Solid State Commun. **1** (1963) 226.

63N1   Nathan, M.I.: Phys. Rev. **130** (1963) 2201.

63S1   Seeger, K.: Z. Phys. **172** (1963) 68.

64P1   Pinson, W.E., Bray, R.: Phys. Rev. **136** (1964) A 1449.

64Z1   Zylbersztejn, A.: Proc. Int. Conf. Physics of Semiconductors, Paris, Hulin, M., (ed.), Paris: Dunod **1964**, 505.

65A1   Asche, M., Boitschenko, B.L., Sarbei, O.G.: Phys. Status Solidi **9** (1965) 323.

65K1   Kästner, P., Röth, E., Seeger, K.: Z. Phys. **187** (1965) 359.

65S1   Schweitzer, D., Seeger, K.: Z. Phys. **183** (1965) 207.

65V1   Vasetskii, V.M., Boitschenko, B.L.: Fiz. Tverd. Tela **7** (1965) 2021.

66I1   Isaacson, R.A., Bridges, F.: Solid State Commun. **4** (1966) 635.

66O1   Onuki, M., Shiga, K.: J. Phys. Soc. Jpn. Suppl. **21** (1966) 427.

67M1   Movchan, E.A., Miselyuk, E.G.: Fiz. Techn. Poluprov. **1** (1967) 1255.

68F1   Ferry, D.K., Heinrich, H.: Phys. Rev. **169** (1968) 670.

68I1   Ivanov-Omskii, V.I., Kolomiets, B.T., Smekalova, K.P., Smirnov, V.A.: Fiz. Techn. Poluprov. **2** (1968) 1197.

68P1   Pozhela, J., Bareikis, V.A., Matulenene, I.B.: Sov. Phys. Semicond. **2** (1968) 503.

68R1   Röth, E., Tschulena, G., Seeger, K.: Z. Phys. **212** (1968) 183.

68R2   Ruch, G., Kino, G.S.: Phys. Rev. **174** (1968) 921.

68Z1   Zavaritskaya, E.I.: Electrical and Optical Properties of Semiconductors, Vol. **37**, Skobel'tsyn, Y., (ed.), 1968, 33.

69B1   Bauer, G., Kuchar, F.: Phys. Lett. **30 A** (1969) 399.

69H1   Hess, K., Seeger, K.: Z. Phys. **218** (1969) 431.

69M1   Maneval, J.P., Zylbersztejn, A., Budd, H.F.: Phys. Rev. Lett. **23** (1969) 848.

69M2   Mansfield, R., Ahmad, I.: J. Phys. C **3** (1970) 423.

69M3   Miyazawa, H.: J. Phys. Soc. Jpn. **26** (1969) 700.

69M4   Movchan, E.A., Miselyuk, E.G.: Sov. Phys. Semicond. **3** (1969) 571.

69N1   Nimtz, G., Seeger, K.: Appl. Phys. Lett. **14** (1969) 19.

69S1   Sigmon, T.W., Gibbons, J.F.: Appl. Phys. Lett. **15** (1969) 320.

69S2   Smith, J.E.: Phys. Rev. **178** (1969) 1364.

69S3	Sze, S.M.: Physics of Semiconductor Devices, New York: Wiley-Interscience, **1969**.
69T1	Tschulena, G.R., Bauer, G.: Solid State Commun. **7** (1969) 1499.
70B1	Bauer, G., Kahlert, H.: Proc. Int. Conf. Physics of Semiconductors, Cambridge, Mass., Keller, S.P., Hensel, J.C., Stern, F., (eds.), Oak Ridge, Tenn.: USAEC **1970**, 65.
70D1	Dalal, V.L.: Appl. Phys. Lett. **16** (1970) 489.
70F1	Fang, F., Fowler, A.B.: J. Appl. Phys. **41** (1970) 1825.
70H1	Heinrich, H., Jantsch, W.: Phys. Status Solidi **38** (1970) 225.
70H2	Heinrich, H., Kriechbaum, M.: J. Phys. Chem. Solids **31** (1970) 927.
70H3	Heinrich, H., Lischka, K., Kriechbaum, M.: Phys. Rev. **B2** (1970) 2009.
70H4	Heinrich, H.: Phys. Lett. **32 A** (1970) 331.
70N1	Neukermans, A., Kino, G.S.: Proc. Int. Conf. Physics of Semiconductors, Cambridge, Mass., Keller, S.P., Hensel, J.C., Stern, F., (eds.), Oak Ridge, Tenn.: USAEC **1970**, 40.
70N2	Neukermans, A., Kino, G.S.: Appl. Phys. Lett. **17** (1970) 102.
70N3	Nimtz, G.: Proc. Int. Conf. Physics of Semiconductors, Cambridge, Mass., Keller, S.P., Hensel, J.C., Stern, F., (eds.), Oak Ridge, Tenn.: USAEC **1970**, 396.
70S1	St.Onge, H., Walpole, J.N., Rediker, R.H.: Proc. Int. Conf. Physics of Semiconductors, Cambridge, Mass., Keller, S.P., Hensel, J.C., Stern, F., (eds.), Oak Ridge, Tenn.: USAEC **1970**, 391.
70S2	Szymanska, W., Maneval, J.P.: Solid State Commun. **8** (1970) 879.
71A1	Asche, M., Zav'yalov, Yu.G., Sarbei, O.G.: JETP Lett. **13** (1971) 285.
71B1	Bastida, E.M., Fabri, G., Svelto, V., Vaghi, F.: Appl. Phys. Lett. **18** (1971) 28; Appl. Phys. Lett. **19** (1971) 122.
71B2	Boers, P.M.: Electron. Lett. **7** (1971) 625.
71C1	Canali, C., Martini, M., Ottaviani, G., Zanio, K.R.: Phys. Rev. **B4** (1971) 422.
71H1	Heinrich, H.: Phys. Rev. **B3** (1971) 416.
71H2	Heinrich, H., Jantsch, W.: Phys. Rev. **B4** (1971) 2504.
71H3	Hess, K., Kahlert, H.: J. Phys. Chem. Solids **32** (1971) 2262.
71J1	Jantsch, W., Heinrich, H.: Phys. Rev. **B3** (1971) 420.
71K1	Kahlert, H., Bauer, G.: Phys. Status Solidi (b) **46** (1971) 535.
71P1	Persky, G., Bartelink, D.J.: J. Appl. Phys. **42** (1971) 4414.
71T1	Tschulena, G.: Acta Phys. Austr. **33** (1971) 42.
72B1	Bauer, G., Kahlert, H.: Phys. Rev. **B5** (1972) 566.
72B2	Bauer, G., Kuchar, F.: Phys. Status Solidi (a) **13** (1972) 169.
72D1	Dargys, A., Banys, T.: Phys. Status Solidi (b) **52** (1972) 699.
72D2	Dick, C.L., Ancker-Johnson, B.: Phys. Rev. **B5** (1972) 526.
72K1	Kriechbaum, M., Heinrich, H., Wajda, J.: J. Phys. Chem. Solids **33** (1972) 829.
72K2	Kriechbaum, M., Lischka, K., Kuchar, F., Heinrich, H.: Proc. Int. Conf. Physics of Semiconductors 1972, Miasek, M., (ed.), Warzawa: Polish Scientific Publishers **1972**, 615.
72K3	Kuchar, F., Philipp, A., Seeger, K.: Solid State Commun. **11** (1972) 965.
72P1	Prew, B.A.: Electron. Lett. **8** (1972) 592.
72V1	Vorob'ev, L.E., Stafeef, V.I., Ushakov, A.V.: Phys. Status Solidi (b) **53** (1972) 431.
72V2	Vorob'ev, L.E., Komissarow, V.S., Stafeef, V.I.: Phys. Status Solidi (b) **52** (1972) 25.
72V3	Vorob'ev, L.E., Komissarow, V.S., Stafeef, V.I.: Phys. Status Solidi (b) **54** (1972) K 61.
73B1	Bauer, G., Kahlert, H.: J. Phys. C **6** (1973) 1253.
73C1	Christensen, O.: Phys. Rev. **B7** (1973) 763.
73C2	Costato, M., Reggiani, L.: Phys. Status Solidi (b) **58** (1973) 47.
73G1	Glover, G.H.: J. Appl. Phys. **44** (1973) 1295.
73H1	Hess, K., Vana, H.: J. Phys. C **6** (1973) L 150.
73J1	Jantsch, W., Heinrich, H.: Solid State Commun. **13** (1973) 715.
73K1	Kahlert, H., Bauer, G.: Phys. Rev. **B7** (1973) 2670.
73N1	Nougier, J.P., Rolland, M.: Phys. Rev. **B8** (1973) 5728.
73V1	Vorob'ev, L.E., Komissarow, V.S., Stafeef, V.I.: Sov. Phys. Semicond. **7** (1973) 59.
74B1	Beneslavskii, S.D., Ivanov-Omskii, V.I., Kolomiets, B.T., Smirnov, V.A.: Fiz. Tverd. Tela **16** (1974) 1620.
74D1	Dornhaus, R., Happ, K., Müller, K.H., Nimtz, G., Schlabitz, W., Zaplinski, P., Bauer, G.: Proc. Int. Conf. Physics of Semiconductors, Stuttgart 1974, Pilkuhn, H.J., (ed.), Stuttgart: Teubner **1974**, 1157.
74E1	Elliot, C.T., Spain, I.L.: J. Phys. C **7** (1974) 727.
74H1	Hess, K., Sah, C.T.: J. Appl. Phys. **45** (1974) 1254.

74I1   Ivanov-Omskii, V.I., Kolomiets, B.T., Smirnov, V.A.: Fiz. Techn. Poluprov. **8** (1974) 620.

74K1   Kajita, K., Masumi, T.: Proc. Int. Conf. Physics of Semiconductors, Stuttgart 1974, Pilkuhn, H.J., (ed.), Stuttgart: Teubner **1974**, 844.

74N1   Nimtz, G., Bauer, G., Dornhaus, R., Müller, K.H.: Phys. Rev. **B10** (1974) 3302.

75C1   Canali, C., Jacoboni, C., Nava, F., Ottaviani, G., Alberigi-Quaranta, A.: Phys. Rev. **B12** (1975) 2265.

75C2   Canali, C., Jacoboni, C., Ottaviani, G., Alberigi-Quaranta, A.: Appl. Phys. Lett. **27** (1975) 278.

75J1   Jacoboni, C., Minder, R., Majni, G.: J. Phys. Chem. Solids **36** (1975) 1129.

75K1   Kahlert, H.: Phys. Status Solidi (b) **71** (1975) 151.

75O1   Ottaviani, G., Reggiani, L., Canali, C., Nava, F., Alberigi-Quaranta, A.: Phys. Rev. **B12** (1975) 3318.

76E1   Englert, T., Landwehr, G.: Surf. Sci, **58** (1976) 217.

76N1   Nougier, J.P., Rolland, M.: Proc. Int. Conf. Physics of Semiconductors, Rome 1976, Fumi, F.G., (ed.), Rome: Tipografia Marves **1976**, 1227.

77B1   Bareikis, V.A., Gal'dikas, A.P., Pozhela, J.: Sov. Phys. Semicond, **11** (1977) 210.

77H1   Houston, P.A., Evans, A.G.R.: Solid State Electron. **20** (1977) 197.

77J1   Jacoboni, C., Canali, C., Ottaviani, G., Alberigi-Quaranta, A.: Solid State Electron. **20** (1977) 77.

77R1   Reggiani, L., Canali, C., Nava, F., Ottaviani, G.: Phys. Rev. **B16** (1977) 2781.

78C1   Canali, C., Jacoboni, C., Nava, F.: Solid State Commun. **26** (1978) 889.

78C2   Canali, C., Jacoboni, C., Nava, F., Reggiani, L., Kozlov, S.F.: Proc. Int. Conf. Physics of Semiconductors, Edinburgh 1978, Wilson, B.L.H., (ed.), London, Bristol: The Institute of Physics and the Physical Society **1978**, 327.

78F1   Ferry, D.K.: Solid State Electron. **21** (1978) 115.

78H1   Hess, K.: Solid State Electron. **21** (1978) 123.

78J1   Jantsch, W., Rozenbergs, J., Heinrich, H.: Solid State Electron. **21** (1978) 103.

78M1   Müller, W., Kohl, F., Partl, H., Gornik, E.: Solid State Electron. **21** (1978) 235.

78N1   Ning, T.H.: Solid State Electron. **21** (1978) 273.

78R1   Reggiani, L., Canali, C., Nava, F., Alberigi-Quaranta, A.: J. Appl. Phys. **49** (1978) 4446.

79A1   Asche, M., Kostial, H., Sarbei, O.G.: Phys. Status Solidi (b) **91** (1979) 521.

79N1   Nava, F., Canali, C., Reggiani, L., Gasquet, D., Vaissiere, J.C., Nougier, J.P.: J. Appl. Phys. **50** (1979) 922.

79S1   Seeger, K., Vana, H.: Phys. Status Solidi (b) **96** (1979) 605.

80B1   Bauer, G.: Physics of Nonlinear Transport in Semiconductors, Nato Advanced Study Institute, Series B, Vol. **52**, New York, London: Plenum **1980**, 175.

80G1   Gornik, E.: Lecture Notes in Physics, Vol. 133, Zawadzki, W., (ed.), Berlin, Heidelberg, New York: Springer, **1980**, 160.

80K1   Kuchar, F., Fantner, E.J.: Phys. Status Solidi (a) **61** (1980) 531.

80N1   Nicholas, R.J., Portal, J.C.: Physics of Nonlinear Transport in Semiconductors, Nato Advanced Study Institute, Series B, Vol. **52**, New York, London: Plenum, **1980**, 255.

80N2   Nougier, J.P.: Physics of Nonlinear Transport in Semiconductors, Nato Advanced Study Institute, Series b, Vol. **52**, New York, London: Plenum, **1980**, 415.

84P1   Pfluger, P., Zeller, H.R., Bernasconi, J.: Phys. Rev. Lett. **53** (1984) 94.

15 Electron-hole liquids

15.0 Introduction

15.0.1 General remarks

Three types of condensation of electrons and holes in semiconductors can be imagined to occur below a critical temperature T_c. The subject of this section is a Fermi-Dirac condensation of electron-hole (e-h) pairs into a metallic liquid. The liquid exists in the macroscopic form of e-h drops showing a strong decrease of the e-h density n at their surface. Other condensations not dealt with here – but possible under certain circumstances – are the Bose-Einstein condensation of excitons or excitonic molecules and the van der Waals condensation of excitons.

An electron-hole liquid (EHL) is a collective state of electrons and holes created for example by optical pumping at a photon energy greater than the band gap. The ground state energy per pair $E_G(n)$ is:

$$E_G(n) = E_{kin}^e(n) + E_{kin}^h(n) + E_{ex}(n) + E_{cor}(n). \tag{1}$$

Here $E_{kin}^{e,h}$ is the kinetic energy of the electrons or holes, respectively, E_{ex} is the exchange energy, and E_{cor} is the correlation energy. $E_G(n)$ has a minimum value at an equilibrium density n_0 for $T = 0$ K (Fig. 1). The EHL exists and is stable with respect to decay into excitons or into an e-h plasma if $E_G(n_0)$ is less than the binding energy of the free exciton E_x. The energy difference $\Phi = E_G(n_0) - E_x$ is called the binding energy of the liquid or the work function. The equilibrium density n_0 and the binding energy Φ can be directly determined by experiments: The EHL represents an excited state of the crystal and has a finite lifetime τ_0. Both, the EHL and the free exciton can recombine radiatively. Fig. 2 explains the origin of the EHL luminescence band. The luminescence intensity emitted at a photon energy $h\nu$ can be simply described as the integral over a joint density of states:

$$I(h\nu) = I_0 \int_0^{E'} D_e(E) D_h(E' - E) f(E, E_F^e) f((E' - E), E_F^h) \, dE. \tag{2}$$

Where $E' = h\nu - E_g'$ is the photon energy relative to the band gap energy E_g' within the liquid, which is smaller than that of the crystal in its ground state. $D_{e,h}(E)$ are the densities of states and $f(E, E_F^{e,h})$ are the Fermi functions for electrons and holes, respectively. The densities of states depend solely on renormalized effective masses. The band masses have to be renormalized because of the many-particle interactions [77R]. The Fermi energies depend on the masses and the e-h density. The difference between the high energy edge of the EHL luminescence band – the chemical potential μ – and the energy of the free exciton is the binding energy Φ of the liquid (Fig. 3). The density of the EHL is determined from the Fermi energies fixing the width of the luminescence line (Figs. 2, 3). A simple verification of the existence of a liquid is thus possible: The shape of the EHL luminescence line is time independent at $T = 0$ K demonstrating the time independence of the density (Fig. 4).

The observation of the temperature evolution of the luminescence line shape permits the construction of the phase diagram in a density-temperature space (Fig. 5). At low densities (to the left) a single gaseous phase exists. At high densities (to the right) there exists also a single phase, *a metallic plasma*. For intermediate densities and temperatures below the critical temperature T_c there is a region of coexistence of two phases: gas and liquid. A phase separation into these two phases occurs, following the creation of an e-h plasma with an "intermediate" density. The experimental determination of the critical density n_c is more difficult than that of T_c. There is no universal phase diagram for EHL's in different materials. The phase diagrams depend on particularities of the band structure e.g. the non-parabolic dispersion curves of the camel's back in GaP and AlAs [80R]. A characterization of the EHL includes besides the microscopic properties the macroscopic parameters of drops: the radius, the surface energy and the droplet net charge Q. Little information about the size of these parameters in different materials has been gathered to date, with the exception of Ge, where the drops are large enough to apply e.g. light scattering techniques (Mie scattering) (Fig. 6).

The EHL is an ideal testing ground for many-body theories and thermodynamic theories [74V, 74B3], since upon application of a uniaxial stress a given band structure and the degeneracies of its conduction and valence bands can be modified in a defined way. A number of results for the ground state and the critical parameters T_c, n_c of the EHL will be given for Si and Ge under uniaxial stress. The nomenclature Si(n, m) indicates that the conduction band of Si is n-fold and that the valence band is m-fold degenerate. Reviews covering mainly the properties of Ge and Si are [72P, 75B2, 75J, 75V2] and in particular [77H3] and [77R]. [85B] reviews the more recent work on the EHL in magnetic fields.

In the Tables 1···8 only the properties of the EHL in materials with **indirect** band gap are given. There is clear evidence from experiments **and** theory for the existence of the EHL in these materials. The distribution of electrons (holes) among a number of equivalent conduction (valence) band valleys reduces the repulsive kinetic energy sufficiently to lower the ground state energy E_G of the EHL below the exciton binding energy, which is the necessary prerequisite for the existence of an EHL. Ground state calculations of the EHL for a number of direct semiconductors are strongly controversial [78B1, 78M1, 77R]. There is **no** clear experimental evidence for the existence of a stable EHL in these materials. Recent reviews on high excitation effects in direct gap materials are [79G] and [81K].

15.0.2 Frequently used abbreviations and symbols

Abbreviations

BE	bound exciton
EHD	electron hole drop
EHL	electron hole liquid
EHP	electron hole plasma
FE	free exciton
FIR	far infrared
LA	longitudinal acoustic (phonon)
LO	longitudinal optic (phonon)
NP	no-phonon (spectral line)
TA	transverse acoustic (phonon)
TO	transverse optic (phonon)

Symbols

B	T	magnetic induction
D	$eV^{-1} cm^{-3}$	density of states
e		electron
E	eV	energy
f		Fermi function
g		factor
h		hole
I	arb. units	intensity
k	cm^{-1}	wave vector
k_B		Boltzmann's constant
n	cm^{-3}	density of e-h pairs
Q	C	electric change
r	µm	radius of e-h drops
t	s	time
T	K, °C	temperature
α	dB/cm	attenuation constant
$\Delta\varphi$	eV	dipole layer energy
λ	cm	wavelength
μ	eV	chemical potential
ν	Hz	optical frequency
Φ	eV	binding energy
σ	kg/mm^2	stress
σ	erg/cm^2	surface energy or surface tension
τ	s	lifetime
θ	°	angle

Indices

b(ath)	bath (temperature)
c	critical
cor	correlation
D	diffusion
e-h	electron-hole
ex	exchange
exc	excitation
exp	experimental
F	Fermi
g	band gap
G	ground state
kin	kinetic
o	zero, equilibrium value
s	scattered (light)
tot	total
T	transmitted
X	exciton

15.1 Electron-hole liquid binding energy

The EHL binding energy Φ is related to the ground state energy E_G by $E_G = \Phi + E_X$. The energy E_X is the lowest energy state of the free exciton. For values E_X of the relevant semiconductor materials, see Subvolumes III/17a, b.

Table 1. Numerical value of the EHL binding energy Φ. (n = degeneracy of conduction band, m = degeneracy of valence band, FIR = far infrared, LLSA = luminescence line shape analysis.)

Material (n, m)	Φ meV	T K	Experimental method, remarks	Ref.
Ge (4, 2)	1.65 (10)	2	onset of FIR-absorption	76G
	1.55	4.9	onset of FIR absorption	76T2
	1.54 (25)	2.5	onset of luminescence	73L
	2.0	2	LLSA	73B
	1.8 (2)	3.5	LLSA	76T1
	1.8 (2)	1.08	value recommended in [77H1]	73T
Ge (4, 1)	1.05	1.5	cyclotron resonance	74O
Ge (1, 1)	0.74	1.71	LLSA	78F
	0.40		theoretical value, derived from $E_G = \Phi + E_X$ with $E_X = 2.655$ meV	74V
$Ge_{0.85}Si_{0.15}$	3	2	photoluminescence	74B2
Si (6, 2)	8.3⋯8.8		LLSA	75V1
	8.2	4.2	value recommended in [77H1], Fig. 7	76H
	9.3 (2)	2	LLSA	81F
	8.0 (2)	1.8	LLSA	78K2
Si (6, 1)	5.3 (2)	2	LLSA	81F
	5.5 (2)	1.8	LLSA	78K2
Si (4, 1)	4.8 (2)	2	LLSA	81F
	3.1 (2)	1.8	LLSA	
Si (2, 1)	2.1 (2)	2	LLSA	81F
	0.5 (2)	1.8	LLSA	78K2
	1.76		theoretical value, derived from $E_G = \Phi + E_X$ with $E_X = 12.85$ meV	74V
GaP	14	2	LLSA	77S1
	17.5 (30)	1.8	LLSA, Fig. 8	79B1
	11		theoretical value	80R
3C—SiC	17 (3)	1.8	LLSA, Fig. 9	79B2
15R—SiC	20 (5)	1.8	LLSA	78S3
AlAs	20.3		theoretical value, taking into account the camel's back in the conduction band and using the ε_0^*-approximation of the electron—phonon-interaction	81B
$Ga_{0.08}Al_{0.92}As$	16 (4)	1.8	LLSA, Fig. 10	81B
AgBr	55	18.5⋯45	LLSA; the EHL luminescence was detected before the system had reached thermal equilibrium and the analysis is thus thought to be somewhat doubtful (79 K2).	77H1
	60	1.9	time-resolved luminescence	79K2
	30		theoretical value	78B1
TlBr	2.2		theoretical value	78B1
TlCl	0		theoretical value	78B1

15.2 Equilibrium density n_0

Most of the experimental data on the equilibrium density n_0 at $T=0$ K are derived from analysis of luminescence line shapes. The Fermi energy is directly determined and relates in the most simple case of a parabolic conduction band to the density via

$$E_F = \hbar^2 (3\pi^2 n)^{2/3} / 2 m_e.$$

An uncertainty of $\approx 15\%$ is usually introduced by using unrenormalized masses instead of renormalized ones in this analysis, besides a number of other uncertainties which are a consequence of approximations used in the fitting procedure. For a discussion of these approximations, see [79B1]. In an external magnetic field, the density oscillates and changes as a function of the direction and strength of the field (Figs. 11, 12) [79S1, 80S].

Table 2. Numerical values of the equilibrium density n_0 at $T \approx 0$ K. (n = degeneracy of conduction band, m = degeneracy of valence band, LLSA = luminescence line shape analysis.)

Material (n, m)	n_0 cm^{-3}	T K	Experimental method, remarks	Ref.
Ge (4, 2)	$2.6 \cdot 10^{17}$	2	LLSA	73B
	2.38 (5)	1.08	LLSA	73T
	$2.2 \cdot 10^{17}$	1.8	lifetime broadening included in the analysis	77M1
	$2.57 \cdot 10^{17}$	1.8	lifetime broadening and mass renormalization included in the analysis	79S1
	$2.45 \cdot 10^{17}$	2	magnetooscillatory experiments, mass renormalization included in the analysis	80S
Ge (1, 1)	$1.05 \cdot 10^{16}$	1.71	LLSA	78F
	$1.0 (2) \cdot 10^{16}$	2	LLSA	78T1
	$1.11 \cdot 10^{16}$		theoretical value	74V
Si (6, 2)	$3 \cdots 3.5 \cdot 10^{18}$	2	LLSA	75V1
	$3.33 (5) \cdot 10^{18}$	4.2	LLSA	76H
	$3.5 (1) \cdot 10^{18}$	2	LLSA	81F, 78K2
Si (6, 1)	$1.13 (2) \cdot 10^{18}$	2	valence band splitting 31 meV, conduction band splitting 0 meV	81F
	$0.98 (3) \cdot 10^{18}$	1.8	LLSA	78K2
Si (4, 1)	$0.93 (2) \cdot 10^{18}$	2	valence band splitting 29 meV, conduction band splitting 25 meV	81F
	$0.90 (3) \cdot 10^{18}$	1.8	LLSA	78K2
Si (2, 1)	$0.48 (1) \cdot 10^{18}$	2	valence band splitting 17 meV, conduction band splitting 30 meV	81F
	$0.48 (2) \cdot 10^{18}$	1.8	LLSA	78K2
	$0.35 \cdot 10^{18}$	1.8		80G
	$0.447 \cdot 10^{18}$		theoretical value	74V
Ge$_{0.85}$Si$_{0.15}$	$5 \cdot 10^{17}$	2	LLSA	74B2
GaP	$6 \cdot 10^{18}$	2	LLSA	77S1
	$8.6 (15) \cdot 10^{18}$	1.8	LLSA	79B1
	$7.9 \cdot 10^{18}$		theoretical value	80R
3C—SiC	$10.0 (17) \cdot 10^{18}$	1.8	LLSA	79B2
	$10.3 \cdot 10^{18}$	1.8	theoretical value	80R
15R—SiC	$2.8 \cdot 10^{19}$	1.8	estimate from luminescence experiments	78S3
AlAs	$2.46 \cdot 10^{19}$		theoretical value	81B
Ga$_{0.08}$Al$_{0.92}$As	$1.6 \cdot 10^{19}$	1.8	LLSA	81B
AgBr	$8 \cdot 10^{18}$	18.5	LLSA	77H1
	$1.0 \cdot 10^{19}$		theoretical value	78B1
TlBr	$1.4 \cdot 10^{19}$		theoretical value	78B1
TlCl	$2.2 \cdot 10^{19}$		theoretical value	78B1

15.3 Critical temperature T_c

The theoretical values of the critical temperature T_c for GaP, SiC, and AlAs are calculated based on the uniform plasma approach [74C] and scaled by 20% to account for the effect of fluctuations.

Table 3. Numerical values of the critical temperature T_c. (n = degeneracy of the conduction band, m = degeneracy of the valence band.)

Material (n, m)	T_c K	Experimental method, remarks	Ref.
Ge (4, 2)	6.5	photoluminescence, complete phase diagram: Fig. 13	74T
	6.7	photoluminescence	78T2
	7.0	photoluminescence	77M2
	6.73	theory, droplet fluctuation model	79R1
Ge (1, 2)	4.89	theory, droplet fluctuation model	79R1
Ge (1, 1)	2.91	theory, droplet fluctuation model	79R1
	3.5	photoluminescence, the valence bands were not completely decoupled at the stress values which were employed.	77F
Si (6, 2)	23 (1)	photoluminescence	81F
	25 (5)	photoluminescence	76H
	27 (1)	photoluminescence	77S2
	23.5	theory, droplet fluctuation model	79R1
Si (6, 1)	16.9 (5)	photoluminescence, phase diagram: Fig. 14	81F
	16.5	theory, droplet fluctuation model	81F
Si (4, 1)	16.4 (5)	photoluminescence, phase diagram: Fig. 14	81F
	15.8	theory, droplet fluctuation model	81F
Si (2, 2)	18.8	theory, droplet fluctuation model	79R1
Si (2, 1)	14.2	theory, droplet fluctuation model	79R1
	14.0 (5)	photoluminescence, phase diagram: Fig. 14	81F
GaP	45	photoluminescence	79B1
	40	photoluminescence	77S1
	44	theory	80R
3C—SiC	$\geqq 41$	photoluminescence	78B2
	45	theory	80R
15R—SiC	76	estimate from experiment based on scaling rules	78S3
AlAs	56	theory	80R
$Ga_{0.08}Al_{0.92}As$	52	photoluminescence	81B
AgBr	> 100?	photoluminescence	77H1

15.4 Critical density n_c

Experimental determinations of the critical density n_c were performed only for a few materials. The error of the n_c-values can be regarded as rather large (see Figs. 13, 14). The theoretical values for GaP, SiC and AlAs are calculated using the uniform plasma approach [74C].

Table 4. Numerical values of the critical density n_c. (n = degeneracy of the conduction band, m = degeneracy of the valence band.)

Material (n, m)	n_c cm^{-3}	Experimental method, remarks	Ref.
Ge (4, 2)	$6 (1) \cdot 10^{16}$	photoluminescence	78T2
	$8.9 (5) \cdot 10^{16}$	photoluminescence	77M2
	$8 (2) \cdot 10^{16}$	photoluminescence	74T
	$6.56 \cdot 10^{16}$	theory, droplet fluctuation model	79R1

(continued)

Table 4 (continued)

Material (n, m)	n_c cm^{-3}	Experimental method, remarks	Ref.
Ge (1, 2)	$2.04 \cdot 10^{16}$	theory, droplet fluctuation model	79R1
Ge (1, 1)	$3.2 \cdot 10^{15}$	theory, droplet fluctuation model	79R1
	$7.7 (20) \cdot 10^{15}$	photoluminescence, valence bands not completely decoupled at the stress values which were employed	77F
Si (6, 2)	$1.1 \cdot 10^{18}$	photoluminescence	77S2
	$1.2 (5) \cdot 10^{18}$	photoluminescence	77H3, 81F
	$0.96 \cdot 10^{18}$	theory, droplet fluctuation model	79R1
Si (6, 1)	$2.1 \cdot 10^{17}$	theory, droplet fluctuation model	81F
	$3.6 \cdot 10^{17}$	photoluminescence	81F
Si (4, 1)	$2.9 \cdot 10^{17}$	photoluminescence	81F
	$2.04 \cdot 10^{17}$	theory, droplet fluctuation model	
Si (2, 2)	$4.43 \cdot 10^{17}$	theory, droplet fluctuation model	79R1
Si (2, 1)	$1.44 \cdot 10^{17}$	theory, droplet fluctuation model	81F
	$1.8 \cdot 10^{17}$	photoluminescence	81F
GaP	$2.2 \cdot 10^{18}$	theory	80R
3C—SiC	$0.53 \cdot 10^{18}$	theory	80R
AlAs	$1.72 \cdot 10^{18}$	theory	80R

15.5 Lifetime τ_0 of the EHL

The lifetime τ_0 of the EHL is, in general, a function of the doping level of the crystal, and of the temperature of the electrons and holes which are in thermal equilibrium with each other but not necessarily with the lattice. In "high-purity" crystals – the nonappearance of a no-phonon EHL luminescence line can be taken as indication of "high-purity" – τ_0 is independent of the doping level. At very low temperatures of the liquid, thermionic emission from the EHL becomes unimportant and τ_0 is also independent of T, probably with the exception of Ge, where at low temperatures of 0.65 K and low excitation levels, lifetimes were found smaller than those at 2 K [78K1]. In an external magnetic field the lifetime τ_0 of the EHL shows oscillations and a strong decrease as a function of the direction and the strength of the magnetic field. Results of magnetic field experiments are summarized in [80S] and [85B]. See also Fig. 15. Upon application of a uniaxial stress, the lifetimes also change considerably. A great uncertainty exists about the reliability of the relevant experimental data since a slightly inhomogeneous application of the stress strongly influences the life-time via enhanced surface recombination (see e.g. the discussion in [78K2]).

Table 5. Numerical values of the lifetime τ_0 of the EHL. (n=degeneracy of the conduction band, m= degeneracy of the valence band.)

Material (n, m)	τ_0 ns	T K	Experimental method, remarks	Ref.
Ge (4, 2)	$40 \cdot 10^3$	2	photoluminescence	72B, 80S
	$33 \cdot 10^3$	0.65		78K1
Ge (1, 1)	$600 \cdot 10^3$	2	photoluminescence, stress of 15 kg/mm^2, for the variation with stress, see Fig. 16	78C
Ge$_{0.85}$Si$_{0.15}$	$4.4 \cdot 10^3$	2	photoluminescence	74B2
Si	150	1.8···4.2	photoluminescence	70C
	250	2···4.2	photoluminescence	78K2
GaP	35	2	photoluminescence	77S1
	30	20	photoluminescence	77H2
	37(2)	1.8	photoluminescence	79B1

(continued)

Table 5 (continued)

Material (n, m)	τ_0 ns	T K	Experimental method, remarks	Ref.
3C—SiC	57(3)	1.8	photoluminescence	78B2
15R—SiC	6(2)	1.8	photoluminescence	78S3, 79S2
$Ga_{0.08}Al_{0.92}As$	35(5)	5	lifetime shows wavelength dependence	81B
AgBr	15		photoluminescence	77H1

15.6 Effect of surface tension

Only for Ge and Si, investigations of the surface properties of electron-hole droplets exist. The values of the surface tension σ obtained from different experiments scatter considerably. This scatter is not necessarily due to experimental uncertainties but it might display a variation of surface properties with drop size. Specifically the surface energy for large droplets appears to be roughly three times larger than that obtained by fitting the phase curve. For a detailed discussion, see [79R1]. The surface tension is temperature dependent (see Fig. 20 [77W1]).

Table 6. Numerical values of the surface tension σ at $T=0$ K. (n = degeneracy of the conduction band, m = degeneracy of the valence band, HK theory = theory based on density-functional formalism of Hohenberg and Kohn.)

Material (n, m)	σ 10^{-4} erg cm^{-2}	T K	Experimental method, remarks	Ref.
Ge (4, 2)	1.84		theory, non-interacting droplet fluctuation model (small drops)	79R1
	3.7		HK theory	78K3
	1.8	2···3.2	temperature dependence of drop concentration (Fig. 17)	75B1, 77H3
	2.6 (3)	1.3···2.5	fit of phase diagram	76W
	3.8	2.4	hysteresis in luminescence thresholds	76S1
	1.8···2.6	1.8	differential detection of the shift of the luminescence edge	76E1
	3	1.8	attenuation of 1.5 GHz ultrasound by capillary modes (Fig. 19)	76E2
Ge (1, 2)	0.452		theory, non-interacting droplet fluctuation model (small drops)	79R1
	1		HK theory	78K3
Ge (1, 1)	0.0679		theory, non-interacting droplet fluctuation model (small drops)	79R1
	0.2		HK theory	78K3
Si (6, 2)	125 (60)	6.6	hysteresis effect of luminescence	77C
	87.4		HK theory	78K3
	32.0		theory, non-interacting droplet fluctuation model (small drops)	79R1
	≈100		estimate for large drops	79R1
Si (2, 2)	32.8		HK theory	78K3
	11.6		theory, non-interacting droplet fluctuation model (small drops)	79R1
Si (2, 1)	11.4		HK theory	78K3
	3.53		theory, non-interacting droplet fluctuation model (small drops)	79R1

15.7 Charge on drops and dipole layer at surface

Electrons and holes in the EHL possess different chemical potentials. Thus, the thermionic emission at a low but finite temperature will be different for electrons and holes. The drops will begin to charge until, at equilibrium, the work functions for electrons and holes are equal [74R1]. A surface dipole layer results. The size of the charge Q is supposed to depend on the size of the drops.

Table 7. Sign of charge on the drops and numerical value of the dipole layer energy $\Delta\varphi$ at the surface. (n = degeneracy of the conduction band, m = degeneracy of the valence band, e = elementary charge, HK theory = theory based on density-functional formalism of Hohenberg-Kohn.)

Material (n, m)	$\Delta\varphi$ meV	Sign or size of charge Q	Experimental method, remarks	Ref.
Ge (4, 2)	0.39	negative	HK theory	78K3
Ge (1, 2)	−0.49	positive	HK theory	
Ge (1, 1)	0.6	negative	HK theory	
Si (6, 2)	1.67	negative	HK theory	
Si (2, 2)	−0.5	neutral	HK theory	
Si (2, 1)	1.93	negative	HK theory	
Ge (4, 2)	–	−100 |e|	shift of space distribution of emission intensity in an electric field	74P
	–	−400 |e|		77N
Ge (1, 2)	–	positive	shift of space distribution of emission intensity in an electric field	74P

15.8 Size of EHL drops

The size of the drops depends on the nucleation conditions: the temperature, the excitation intensity, the risetime of the exciting light, as well as on the history of excitation [77R, 77H3]. For detailed calculations of drop size distributions in Ge as a function of the above mentioned conditions, see [79K3, 76W] and Fig. 18. Under precisely defined excitation conditions, the distribution of drop sizes is sharply peaked around the stable radius. Light scattering (Mie scattering) experiments are an elegant way to experimentally determine the size of drops. For experiments performed on Ge under different conditions, see [71P, 74B1, 76B] and Figs. 21 and 22. Other experiments include investigations of the line shape and hysteresis effects as a function of excitation conditions and temperature.

15.9 Renormalization of electron and hole masses

A renormalization of the masses of electrons and holes inside of the EHL results from many-particle interactions [74R2]. Theoretical predictions and experimental results exist only for Ge. Evidence for mass renormalization is based on experiments in an external magnetic field.

Table 8. Percentage of the mass increase for Ge. (m_{el} = longitudinal electron mass, m_{et} = transverse electron mass, m_{lh} = light hole mass, m_{hh} = heavy hole mass.)

Mass m	Mass increase %	T K	Experimental method, remarks	Ref.
m_{el}	±0		theory	74R2
	±0	1.6	magneto-oscillation of integrated luminescence	80S
	−4	2 (?)	FIR magneto-plasma absorption	80N
m_{et}	+10		theory	74R2
	+10	1.6	magneto-oscillation of integrated luminescence	80S
	+15	2 (?)	FIR magneto-plasma absorption	80N

(continued)

Table 8 (continued)

Mass m	Mass increase %	T K	Experimental method, remarks	Ref.
m_{lh}	+10		theory	74R2
	+10	1.6	magneto-oscillation of integrated luminescence	80S
	+20	2 (?)	FIR magneto-plasma absorption	80N
m_{hh}	+14		theory	74R2
	+10	1.6	magneto-oscillation of integrated luminescence	80S
	+15	2 (?)	FIR magneto-plasma absorption	80N

15.10 EHL in doped semiconductors

Impurities influence the EHL via nucleation processes even at lowest doping levels of $\approx 10^{10}$ cm^{-3}. Drops bind to crystal defects. Detailed theoretical predictions were made [78S1, 78S2] and experimentally verified for Ge [79W], where a binding energy of a drop to an impurity of $\geqq 5$ meV was found.

At intermediate impurity concentrations, EHL recombination lines without phonon participation appear.

At impurity concentrations of e.g. $\approx 10^{16}$ cm^{-3} in Ge and $\approx 10^{17}$ cm^{-3} in Si, the character of the EHL luminescence and the energetics change significantly (Fig. 23). For a review of older work, see [77H3]. Detailed investigations of the influence of nitrogen doping on the EHL in GaP are presented in [79S3, 79K1].

15.11 γ-drops

Upon application of a suitable contact stress to a cubic crystal, one or more electronic potential valleys inside the crystal are created, depending on the direction of stress relative to the crystal axes. These potential valleys are regions where the band gap is smaller than in the neighbouring regions. The potential valleys attract charge carriers, excitons and electron-hole drops which are created e.g. by photoexcitation. These particles coalesce inside the potential valley to create a large e-h liquid, 300···500 μm in diameter in Ge, called γ-drop. The γ-drops were discovered by Alfvén-wave resonance [74M] and were studied theoretically and experimentally by various methods. The most direct method is the photographic method. For reviews, see [76J, 85B]. γ-drops deform strongly upon application of an external magnetic field in a direction perpendicular to the field [76S2, 77B, 77W2, 78M2].

Figures for chapter 15

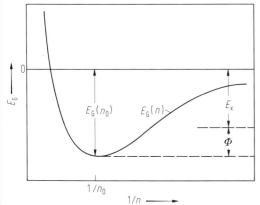

Fig. 1. Schematics of the ground state energy E_G of an e-h condensate vs. reciprocal e-h pair density $1/n$ at $T=0$ K [77H3]. $\Phi = E_G(n_0) - E_X$ is the binding energy of the liquid with respect to the free exciton energy E_X.

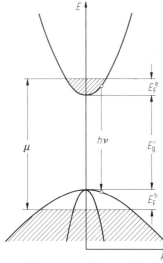

Fig. 2. Energy diagram E vs. k to explain the origin of the luminescence from the e-h liquid. A transition is shown representing the recombination of an electron with a hole, each in its respective Fermi sea (for simplicity the indirect nature of the transition is ignored). The characteristic energies of the e-h liquid – the chemical potential μ, the Fermi energies E_F^e and E_F^h and the bottom of the band E_g' – are identified [77H3].

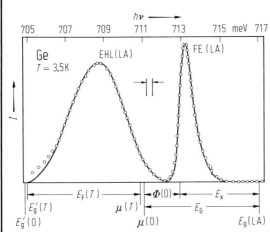

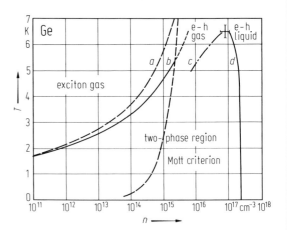

Fig. 3. Ge. Luminescence intensity I vs. photon energy $h\nu$ for germanium. Line shape fits to the LA phonon replicas of the EHL and the free exciton (FE) are shown. The data (circles) were measured at $T = 3.5$ K. The solid curve for the EHL was calculated from Equation (2). $E'_g(T)$ is the reduced gap at a temperature T, E_F is the sum of the Fermi energies, μ is the chemical potential, Φ is the binding energy of the EHL, E_X is the free exciton binding energy, E_G is the ground state energy of the EHL and $E_g(LA)$ is the LA phonon replica of the band gap [76T1].

Fig. 5. Ge. Temperature T vs. e-h pair density n. Phase diagram of the e-h system. The exciton gas and e-h liquid single-phase regions are delineated by curves b and d, respectively. The region enclosed by b and d at intermediate densities is the two-phase region in which exciton gas and e-h drops coexist. The critical point is marked by the large cross. Near the critical point is another less well defined region in which an e-h gas may exist. The solid curves are calculated from empirical parameters. For the sake of comparison with the real exciton gas (curve b) we include curve a for an ideal gas. Curve c is an extrapolation from the critical point based on the "universal" phase diagram. The dashed curve is the Mott criterion for the metal-insulator transition [76T1].

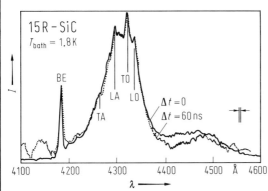

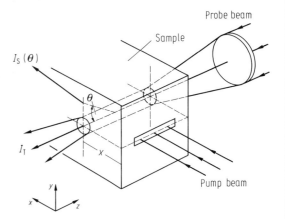

Fig. 4. 15 R—SiC. Luminescence intensity I vs. wavelength λ for silicon carbide. Time delayed luminescence spectra, for times $\Delta t = 0$, and $\Delta t = 60$ns after the exciting laser pulse. The peaks of the various EHL phonon replicas LO, TO, LA, TA are indicated. BE is a bound exciton [78S3].

Fig. 6. Schematic of the typical geometry employed in light scattering and absorption measurements. I_T is the transmitted light intensity and $I_S(\theta)$ is the scattered light intensity of the probe beam.

For Figs. 7 and 8, see next page.

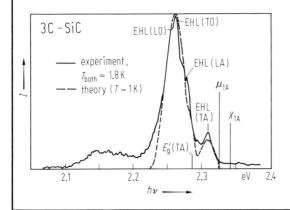

◀

Fig. 9. 3 C—SiC. Luminescence intensity I vs. photon energy $h\nu$. Comparison of theoretical ($T = 1$ K) and experimental line shapes. X_{TA} is the free exciton TA energy and μ_{TA} the EHL—TA chemical potential [79B2].

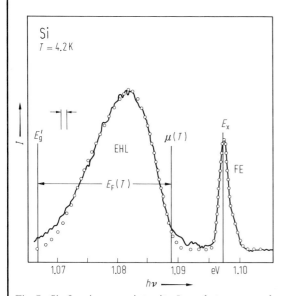

Fig. 7. Si. Luminescence intensity I vs. photon energy $h\nu$. Line shape fits to the unresolved TO/LO phonon replicas of the e-h liquid (EHL) and excitons (FE) are shown. The tracing is the measured spectrum at $T = 4.2$ K upon which is superimposed a fit (points) [76H].

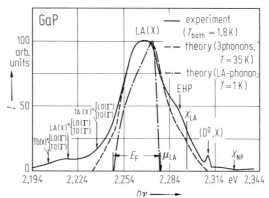

Fig. 8. GaP. Luminescence intensity I vs. photon energy $h\nu$. Comparison of experimental and theoretical line shapes of the EHL. The solid line is an experimental curve for $T_{bath} = 1.8$ K at high excitation intensity ($\simeq 5$ MW/cm^2). (D^0, X) is the zero-phonon line of an exciton bound to a neutral S(or Te) donor. The dashed line is a fit with a $T = 35$ K theoretical curve and the dot-dashed line is the $T = 1$ K LA-phonon component of this line. μ is the chemical potential, X is the free-exciton energy gap, E_F is the Fermi energy and TA, LA, LO, TO designate the various phonon replicas [79B1].

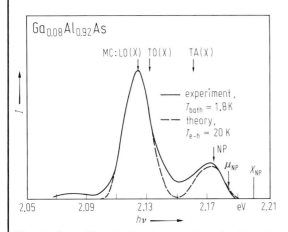

Fig. 10. Ga$_{0.08}$Al$_{0.92}$As. Luminescence intensity I vs. photon energy $h\nu$. Comparison of experimental and theoretical line shapes of the EHL. The solid curve is an experimental curve taken at $T_{bath} = 1.8$ K, $I_{exc.} = 0.5\ I_0$, at a delay $\Delta t = 60$ ns after excitation. The dashed curve is a theoretical curve for $T_{e-h} = 20$ K and a density of $n = 1.6 \cdot 10^{19}$ cm^{-3}. μ_{NP} is the chemical potential. X_{NP} is the free exciton energy gap and TA, TO, etc. designate the various phonon replicas. NP means no-phonon. MC means momentum conserving [81B].

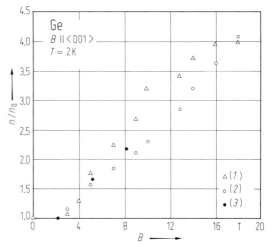

Fig. 11. Ge. Density increase n/n_0 vs. magnetic field $B \parallel \langle 001 \rangle$. The experimental points (1) are obtained from the ratio of (I_{tot}/τ) at each magnetic field value normalized to its value at $B = 1$ T to allow for the anomalous increase in I_{tot} from $B = 0 \cdots 1$ T. The EHL density change from $B = 0 \cdots 1$ T is negligible. The points (2) indicate the experimental values reported in [79S1] for comparison with the (I_{tot}/τ) values. The points (3) represent the density values deduced from Landau oscillations [80S].

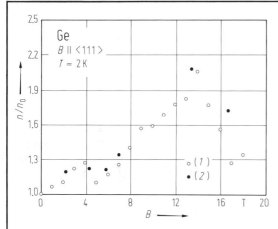

Fig. 12. Ge. Density increase n/n_0 vs. magnetic field $B \parallel \langle 111 \rangle$. The density values (1) are derived from the ratio I_{tot}/τ at each magnetic field. The points (2) indicate density values derived from the hole and electron Landau oscillations. The oscillations in n arise from Landau level crossings of the electron (hole) Fermi energy [80S].

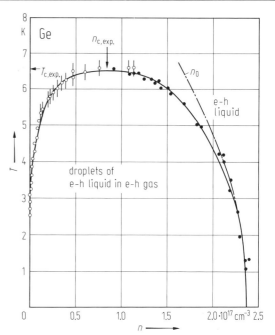

Fig. 13. Ge. Temperature T vs. e-h density n. Comparison between droplet fluctuation model and measurements of EHL phase diagram. Data from [74T]. The reanalysis of the data for the gas side of the phase diagram from [78T2] is not included [79R1].

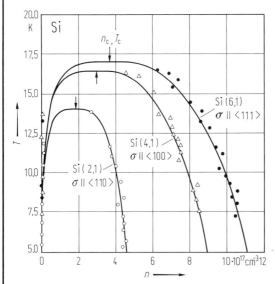

Fig. 14. Si. Temperature T vs. e-h density n. Phase diagrams for Si under high uniaxial stress $\sigma \parallel \langle 111 \rangle$, $\langle 110 \rangle$ and $\langle 100 \rangle$ directions. The size of the symbols corresponds to half of the error bars in the low temperature region. Near the critical point the error bars are about twice as large. The solid curves are from phase diagram fits using the droplet fluctuation model [81F]. The numbers in parentheses (n, m) indicate the degeneracy of the conduction and valence band, respectively.

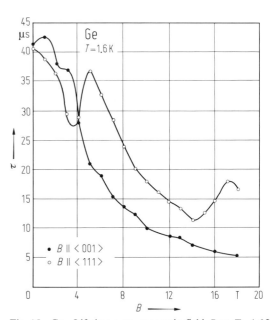

Fig. 15. Ge. Lifetime τ vs. magnetic field B at $T = 1.6$ K for $B \parallel \langle 001 \rangle$ and $B \parallel \langle 111 \rangle$. The experimental points are taken at 1 T-intervals [80S].

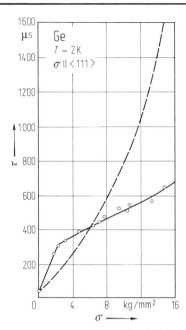

Fig. 16. Ge. Lifetime τ vs. stress $\sigma \parallel \langle 111 \rangle$. The solid curve through the data points is drawn for visual guide. The dashed curve is calculated by assuming that $Ag_{eh}(0)$ and $Bg_{eh}^2(0)$ are independent of stress. $g_{eh}(0)$ is the enhancement factor. A and B are constants proportional to the radiative and nonradiative decay rates [78C].

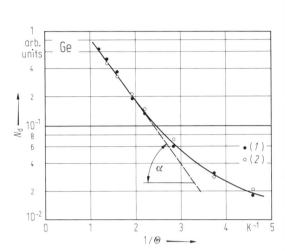

Fig. 17. Ge. Drop concentration N_d vs. reciprocal reduced temperature $1/\Theta = (1/T)(\ln(T_0/T))^2$ with $T_0 = 4.4$ K (threshold temperature at a given generation rate). The surface tension σ is deduced from the angle α according to the formula $\tan \alpha = (16\pi \sigma^3)/(6.9 \, k_B \, n_0^2 \, \Phi^2)$. ($n_0 =$ carrier density in the liquid, $\Phi =$ binding energy per pair of e-h particles in a drop relative to the exciton level.) N_d was measured by optical absorption (scattering) (2) and luminescence (1) [75B1].

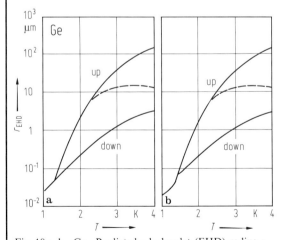

Fig. 18 a, b. Ge. Predicted e-h droplet (EHD) radius r_{EHD} near threshold vs. temperature T. a) homogeneous and b) inhomogeneous nucleation. The curves labelled up and down are for monotonically increasing and decreasing the excitation. Above $T \approx 2.5$ K the diffusion-limited radius r_D, shown as the dashed curves, should be observed rather than the further increasing radius [76W].

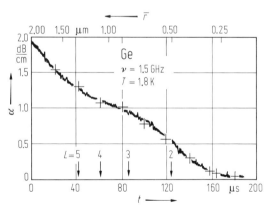

Fig. 19. Ge. Ultrasonic attenuation α vs. time t and droplet mean radius $\bar{r}$ after the light is cut off. The solid line corresponds to the experimental results and the crosses correspond to a fit. It is summed from $L=1$ to $L=10$ (different oscillation shapes of the droplets) with $\sigma = 3 \cdot 10^{-4}$ erg/cm^2 (surface tension), $\gamma = 1.4 \cdot 10^8$ s^{-1} (damping constant for the modes), $r_0 = 2.0$ μm and $\Delta r_0 = \pm 0.5$ μm (droplet radius and mean variation at $t=0$, respectively). For clarity only a few representative points are shown. The arrows indicate the position of some capillary modes [76E2].

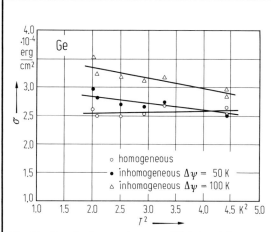

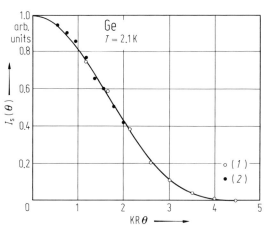

Fig. 20. Ge. Surface tension σ vs. T^2 deduced from measured threshold ratios P_+/P_- (P_+ and P_-: increasing and decreasing threshold excitation power, respectively) for three separate assumptions on the nucleation as indicated in the figure. $\Delta\psi$ is the energy barrier lowering introduced by the neutral impurity. The straight lines are least square fits to the data [77W1].

Fig. 21. Ge. Scattered intensity $I_s(\theta)$ vs. the argument $u \approx K R \theta$ of the Rayleigh-Gans formula. The droplet radius in Ge is determined from this light scattering experiment at $\lambda = 3.39\,\mu m$ wavelength. Two sets of data taken at different "depths" x are superimposed to the universal scattering curve and yield $r = 7.6\,\mu m$ (1) and $r = 3.4\,\mu m$ (2) [71P].

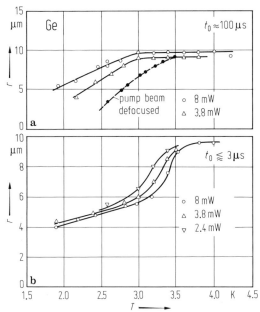

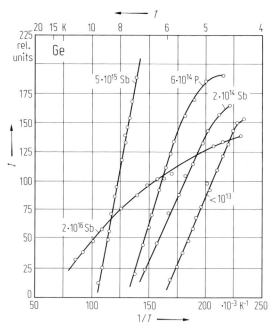

Fig. 22a, b. Ge. Droplet radius r vs. temperature T as measured by light scattering at $\lambda = 3.39\,\mu m$, parameter is the excitation power. The sample was bulk excited by focused $\lambda = 1.52\,\mu m$ radiation (He—Ne laser) chopped with two different rise times t_0: a) $t_0 = 100\,\mu s$, b) $t_0 \lesssim 3\,\mu s$ [76B].

Fig. 23. Ge. Luminescence intensity I vs. reciprocal temperature $1/T$. Effect of doping on the temperature threshold for condensation of the e-h liquid in Ge. Data points represent the e-h liquid luminescence intensity (LA phonon replica) measured at a fixed pumping power. Dopant species and concentration given in [cm^{-3}] are indicated along each curve [71A].

References for 15

70C   Cuthbert, J.C.: Phys. Rev. **B1** (1970) 1552.

71A   Alekseev, A.S., Bagaev, V.S., Galkina, T.I., Gogolin, O.V., Penin, A.N.: Sov. Phys. Solid State **12** (1971) 2855.

71P   Pokrovskii, Ya.E., Svistunova, K.I.: JETP Lett. **13** (1971) 212.

72B   Benoit à la Guillaume, C., Voos, M., Salvan, F.: Phys. Rev. **B5** (1972) 3079.

72P   Pokrovskii, Ya.E.: Phys. Status Solidi (a) **11** (1972) 385.

73B   Benoit à la Guillaume, C., Voos, M.: Phys. Rev. **B7** (1973) 1723.

73L   Lo, K., Feldman, B.J., Jeffries, C.D.: Phys. Rev. Lett. **31** (1973) 224.

73T   Thomas, G.A., Phillips, T.G., Rice, T.M., Hensel, J.C.: Phys. Rev. Lett. **31** (1973) 386.

74B1 Bagaev, V.S., Penin, N.A., Sibeldin, N.N., Tsvetkov, V.A.: Sov. Phys. Solid State **15** (1974) 2179.

74B2 Benoit à la Guillaume, C., Voos, M., Petroff, Y.: Phys. Rev. **B10** (1974) 4995.

74B3 Bhattacharyya, P., Massida, V., Singwi, K.S., Vashishta, P.: Phys. Rev. **B10** (1974) 5127.

74C   Combescot, M.: Phys. Rev. Lett. **32** (1974) 15.

74M   Markiewicz, R.S., Wolfe, J.P., Jeffries, C.D.: Phys. Rev. Lett. **32** (1974) 1357; Phys. Rev. Lett. **34** (1976) 59(E).

74O   Ohyama, T., Sanada, T., Otsuka, E.: Phys. Rev. Lett. **33** (1974) 647.

74P   Pokrovskii, Y.E., Svistunova, K.I.: Proc. 12th Intern. Conf. on the Physics of Semiconductors, Stuttgart **1974**; Pilkuhn, M.H., (ed.) Stuttgart: Teubner Verlag, **1975**, p.71.

74R1 Rice, T.M.: Phys. Rev. **B9** (1974) 1540.

74R2 Rice, T.M.: Nuovo Cimento **B23** (1974) 226.

74T   Thomas, G.A., Rice, T.M., Hensel, J.C.: Phys. Rev. Lett. **33** (1974) 219.

74V   Vashishta, P., Bhattacharyya, P., Singwi, K.S.: Phys. Rev. **B10** (1974) 5108.

75B1 Bagaev, V.S., Sibeldin, N.N., Tsvetkov, V.S.: JETP Lett. **21** (1975) 80.

75B2 Bagaev, V.S.: Springer Tracts Mod. Phys. **73** (1975) 72.

75J   Jeffries, C.D.: Science **189** (1975) 955.

75V1 Vouk, M.A., Lightowlers, E.C.: J. Phys. **C8** (1975) 3695.

75V2 Voos, M., Benoit à la Guillaume, C.: Optical Properties of Solids: New Developments; Seraphin, B.O., (ed.), Amsterdam: North Holland, **1975**, p. 143.

76B   Bagaev, V.S., Zamkovets, N.V., Keldysh, L.V., Sibeldin, N.N., Tsvetkov, V.S.: Sov. Phys. JETP **43** (1976) 783.

76E1 Etienne, B., Sander, L.M., Benoit à la Guillaume, C., Voos, M.: Phys. Rev. **B14** (1976) 712.

76E2 Etienne, B., Sander, L.M., Benoit à la Guillaume, C., Voos, M.: Phys. Rev. Lett. **37** (1976) 1299.

76G   Gershenzon, E.M., Goltsman, G.N., Ptitsina, N.G.: Sov. Phys. JETP **43** (1976) 116.

76H   Hammond, R.B., McGill, T.C., Mayer, J.W.: Phys. Rev. **B13** (1976) 3566.

76J   Jeffries, C.D., Wolfe, J.P., Markiewicz, R.S.: Proc. 13th Intern. Conf. on the Physics of Semiconductors, Rome 1976; Fumi, F., (ed.), **1976**, p. 879.

76S1 Staehli, J.L.: Phys. Status Solidi (b) **75** (1976) 451.

76S2 Störmer, H.L., Bimberg, D.: Commun. Phys. **1** (1976) 131.

76T1 Thomas, G.A., Frova, A., Hensel, J.C., Miller, R.E., Lee, P.E.: Phys. Rev. **B13** (1976) 1692.

76T2 Timusk, T.: Phys. Rev. **B13** (1976) 3511.

76W   Westervelt, R.M.: Phys. Status Solidi (b) **74** (1976) 727; Phys. Status Solidi (b) **76** (1976) 31.

77B   Bimberg, D., Störmer, H.L.: Nuovo Cimento **B39** (1977) 615.

77C   Collet, J., Pugnet, M., Barrau, J., Brousseau, M., Maaref, H.: Solid State Commun. **24** (1977) 335.

77F   Feldman, B.J., Chou, H.-H., Wong, G.K.: Solid State Commun. **24** (1977) 521.

77H1 Hulin, D., Mysyrowicz, A., Combescot, M., Pelant, I., Benoit à la Guillaume, C.: Phys. Rev. Lett. **39** (1977) 1169.

77H2 Hulin, D., Combescot, M., Bontemps, N., Mysyrowicz, A.: Phys. Lett. **A61** (1977) 349.

77H3 Hensel, J.C., Phillips, T.G., Thomas, G.A.: Solid State Phys. Vol. **32**, p. 87, Ehrenreich, H., Seitz, F., Turnbull, D., (eds.), New York: Academic Press, 1977.

77M1 Martin, R.W., Störmer, H.L.: Solid State Commun. **22** (1977) 523.

77M2 Miniscalco, W., Huang, C.-C., Salamon, M.B.: Phys. Rev. Lett. **39** (1977) 1356.

77N   Nakamura, A.: Solid State Commun. **21** (1977) 1111.

77R   Rice, T.M.: Solid State Phys. Vol. **32**, p. 1; Ehrenreich, H., Seitz, F., Turnbull, D., (eds.), New York: Academic Press, 1977.

77S1 Shah, J., Leheny, R.F., Harding, W.R., Wight, D.R.: Phys. Rev. Lett. **38** (1977) 1164.

77S2	Shah, J., Combescot, M., Dayem, A.H.: Phys. Rev. Lett **38** (1977) 1497.
77W1	Westerveldt, R.M.: Ph.D. Thesis, University of California Berkeley **1977**.
77W2	Wolfe, J.P., Markiewicz, R.S., Furneaux, J.E., Kelso, S.M., Jeffries, C.D.: Phys. Status Solidi (b) **83** (1977) 305.
78B1	Beni, G., Rice, T.M.: Phys. Rev. **B18** (1978) 768.
78B2	Bimberg, D., Skolnick, M.S., Choyke, W.J.: Phys. Rev. Lett. **40** (1978) 56.
78C	Chou, H.-H., Wong, G.K.: Phys. Rev. Lett. **41** (1978) 1677.
78F	Feldman, B.J., Chou, H.-H., Wong, G.K.: Solid State Commun. **28** (1978) 305.
78K1	Karuzskii, A.C., Zhurkin, B.G., Fradkov, V.A.: Solid State Commun. **25** (1978) 177.
78K2	Kulakovskii, V.D., Timofeev, V.B., Edel'shtein, V.M.: Sov. Phys. JETP **47** (1978) 193.
78K3	Kalia, R.K., Vashishta, P.: Phys. Rev. **B17** (1978) 2655.
78M1	Müller, G.O., Zimmermann, R.: Proc. 14th Intern. Conf. on the Physics of Semiconductors; Wilson, B.L.H., (ed.), Inst. of Phys. Conf. Series **43** (1978) 165.
78M2	Markiewicz, R.S.: Phys. Rev. **B17** (1978) 4788.
78S1	Smith, D.L.: Solid State Commun. **18** (1978) 637.
78S2	Sander, L.M., Shore, H.B., Rose, J.H.: Solid State Commun. **27** (1978) 331.
78S3	Skolnick, M.S., Bimberg, D., Choyke, W.J.: Solid State Commun. **28** (1978) 865.
78T1	Thomas, G.A., Pokrovskii, Ya.E.: Phys. Rev. **B18** (1978) 864.
78T2	Thomas, G.A., Mock, J.B., Capizzi, M.: Phys. Rev. **B18** (1978) 4250.
79B1	Bimberg, D., Skolnick, M.S., Sander, L.M.: Phys. Rev. **B19** (1979) 2231.
79B2	Bimberg, D., Sander, L.M., Skolnick, M.S., Rössler, U., Choyke, W.J.: J. Lumin. **18/19** (1979) 542.
79G	Göbel, E., Mahler, G.: Advances in Solid State Physics **XIX**; Treusch, J., (ed.) Braunschweig: Vieweg Verlag, **1979** p. 105.
79K1	Kardontchik, J.E., Cohen, E.: Phys. Rev. **B19** (1979) 3181.
79K2	Kleinefeld, Th., Stolz, H., von der Osten, W.: Solid State Commun. **31** (1979) 59.
79K3	Koch, S.W., Haug, H.: Phys. Status Solidi (b) **95** (1979) 155.
79R1	Reinecke, T.L., Lega, M.C., Ying, S.C.: Phys. Rev. **B20** (1979) 1562.
79R2	Reinecke, T.L., Lega, M.C., Ying, S.C.: Phys. Rev. **B20** (1979) 5404.
79S1	Störmer, H.L., Martin, R.W.: Phys. Rev. **B20** (1979) 4213.
79S2	Skolnick, M.S., Bimberg, D.: Solid State Commun. **29** (1979) 633.
79S3	Schwabe, R., Thuselt, F., Weinert, H., Bindemann, R.: Phys. Status Solidi (b) **95** (1979) 571.
79W	Westerveldt, R.M., Culbertson, J.C., Black, B.S.: Phys. Rev. Lett. **42** (1979) 267.
80G	Gourley, P.L., Wolfe, J.P.: Bull. Am. Phys. Soc. **25** (1980) abstract EI6, March Meeting.
80N	Narita, S., Muro, K., Yamanaka, M.: Physics of Semiconductors: Proc. of the Oji Intern. Seminar on the Application of High Magnetic Fields 1980, Miura, N., (ed.), Heidelberg, Berlin, New York: Springer Verlag, **1981**, p. 216.
80R	Reinecke, T.L., Bimberg, D.: Proc. 15th Intern. Conference on the Physics of Semiconductors Tanaka, S., Toyozawa, Y., (eds.), J. Phys. Soc. Jpn. **49A** (1980) 499.
80S	Skolnick, M.S., Bimberg, D.: Phys. Rev. **B21** (1980) 4624.
81B	Bimberg, D., Bludau, W., Linnebach, R., Bauser, E.: Solid State Commun. **37** (1981) 987.
81F	Forchel, A., Laurich, B., Moersch, G., Schmid, W., Reinecke, T.L.: Phys. Rev. Lett. **46** (1981) 678.
81K	Klingshirn, C., Haug, H.: Phys. Rep. **70** (1981) 316.
85B	Bimberg, D.: Infrared and Millimeter Waves, Button, K.J., (ed.), New York: Academic Press, **1985**, to be published.

C. Comprehensive index for III/17a···i

The comprehensive index of all Landolt-Börnstein subvolumes III/17a···i on semiconductor data includes:

1. an index of substances for all substances referred to in subvolumes 17a, b, e···i (except organic semiconductors)
2. an index of all binary and pseudobinary phase diagrams presented in subvolumes 17a, b, e···i
3. an index of mineral and common names quoted in the tables
4. an index of the names of the organic semiconductors presented in chapter 12 of this subvolume 17i
5. a subject index for the technological volumes 17c and 17d.

The chapters in subvolumes 17c, d containing information on the technology of semiconductor materials used in applications or devices differ from the other subvolumes in the form of data presentation. For data retrieval in these subvolumes a subject index seemed to be more useful for the reader than indexes of substances. On the other hand a subject index for the chapters on basic physical data of semiconductors (subvolumes a, b, e···i) seems not necessary since these data are arranged for each substance according to a scheme which will be found on the inside of the front cover of these subvolumes.

For indexes 1)···4) the page numbers are given with the respective subvolumes (a, b, e···i).

Arrangement of substances in the index of substances:

The compounds are arranged according to their chemical formula as given in the tables and figures (cf. below). For a compound $A_n B_m C_p$... the sequence A, B, C... is usually given by the position of the elements A, B, C in the Periodic Table from left to right.

The substances are listed strictly in alphabetical order of the first element (A). Within the listing of all substances beginning with A_n, the sequence is given by the alphabetical order of the second element B etc. For substances within the series $A_n B_m$... the order is according to the magnitude of the quotient n/m. If it was possible to summarize specific compositions of a solid solution series in a more general formula by using subscripts x, y, z, this was performed in many cases.

Substances differing by doping are *not* listed as well as those differing in modifications or phases. Deviations from stoichiometry are given if particular attention to this non-stoichiometry is expressed in the tables.

Arrangement of systems in the index of phase diagrams:

The systems are represented by their elements or compounds as given in the tables and figures. The arrangement of the systems is in alphabetical order. Binary systems are listed *before* quasi-binary systems. Each system is listed only once. – For further references to phase diagrams, see the subject index below.

Arrangement of mineral and common names:

Mineral and common names are listed strictly alphabetically together with the chemical formula and the page number, where the substance is cited first.

Arrangement of substances in the index of organic semiconductors:

In this index of substances the compounds are listed alphabetically under the names which are used in the tables and figures of chap. 12 of subvolume 17i. In some cases the names (or abbreviations) frequently used for the same substance in the literature are added in parantheses. References to radical ions are listed separately.

Arrangement of the subject index:

The subject index is arranged in alphabetical order. The context in which each subject occurs is specified. The respective subvolume (c or d) and the page number are given. Only the first page, where the subject appears, is cited. Further information will be usually found on the pages following the quoted page. Page numbers referring to text or tables are printed in roman, those referring to figures are printed in italic numbers and letters.

Madelung

2. Index of binary and quasi-binary phase diagrams for subvolumes 17a, b, e···i

(T-x-diagrams)

2. Index of binary and quasi-binary phase diagrams for subvolumes 17a, b, e···i

Substance	Page	Substance	Page
Tetranitromethane	158	cyanoquinodimethane (TTF:TCNQ)	
1,3,6,8-Tetranitropyrene	155	Tetrathiatetracene (TTT)	154, 156
Tetraselenafulvalene (TSF)	152, 155, 156, 203	(Tetrathiatetracene)$_2$:I$_3$ ((TTT)$_2$:I$_3$)	146, 190
Tetraselenatetracene (TST)	154		
Tetrathiafulvalene (TTF)	155, 156, 199	Tetrathiomethoxytetra-thiafulvalene (TTMTTF)	156
Tetrathiafulvalene (TTF) (radical ion)	153	Toluene	156
Tetrathiafulvalene:bromine (TTF:Br$_{0.7}$)	146, 191	2,4,7-Trinitrofluoro-9-one	158
Tetrathiafulvalene:tetra-chloro-p-benzoquinone (TTF:chloranil)	150, 196	Triphenylene	151, 155
Tetrathiafulvalene:tetra-	148, 193ff.	Violanthrene A	154, 156
		Violanthrene B	156

5. Subject index for subvolumes 17c, d

Subject Context	page text *page figure*	Subject Context	page text *page figure*
1:1 etch		acceptors	
III-V compounds, wafer prepara-tion	d44	III-V compounds, ion implantation	d94
2:1 etch		IV-VI compounds	d273
III-V compounds, wafer prepara-tion	d44	acoustic amplifier	
7:7:5 etch		II-VI wide gap compounds	*d380*
diagnostic techniques, Si, Ge	c81	II-VI wide gap compounds, devices	d156
A swirl defects		acoustic attenuation	
float-zone Si	c72	II-VI wide gap compounds	*d379*
AAS		acoustic gain	
HgI$_2$ material	d302	II-VI wide gap compounds, devices	d156
III-V compounds, characterization	d58	activation analysis	
AB etch		diagnostic techniques, Si, Ge	c80
etching processes	c282	evaluation of ion implantations	c169
III-V compounds, defect detection	d55	activation energy	
aberration		silicides, metal films	c608
ion-beam lithography	c273	activation enthalpy	
aberration figure		IV-VI compounds, interdiffusion	*d416*
ion-bcam lithography	*c555*	activation, electrical	
abrasion		ion implantation, Si	*c508*
gettering method in Si	c144	activator	
absorption		plating of lead frame	c379
CdTe detector efficiency	d159	active dopants	
HgI$_2$ nuclear particles	*d423*	III-V compounds, implantation	*d362*
absorption coefficient		active layer	
Se material	*d427*	IV-VI compounds, diode fabrica-tion	d264
	d313	activity	
X-ray	*c552*	III-V compounds, ion implantation	d94
absorption, optical		additives	
III-V compounds, characterization	d62	photoresist	c253
Se material	*d426*	ZnO ceramics	d158
abundance, natural		adhesion	
isotopes, ion implantation	*c512*	metal films to substrates	c611
nuclear transmutation doping, Si	c186	packaging plastics	c381
		plating, metal films, Si	c602

5. Subject index for subvolumes 17c, d

5. Subject index for subvolumes 17c, d

5. Subject index for subvolumes 17c, d

Schulz

Subject Context	page text *page figure*	Subject Context	page text *page figure*
rhombohedral angle		SAW amplifier	
IV-VI compounds	*d417*	II-VI wide gap compounds	*d381*
ribbon against drop		SAW convolver	
unconventional growth method, Si	c54	II-VI wide gap compounds, devices	d157
	c444	SAW transducer	
ribbon growth		II-VI wide gap compounds	*d380*
unconventional growth method, Si	c52	sawing	
ribbon to ribbon		final device preparation	c387
unconventional growth method, Si	*c442*	III-V compounds	d142
RIBE apparatus		SBD etch	
reactive dry etching	c326	etching processes	c283
Richards Crockers		scanning	
etching processes	c282	ion implantation	c178
RIE apparatus		lithography	*c550*
reactive dry etching	c326	scanning projection	
rod		lithography	*c546*
poly-Si	c25	scanning system	
roller quenching		e-beam lithography	c260
unconventional growth method, Si	c53	scattering coefficient	
	c443	HgI$_2$ nuclear particles	*d423*
rotating technique		Si, Ge, nuclear particles	*c482*
III-V compounds, LPE	d120	Schell etch	
	d373	etching processes	c283
rotation		Schimmel I etch	
crucible, Czochralski growth, Si	c31	wafer preparation, Si	c55
crystal, Czochralski growth, Si	c31	Schimmel II etch	
roughness		wafer preparation, Si	c55
SiO$_2$	c226	Schottky barrier	
Russian etch		basic device structures, Si	c90
etching processes	c283	device fabrication, SiC	c409
Rutherford backscattering		II-VI wide gap compounds, electro-	
evaluation of ion implantations	c166	luminescent system	d165
		II-VI wide gap compounds, photo-	
S-pits		detectors	d163
delineation by etchants	c300	III-V compounds, photodetector	d77
safelight zone		IV-VI compounds, epitaxy	d282
photoresist	c253	silicides, metal films	c609
Sailor's etch		Schottky contacts	
etching processes	c283	III-V compounds	d136
SAMOS memory transistor		Schottky diode	
basic device structures, Si	c101	II-VI narrow gap compounds,	
sandblasting		devices	d229
gettering method in Si	c144	II-VI wide gap compounds	*d382*
sandwich		III-V compounds, microwave devices	d81
magnetron sputter, metal layers, Si, Ge	c600		*d357*
sandwich gate electrode		SCIM method	
metal films	*c628*	Si, unconventional growth method	*c444*
sapphire		scission of chains	
substrate, Si VPE	c201	e-beam lithography	c260
saturated pressure		screened technique	
Se material	*d424*	HgI$_2$, crystal growth	*d420*
saturation carrier concentration		screening radius	
IV-VI compounds	*d419*	dry etching	c307
saturation velocity		screw dislocation	
CdTe detector	d159	III-V compounds, characterization	d52
nuclear detector, Ge, Si, CdTe, HgI$_2$	c108	III-V compounds, LPE	d116

5. Subject index for subvolumes 17c, d